ADVANCES IN ELECTRONICS AND ELECTRON PHYSICS

VOLUME 41

Contributors to This Volume

M. Campagna
Richard G. Fowler
Henning F. Harmuth
George Karady
Steven T. Manson
F. Meier
D. T. Pierce
K. Sattler
H. C. Siegmann
J. F. Verwey

Advances in Electronics and Electron Physics

EDITED BY
L. MARTON

Smithsonian Institution, Washington, D.C.

Assistant Editor
CLAIRE MARTON

EDITORIAL BOARD

T. E. Allibone   E. R. Piore
H. B. G. Casimir   M. Ponte
W. G. Dow   A. Rose
A. O. C. Nier   L. P. Smith
      F. K. Willenbrock

VOLUME 41

1976

ACADEMIC PRESS   New York San Francisco London
A Subsidiary of Harcourt Brace Jovanovich, Publishers

QC
501
A36
V.41
ENG1

COPYRIGHT © 1976, BY ACADEMIC PRESS, INC.
ALL RIGHTS RESERVED.
NO PART OF THIS PUBLICATION MAY BE REPRODUCED OR
TRANSMITTED IN ANY FORM OR BY ANY MEANS, ELECTRONIC
OR MECHANICAL, INCLUDING PHOTOCOPY, RECORDING, OR ANY
INFORMATION STORAGE AND RETRIEVAL SYSTEM, WITHOUT
PERMISSION IN WRITING FROM THE PUBLISHER.

ACADEMIC PRESS, INC.
111 Fifth Avenue, New York, New York 10003

United Kingdom Edition published by
ACADEMIC PRESS, INC. (LONDON) LTD.
24/28 Oval Road, London NW1

LIBRARY OF CONGRESS CATALOG CARD NUMBER: 49-7504

ISBN 0-12-014541-3

PRINTED IN THE UNITED STATES OF AMERICA

CONTENTS

CONTRIBUTORS TO VOLUME 41 vii
FOREWORD viii

Nonlinear Electron Acoustic Waves, Part II
RICHARD G. FOWLER

I. Introduction 1
II. Natural Phenomena 2
III. Laboratory Experimentation 31
IV. Theories of the Electron Waves 50
V. Conclusion 68
References 70

Atomic Photoelectron Spectroscopy, Part I
STEVEN T. MANSON

I. Introduction 73
II. Theoretical Description of the Photoionization Process 75
III. Experimental Techniques 106
References 108

Emission of Polarized Electrons from Solids
M. CAMPAGNA, D. T. PIERCE, F. MEIER,
K. SATTLER, AND H. C. SIEGMANN

I. Introduction 113
II. Techniques and Apparatus 115
III. Results of Spin-Polarized Photoemission 130
IV. Results of Other Techniques 157
V. Conclusion 161
References 162

Generation of Images by Means of Two-Dimensional, Spatial Electric Filters
HENNING F. HARMUTH

I. Introduction 168
II. Image Generation by Linear Transformations 172
III. Focusing for Spherical Wavefronts 197
IV. Reduction of the Array Size 200
V. Implementation of Filters by Digital Circuits 216
VI. Beyond the Capabilities of Photography and Holography 223
VII. Experimental Equipment and Test Results 242
Appendix 246
References 247

Nonvolatile Semiconductor Memories
J. F. Verwey

I. Introduction	249
II. Some Nonvolatile Memory Devices	250
III. Semiconductor Memory Devices	255
IV. Reprogrammable Read-Only Memory (RePROM) Devices	262
V. Injection and Conduction Mechanisms	264
VI. MIOS Devices	274
VII. Floating-Gate Devices	294
VIII. Discussion and Conclusions	302
Glossary of Symbols Used in Text	304
References	306

High-Power Electronic Devices
George Karady

I. Introduction	311
II. Semiconductor Devices	313
III. Thyristor Systems	336
IV. Development of Calculation Methods	341
V. Field of Application	354
VI. Conclusion	367
References	368

Author Index	371
Subject Index	380

CONTRIBUTORS TO VOLUME 41

Numbers in parentheses indicate the pages on which the authors' contributions begin.

M. CAMPAGNA, Bell Telephone Laboratories, Murray Hill, New Jersey (113)

RICHARD G. FOWLER, University of Oklahoma, Norman, Oklahoma (1)

HENNING F. HARMUTH, Department of Electrical Engineering, The Catholic University of America, Washington, D.C. (167)

GEORGE KARADY, Hydro-Quebec Institute of Research, Varennes, Quebec, Canada (311)

STEVEN T. MANSON, Department of Physics, Georgia State University, Atlanta, Georgia (73)

F. MEIER, Laboratorium für Festkörperphysik der ETH, Zurich, Switzerland (113)

D. T. PIERCE, National Bureau of Standards, Washington, D.C. (113)

K. SATTLER, Laboratorium für Festkörperphysik der ETH, Zurich, Switzerland (113)

H. C. SIEGMANN, Laboratorium für Festkörperphysik der ETH, Zurich, Switzerland (113)

J. F. VERWEY, Philips Research Laboratories, Eindhoven, The Netherlands (249)

FOREWORD

The present volume begins with the second part of R. G. Fowler's review on "Nonlinear Electron Acoustic Waves," the first part of which was published in Volume 35 of these Advances (1974). In the first part the author presented the relevant theories and forecast an examination of these theories with respect to various natural phenomena, such as lightning. This is done in the second part, with added laboratory experimentation and a renewed look at the theories of electron waves.

Atomic photoelectron spectroscopy has become a very important subject in recent years. Its importance is indicated by the great number of relevant publications, which obliged S. T. Manson to split his review of the subject into two parts. In this volume we present the first part, containing the theoretical aspects of the photoionization process, namely, photoionization cross sections, angular distributions of photoelectrons, and the ionization of atoms by fast charged particles. This is followed by a brief review of the experimental methods. The second part, which we expect to publish shortly, will discuss a comparison of the experimental and theoretical results.

Electron polarization was reviewed in 1965 in these Advances by P. S. Farago. Since that time considerable progress has been achieved, particularly through the efforts of H. C. Siegmann and his co-workers M. Campagna, D. T. Pierce, F. Meier, and K. Sattler. Their joint review retraces the early work with a succinct presentation of recent results in the emission of polarized electrons from solids.

Image generation is not usually considered to be a subject fitting within the framework of electronics or of electron physics. H. F. Harmuth demonstrates in his review, entitled "Generation of Images by Means of Two-Dimensional, Spatial Electric Filters," that besides lenses, echoes, and holography, there exists a useful new method for generation of images by means of electromagnetic or acoustic waves. The review covers both the principles and the applications of this new field.

More than ten years have elapsed since we published our last review of memory devices (in Volume 21). J. F. Verwey, although mentioning briefly such nonvolatile devices as magnetic cores, bubbles, and ovonics, devotes his review in this issue to nonvolatile semiconductor memories. One class, bipolar transistors, has been the subject of an even earlier review, in Volume 18. Most of the discussion in Verwey's review centers around various forms of the metal–oxide–semiconductor transistors used as storage devices.

FOREWORD

The last review, by G. Karady, is on high power electronic devices. High power, for the purposes of his discussion, may be defined as power exceeding 10 amperes and 100 volts. The two main representatives of such devices are silicon diodes and thyristors. Present-day examples can handle as much as kiloamperes and kilovolts. Karady reviews the manufacturing of such devices, their operating principles, major parameters, and failure modes, as well as their development.

As in the past, we list here the critical reviews scheduled for forthcoming volumes of *Advances in Electronics and Electron Physics:*

Time Measurements on Radiation Detector Signals	S. Cova
The Photovoltaic Effect	Joseph J. Loferski
In Situ Electron Microscopy of Thin Films	A. Barna, P. B. Barna, J. P. Pócza, and I. Pozsgai
Physics and Technologies of Polycrystalline Si in Semiconductor Devices	J. Kobayashi
Charged Particles as a Tool for Surface Research	J. Vennik and L. Fiermans
Electron Micrograph Analysis by Optical Transform	G. Donelli and L. Paoletti
X-Ray Image Intensifiers	J. Houston, K. H. Vosburgh, and R. K. Swank
Electron Bombardment Semiconductor Devices	D. J. Bates, R. Knight, and S. Spinella
Thermistors	G. H. Jonker
Atomic Photoelectron Spectroscopy. II	S. T. Manson
Electron Spectroscopy for Chemical Analysis	D. Berényi
Laboratory Isotope Separators and Their Application	S. B. Karmohapatro
Recent Advances in Electron Beam Addressed Memories	J. Kelly
Light-Emitting Devices, Methods and Applications. I and II	H. F. Mataré
Mass Spectroscopy	F. E. Saalfeld, J. J. DeCorpo, and J. R. Wyatt
Nonlinear Atomic Processes	J. Bakos
High Injection in a Two-Dimensional Transistor	W. L. Engl
Semiconductor Microwave Power Devices. II	S. Teszner and J. L. Teszner
Plasma Instabilities	A. Garscadden
Basic Concepts of Minicomputers	L. Kusak
Physics of Ion Beams from a Discharge Source	Gautherin and C. Lejeune
Physics of Ion Source Discharges	Gautherin and C. Lejeune
Auger Electron Spectroscopy	N. C. Macdonald and P. W. Palmberg
High Power Electron Beams as Power Tools	B. W. Schumacher
Terminology and Classification of Particle Beams	B. W. Schumacher and J. H. Fink
On Teaching of Electronics	H. E. Bergeson and G. Cassidy
Wave Propagation and Instability in Thin Film Semiconductor Structures	A. A. Barybin
The Gunn–Hilson Effect	M. P. Shaw

A Review of Applications of Superconductivity	W. B. Fowler
Minicomputer Technology	C. W. Rose
Digital Filters	S. A. White
Physical Electronics and Modeling of MOS Devices	J. N. Churchill, T. W. Collins, and F. E. Holmstrom
Measurement and Application of Precise Time	G. M. R. Winkler
Thin Film Electronics Technology	T. P. Brody
Characterization of MOSFET's Operating in Weak Inversion	R. J. Van Overstraeten
Electron Impact Processes	S. Chung
Sonar	F. N. Spiess
Micro-Channel Electron Multipliers	R. F. Potter
The Negative Hydrogen Ion	R. Geballe
Electron Attachment and Detachment	R. S. Berry

Supplementary Volumes

Sequency Theory	H. F. Harmuth
Computer Techniques for Image Processing in Electron Microscopy	W. G. Saxton
High Voltage and High Power Applications of Thyristors	G. Karady

We wish to express our best thanks to the many friends whose help makes it possible to produce these volumes. As in the past we would be very grateful for further advice and constructive criticism.

L. MARTON
C. MARTON

Nonlinear Electron Acoustic Waves, Part II

RICHARD G. FOWLER

University of Oklahoma,
Norman, Oklahoma

I. Introduction 1
II. Natural Phenomena 2
   A. Lightning 2
   B. Solar Phenomena 22
   C. Ionospheric Effects 26
III. Laboratory Experimentation 31
   A. Primary Breakdown Waves 31
   B. Microwave and Laser Breakdown 34
   C. Charge Injection Waves 36
   D. Avalanche Processes 37
   E. Long Sparks 42
IV. Theories of the Electron Waves 50
   A. Theories Based on Electron Impact Ionization 50
   B. Diffusion Equation Theories 60
   C. Time-Dependent Solutions 62
   D. Photoionization Models 63
   E. Computer Avalanche Simulations 65
V. Conclusion 68
   References 70

I. Introduction

Among a variety of propagating phenomena which are present during the initiation of gas discharges in an electric field, such as, for example, a lightning stroke, there is a class of waves having speeds in the range 10^6 to 10^8 m/sec. This article is the second part of a two-part review of these phenomena intended to emphasize the now widely held opinion that these waves are governed by the fluid dynamical properties of the electron gas.

In the time elapsed since this review was undertaken, a noticeable increase has occurred in the number of publications pertaining to the topic. In fact, in the interval between the preparation of the two parts (*1*) of the review, significant new theory has appeared, so that at present the quality of the theory seems to exceed that of the experimental results available. To the observer, the technical problems of resolution in time at the high speeds present in these phenomena continue to offer a real challenge. It is a challenge, however, which it should be within the possibilities of modern technology to meet.

For many years corona and streamer physicists have insisted that photoionization plays the dominant role in the propagation process, chiefly because the existence of antiforce waves (those propagating against the electric force) seemed to demand a charge-independent agent. Until recently the acceptance of this concept had been hindered in many minds by the inability to reduce it to a mathematical model. A model has now been put forward which involves photoionizing processes, but it also includes electron fluid behavior, and as such is not the pure photoionizing process that was originally envisioned. During this same period the model based on pure electron excitation has been extended and found to be in broad agreement with experiment. It is now a matter for careful quantitative experimentation to decide whether the photoionization process is actually present in all or some of the many types of propagating phenomena discernible as members of this class.

The structural plan of this review is to examine at the outset the experimental evidence, taking up first that relating to natural occurrences (which are presumed to contain waves of the class under study), and second, that relating to laboratory simulations and abstractions, and then to discuss subsequently efforts to provide theoretical models for these observations.

II. Natural Phenomena

A. Lightning

1. General Description

Immersed as we are in a medium which is busily generating electric spark discharges it has been easy to feel that we know "all" about lightning processes after we understand basic electricity and atomic physics. And looking at the laboratory sparks which we can generate with induction coils and electrostatic generators we gain the further feeling that we are seeing here in miniature something which if merely scaled up by a factor of 10^5 would be the same as lightning. Both of these views, often found in beginning textbooks on physics, are incorrect. Lightning is a very complex phenomenon, and laboratory sparks have not as yet been made long enough to show the full range of properties of lightning. In fact, there is some reason to think that such sparks never can be made (2). Accordingly, they will be discussed separately.

Lightning strokes can occur between cloud and cloud, cloud and ground, cloud and open air, or wholly within a cloud. Of course, only those in which the channel comes into view can be explored optically, and any information on others must come from electric and magnetic fields and long wavelength radiations. The modern description of a visible lightning discharge moving away from a cloud with negative charge has changed little from the one built up over the period 1934–1953 in a series of nine papers by Schonland, Malan, and Collens (*3–11*). After a period of time during which a slow buildup of the electric field is observed at ground antennas, luminosity emerges from the cloud and elongates as a cylindrical column several meters in diameter. As it grows in length, the column repeatedly, almost periodically, decreases in brilliance but is abruptly rekindled by a wave or front which propagates without noticeable change down from the top along the entire (available) length of the column, until it nears the end of that portion of the column traversed by the front just previous. Here it undergoes a sudden increase in luminosity which it continues to display as it blazes its way over untrod ground, until it extends the visible column by an almost quantized step of about 50 m, whereupon it seems to cease to advance, and the process is repeated until the available gap is bridged by the column or nearly so. The entire process is termed the stepped leader.

Each step is nearly rectilinear, but need not be collinear with the previous step, although its direction is always down and away from the point of origin of the column. Branching of the column is the rule rather than the exception, and branches always form at the beginning of a bright step phase of the leader process. The rekindling wave from the column head then divides evenly, traversing both branches to form steps in unison, each with normal randomicity of direction. Occasionally the rekindling waves fail one, two, n times to plow out a new bright step. Then the wave which finally breaks through develops an extra long (roughly an $n + 1$ multiple) step.

Two types of leaders have been recognized, a low-luminosity, slow-moving leader with regular short steps and little branching (called type α), and a high-luminosity, fast-moving leader with irregular tortuous steps and much branching (called type β). Subclasses β_1 and β_2 were also distinguished by Schonland, characterized by the existence of an abrupt transition point in flight from type β to type α behavior in the β_1 class, and the occurrence of one or more extra intensity rekindling waves resulting in much branching in the β_2 class.

If the gap is from cloud to ground, after a last step (which may have anomalous behavior itself) a highly luminous wave bearing a large amount of charge moves very quickly from the ground to the cloud. This is called the *return stroke*. The return stroke not only traverses the main column, but at each branch moves synchronously away along the branch to its end. The return stroke is quite clearly a stripping wave which peels off and transports to ground the charge deposited along the column by the leader process.

If the gap being bridged is from cloud to cloud the discharge after closure seems to be a steady current with occasional weak pulses, seemingly return strokes. If the gap is from cloud to air, only a weak steady current is observed.

The last step in forming the bridge is frequently not a cloud-to-ground wave, but a ground-to-cloud wave which completes the connection in a single (unstepped) advance. It has normally been difficult to observe because of intervening structures.

The return stroke wave is followed by the conduction of currents in tens of kiloamperes for durations of milliseconds. It is not the purpose of this review to follow the discharge into this period, during which channel heating and contraction develop the well-known acoustical phenomena, and such interesting peculiarities as bead lightning take place. For recent treatment of these the reader is referred to Uman (*12*).

A period of tens of milliseconds then elapses during which the luminosity dies away, and abruptly a wave (of nearly the same properties of intensity and speed as any one of the rekindling waves in the stepped leader process) propagates smoothly from cloud to ground, followed by a new return stroke. This is called the *dart leader* because it is highly luminous at its leading edge and so resembles a thrown dart. The process may repeat itself several times, as many as 20 repetitions having been observed. (Each pair consisting of a leader wave and a return wave is called a *stroke*. The entire set of strokes is called a *flash*.) The longer the interval between strokes, the slower the dart leader advances, and if the interval is excessive, the lower end of the channel is not reached by the dart leader, but requires several steps of the step process to renew the conductivity of the new channel. The dart leaders almost never reactivate any of the branches, but confine themselves to the trunk channel.

The term dart leader has also been applied to the rekindling waves which come down the column intermittently in the initial stepping process, especially to the extra bright ones which occasionally (about one in 10 times) occur. Quite often, the decaying arc luminosity of the return stroke channel is also abruptly rekindled weakly over its whole length, presumably as the result of a very fast wave traveling downward from the cloud. Schonland called these resurgences M-strokes.

For some reason, perhaps associated with the observing terrain and climate of South Africa, Schonland *et al.* observed cloud-to-ground strokes only from negatively charged clouds, leading to the assumption by an occasional writer, based on the massive nature of the African studies, that only these occur in nature. When, therefore, in the 1939–1947 Empire State Building studies by McEachron (*13*) it was observed that upward-moving stepped leaders existed and some unstepped leaders were also seen, belief was expressed here and there that this might be only an artifact from the use of such a high tower. Accordingly the 1955–1965 work of Berger (*14*) has been of great importance in clarifying this point. Observing both photographically and electrically the discharges between clouds and two instrumented towers on mountains in the Swiss–Italian Alps, he detected discharges in all four permutations and gave rough statistics on their occurrences in that locality: (a) negative cloud to tower, 38%; (b) tower to positive cloud, 38%; (c) tower to negative cloud, 22%; and (d) positive cloud to tower, 2%. In the terminology of Shelton and Fowler (*15*), the leader phases of (a) and (b) involve proforce waves, those of (c) and (d) involve antiforce waves. Berger's observations on case (a) are identical with those of Schonland, but he finds that stepped leaders are also always observed for proforce waves, i.e., are seen with case (b) as well. For antiforce waves the leader is not stepped, but in case (c) the channel is nevertheless intermittently illuminated without the extra luminosity at the ends which gives the stepping its character. Case (d) is remarkable first because it is so rare (a single event was recorded among 100 of all types, but fortunately this lone observation was of excellent quality), but second because it showed neither stepping nor intermittency. The conducting column grew smoothly downward with a slightly greater luminosity at its head than behind. Berger noted that all branching is toward the direction of leader advance, upward for upward-moving leaders, downward for downward ones.

The stepping that exists on the leading edge of the column in the proforce case has intrigued researchers since Schonland *et al.* first observed it in 1935. Numerous suggestions for its cause have been made, which fall roughly into two groups: those taking the intermittency of the step illumination as reflecting an intermittency of the advance of the column, and those which treat this intermittency as secondary and superimposed on an as yet undetected steady growth of the column. Schonland himself advanced the latter view and termed the hypothetical and unseen wave the *pilot leader*. As Uman (*12*) points out, much imprecise usage of this term by both Schonland and other authors since then, who wished either to explain or to discover other evidence for this process, make the literature difficult to read. The word "streamer" is perhaps the most misused in this context, and will be avoided in this review. It is evidently equivalent in some degree to the word "channel" chosen for use here, but often conveys connotations of motion, growth, and filamentation.

The evidence that exists for the reality of the pilot leader is impressive (*11*). (1) If there were no pilot leader, the stepped leader would need to transport currents of order 20 kA, with corresponding intense electromagnetic signals of about 5-μsec duration every 100 μsec. Only small signals were observed, about 1/10 of the steady change of field between the steps. There is some ambiguity in this observation by Malan (5), for the apparatus used is implied to involve the RC coupled amplifiers described before (7), and if so the bandwidth (unstated) was probably, in the state of the art, unequal to passing 5-μsec pulses without considerable attenuation. (2) The even length of steps is more easily achieved with a timing mechanism such as relaxation of the channel coupled with some near constant velocity (the pilot velocity), than by a process which runs out of driving force abruptly after a certain distance. (3) The failure of one or more steps to be blazed out by the rekindling wave, followed by an extra long step which covers the distance that should have been traversed piecewise but was missed, suggests that the steps are already preconditioned in some fashion. (4) The fact that an unstepped leader of this general speed is observed with antiforce waves suggests that it is the presence of a polarity dependent timing mechanism which distinguishes the two cases. (5) A "pilot" leader has been photographed in laboratory sparks (16). However, whether this is the same item as in natural lightning is an open question. (6) The observed speed of traversal of the step does not seem to diminish even at the very end.

The alternative to the pilot leader is that the succession of rekindling waves simply run until they have distributed all the charge that they brought down the column along the sides of the new column and, owing to electron attachment, the flow along the column is shut off, resulting in a rise in potential at the head of the column and a new breakdown. One piece of evidence for this is the fact that the beginning of the extra luminosity of each new step lies partly back into the step just previous, as if the end of each step were not as highly ionized as the beginning. Another is that no similar leader stepping has ever been observed in long laboratory sparks (but it is also unlikely to be), although Aleksandrov (17) has reported intermittent growth of a spark column in steps of 0.20 m, a length which must be due to another cause since both effects are at the same pressure, atmospheric.

We will now review the known data on each of these processes.

2. Pilot Leader

The characteristics of the pilot leader are inferentially derived from the stepped leader. The velocity would be the rate of advance of the termini of the steps. A judicious fit of the data histogram of Schonland's observations to Poisson probability is $\exp(-v/2.5 \times 10^5)$. The extreme range of possible mean values for the data is 1.2×10^5 to 3.6×10^5 m/sec. Schonland states that no projected velocity below 1.0×10^5 m/sec has ever been observed. Since the observations are stereographic, three-dimensional tortuosity will introduce a factor which Schonland estimated empirically at 1.3, raising this minimum to perhaps 1.3×10^5 m/sec, and the judicious average value to 3.3×10^5 m/sec. If the advance of the lightning were a three-dimensional drunkard's walk the factor (given by Lord Rayleigh) would be 1.5. It is, however, a downhill drunkard's walk, related to Smoluchowski's problem in Brownian motion theory, and since the degree of the preference for forward steps is not known as yet, we can only agree that it should reduce the factor toward that suggested by Schonland. The charge per unit length of column in the vicinity of the pilot leader is estimated at 8×10^{-4} C/m. From this an estimate can be made for the radius of the minimum velocity pilot leader, using the laboratory value of minimum breakdown field, $E = 3 \times 10^6$ V/m. It can be computed from the model of the column as a hollow cylinder of charge of radius a capped by a space charge of the same radius and of thickness l:

$$E = \frac{\lambda}{2\pi\varepsilon_0 a}\left[1 + \frac{l}{a} + \frac{1}{2}\left(1 + \frac{l^2}{a^2}\right)^{-1/2} - \left(1 + \frac{l^2}{a^2}\right)^{1/2}\right].$$

For $l > a$, $a = 4.8$ m. For $l < a$, $a = 2.4$ m. The theory of breakdown waves (15) suggests the latter situation as the more reasonable. The convection current is λv, and hence is 100 A for a minimum velocity leader, with a convection current density of -5.5 A/m². The positive sense of current being downward, displacement current density at a point outside the vertex is $-5.5\, l/a$ A/m² for $l < a$, and $-5.5(1 - l/a)$ A/m² for $l > a$. The displacement current density at a point inside the vertex is $+5.5(1 - l/a)$ A/m² for $l < a$ and $+5.5\, l/a$ A/m² for $l > a$. Thus the total current is -5.5 A/m² inside or out for $l > a$, and $-5.5\, l/a$ A/m² for $l < a$. Since the pilot leader speed is not explicitly observed it is not possible to say whether it varies either (a) from place to place, or (b) within the unobserved intervals between steps. There is some evidence for (a). The possibility of (b) remains a free variable for any theoretical models.

3. Stepped Leaders

The lengths of the steps of stepped leaders vary within narrow bounds. Schonland quotes observations ranging from 10 to 200 m on different flashes. He also shows that the length of a step is related to the time elapsed between steps (Fig. 1). If it is assumed that a pilot leader is moving steadily onward at the head of the column, then the variability of the step length can be assigned to three causes: (1) variation of the angle of the step with the vertical in the line of sight (stereographic projection); (2) variation of the timing mechanism which determines the interval between rekindling waves;

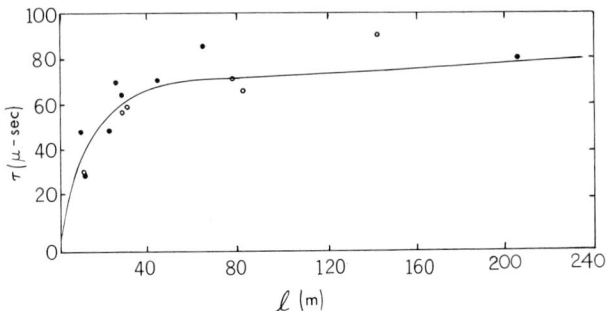

FIG. 1. Leader step length as a function of elapsed time between steps. [After Schonland et al. (4), with the permission of the *Proceedings of the Royal Society*.]

(3) variation in pilot leader speed because of local electric field differences. This last is indicated by the data which can be derived from Fig. 2. For the trunk and all branches above mid-line, steps range from 10 to 24 m. Below midline they range from 20 to 50 m, although the delay intervals are not significantly longer. In both cases the distribution in length is quite compatible with a simple angular dependence of probability such as $(1 + \cos \theta)$. The time intervals observed between successive rekindlings of the column were more variable in the African data than in the Swiss data. Berger found 30 to 52-μsec intervals for 38 proforce cases examined. In the intermittent upward-moving antiforce leader he reported times between recurrences that range more widely, from 40 to 115 μsec.

The velocity of blazing out the steps in the stepped leader is very high. Schonland estimates the value at in excess of 5×10^7 m/sec. No real measurements seem available. Since the weak luminosity which is seen to illuminate the entire column in near unison with the bright step cannot be an instantaneous action at all points, but must have moved from somewhere, presumably from the cloud base, the fastest process that could be invoked would be an electromagnetic one. But because it must move and augment free electrons, it must be an electron acoustic wave, and so should be close kin to the subsequent dart leader preceding secondary strokes. Occasionally a rekindling wave is seen to rise almost to dart leader intensity, so it seems safe to assume that they are essentially weak dart leaders which become brighter when they pass through the electron-rich step newly created by the pilot leader. Since the speed of dart leaders will be seen to become less the less electron-rich (older) the channel, it is probable that the lowest speed dart leaders (2×10^6 m/sec) nearly match the speed of the rekindling wave, while the highest speed ones approach the speed at which the step is traversed (2×10^7 m/sec).

Indirectly Schonland estimated the radii of the pilot leaders at less than

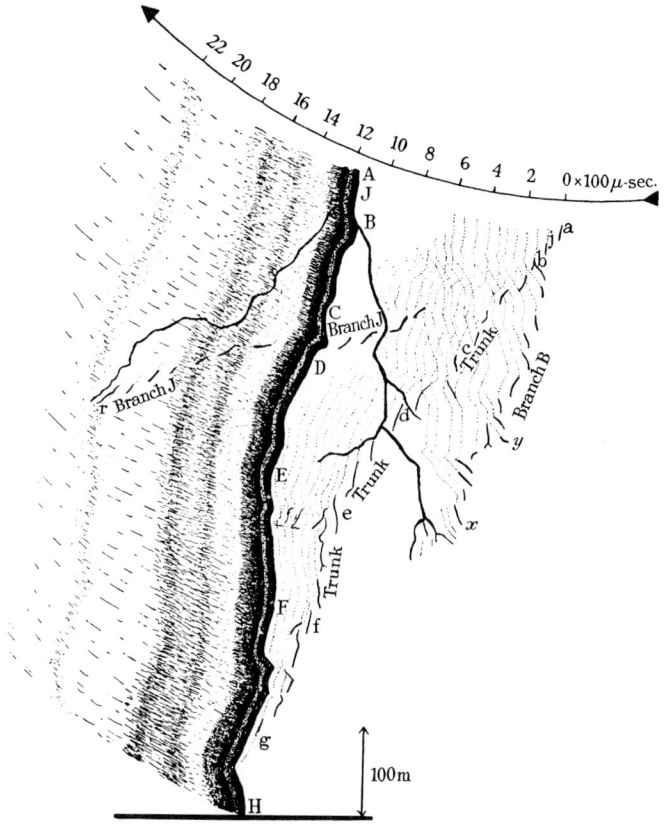

FIG. 2. Characteristic Boys camera record of a stepped leader. [After Schonland *et al.* (4), with permission of the *Proceedings of the Royal Society*.]

5 m. We have shown that a variation in the charge distribution model used could halve this maximum. Schonland argues that the radii of the pilot and the step are essentially equal, and equal also to the overall channel radii. That there is a discrepancy here can be seen by simple measurement of the step diameters which are a clearly resolvable feature of the Boys camera photographs available. Such diameters can only be larger than the true diameters owing to photographic halation, etc. The result in a typical case is a radius of the order of 1.0 m. It is necessary therefore either to assume that there is a central core of photographable luminosity surrounded by an unseen but conducting shell, or that the quantities used in the estimate need reconsideration, or that the girth of the steps is not as great as that of the pilot leader which formed them. This last is perhaps the most likely case, and is further countenanced by the even more shrunken appearance of the

column behind the step, which seems to be reduced by a further factor of 2.

Hodges (18) has given an estimate of the currents in the α and β leader steps based on world-wide average current rise rate measurements made on strokes to tall structures, coupled with his own observations of the low frequency radiation from step processes and return strokes. He concluded that α leader steps carry 600 A and β leader steps carry 2600 A on the average.

4. *First Return Strokes*

The return stroke, being by far the most reliably visible portion of the discharge and the most electrically and acoustically active, has received the bulk of quantitative attention. The Boys camera frequently provides a good pair of images of a return stroke, but never, in Schonland's reported experience, seems to have given a recording of both leader pairs in the few rare cases when a good stepped leader phenomenon was detected on one photo. In Fig. 3 a timetable constructed by Schonland for a typical return stroke 2.5 km long is given. One remarkable feature was that the wave stalled from the sixth microsecond to the tenth at the branch junction marked P. During this time it seemed to explore the branch for about 100 m, and then proceeded upward on the main channel. As a return stroke nears the top of the column, especially after passing the last branch (if any), it weakens in intensity to a level which Schonland called "glow luminosity," a level, however, which is roughly similar to that in the branches themselves, so that it is a paralleling of currents which accounts for the intensity of the lower trunk. It also decreases in velocity by a factor of as much as 4. With about equal frequency, Schonland observed stereographically projected velocities over the lower main trunk between 2.5×10^7 and 1.5×10^8 m/sec, while the upper trunk and branch velocities were also nearly evenly distributed between the limits 1.5×10^7 and 6×10^7 m/sec. Part of the reduction certainly comes from reduced driving field, but part may come from increased probability of inclination of path in the upper portion. This may also account for the greater spread of data in the branches of upper portions *vis à vis* the lower main trunk. Even in the lower main trunk part of the variance must be projectively introduced.

Currents carried by the return stroke are rather well measured and range widely. Berger's data are the most recent, complete and direct, and are given in Fig. 4, where the four possibilities of cloud polarity and wave direction are distinguished. Measurements can also be inferred from the radiation fields if assumptions are made about velocity and orientation of the stroke, as has been discussed by Uman *et al.* (19). In the Florida observations somewhat higher velocities were observed than occurred in the South Africa data,

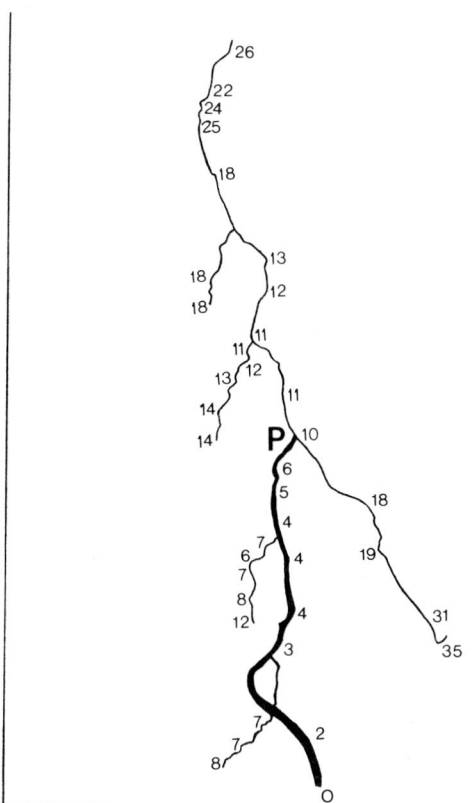

FIG. 3. Timetable for a return stroke. [After Schonland et al. (4), with permission of the Proceedings of the Royal Society.]

i.e., 1.2×10^8 to 2.4×10^8 m/sec. They give a probable occurrence frequency curve for return stroke currents of all kinds as shown in Fig. 5.

The directions of charge motion for the four possibilities are deducible from the moving photographs of Berger. They are given in Table I. The reference sense of the current is positive for positive charge flow away from a positive cloud.

Pruett (20), Uman (21), Uman and Orville (22), and Orville (23) have analyzed spectroscopic emission from return strokes in Arizona, and Uman has found a peak temperature of 24,000°K, an electron density of 4.3×10^{24} m^{-3}, an atomic oxygen density of 2.7×10^{23} m^{-3}, and an atomic nitrogen density of 7.4×10^{23} m^{-3}. Essentially 100% of the ionized atoms are O$^+$ and N$^+$. From these data he has computed a core conductivity of 1.8×10^4 Ω^{-1} m^{-1}.

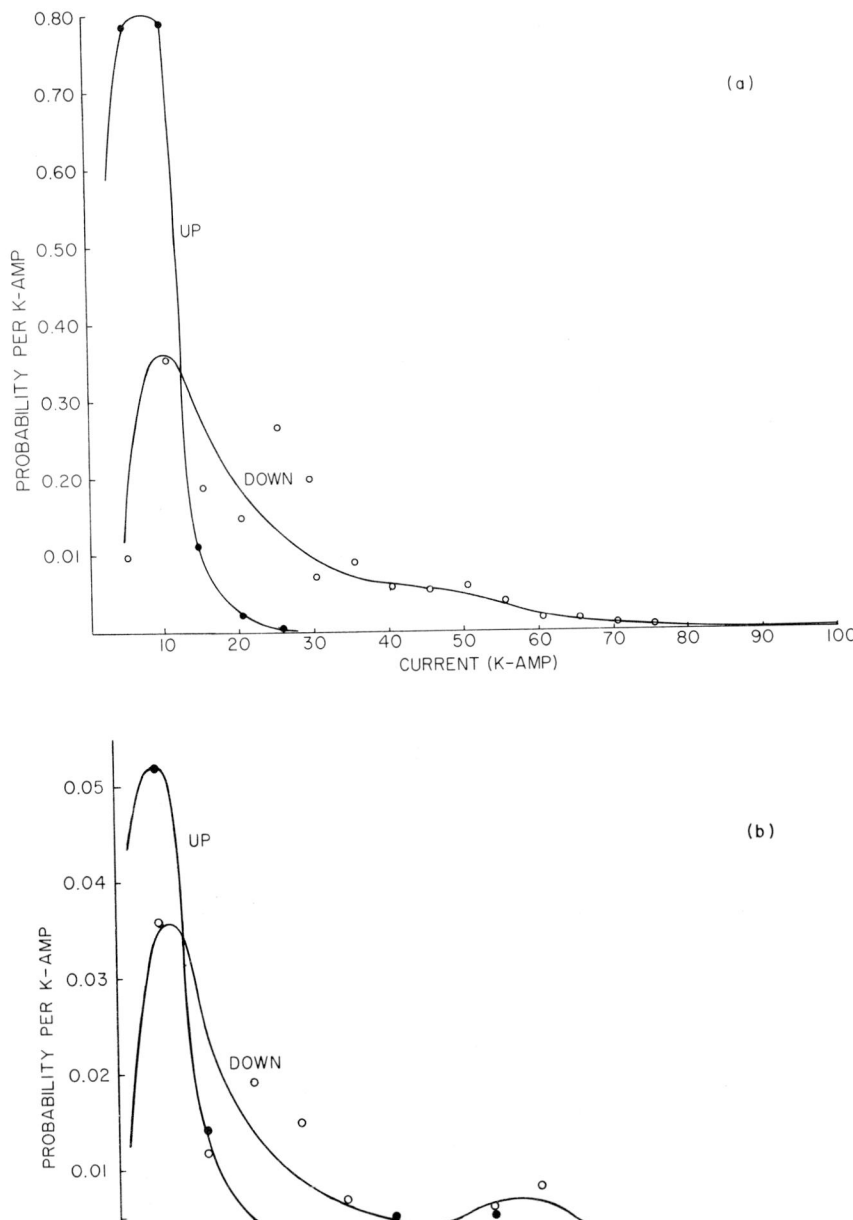

FIG. 4. Probability of stroke current in the range I, dI. (a) cloud positive; (b) cloud negative. [Data by Berger (*14*).]

TABLE I

LIGHTNING PROPAGATION DIRECTIONS[a]

Cloud polarity	Negative	Negative	Positive	Positive
Leader origin	Cloud	Ground[b]	Cloud[d]	Ground
Leader current	Negative	Negative	?	?
Return I origin	Ground	Cloud[c]	Ground	Cloud
Return I current	Negative	Negative	Positive	Positive
Dart origin	Cloud	Cloud	None	None
Return II origin	Ground	Ground	None	None
Return II current	Negative	Negative	None	None

[a] After Berger (14).
[b] Meets downcoming leader.
[c] Repeated twice at least.
[d] Meets upcoming leaders.

The size of the column along which the return stroke moves has been attacked in several ways. Schonland made direct photographic measurements which must at least be upper limits, ranging from 0.075 to 0.115 m radius. Evans and Walker (24) with improved photographic equipment obtained 0.015 to 0.06 m radius in Arizona studies. Uman (25), by causing lightning to pass through fiberglass screens, detected that there are hot cores ranging from milli- to centimeters in radius. Hill (26) and Jones (27) have studied the cratering on electrodes at which lightning strokes have terminated, finding them to be a few millimeters in diameter, but as Uman points out, electrode processes often constrict discharges far below their free column size.

Schonland has observed that his return strokes were wavelike, and had a thickness (column length) of about 500 m. He attributes most of this length determination to the lifetime of the radiating excited states.

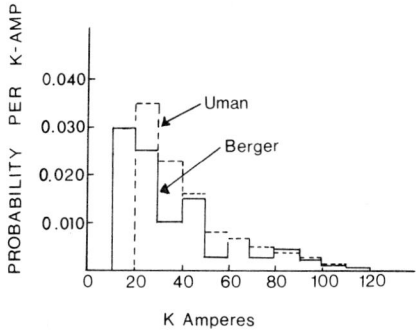

FIG. 5. Probability of a maximum current in the range I, $+dI$ in South Africa and Florida, for negative cloud-to-ground strokes. [Data by Schonland et al. (4) and by Uman et al. (19).]

The primary interest of observers has generally been to obtain practical knowledge of the maximum effects of lightning. In this spirit a most desirable quantity, the charge transported solely during the flight of the return stroke, has not generally been measured but rather the total charge in a stroke or often in entire flashes. The former will be some portion of the latter, and in the case of downward discharges may be a major part of it, but in the case of upward discharges the inclusion of a long continuing current at low levels generally dominates the total charge transported. Examples of such data as taken by Berger are given in Figs. 6 through 9.

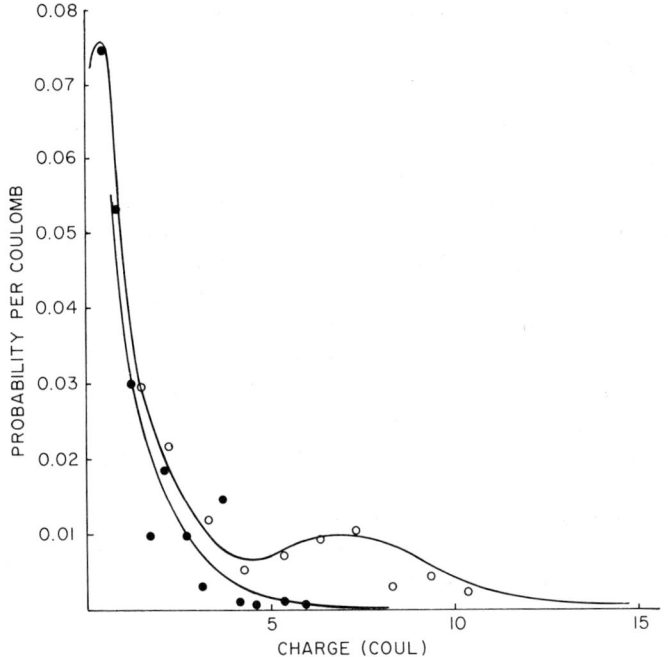

FIG. 6. Probability of charge transported by a single return stroke in the range Q, dQ. Cloud negative. ○, down; ●, up. [Data by Berger (*14*).]

In Figs. 10 and 11 the maximum rate of rise of a stroke current is given for the four possibilities. The cause of a noninstrumental limitation in the rise of current is not precisely known. Clearly there is one, since the four different configurations display different rates. Equally clearly, conventional linear conductor inductance cannot be involved, since the indicated values are less than 10^{-12} H. That the limitation is not in the motion of the return stroke wave can be seen from a calculation of the very small amount of charge transported during the current rise, i.e., $\frac{1}{2}I^2/(dI/dt)$, which ranges

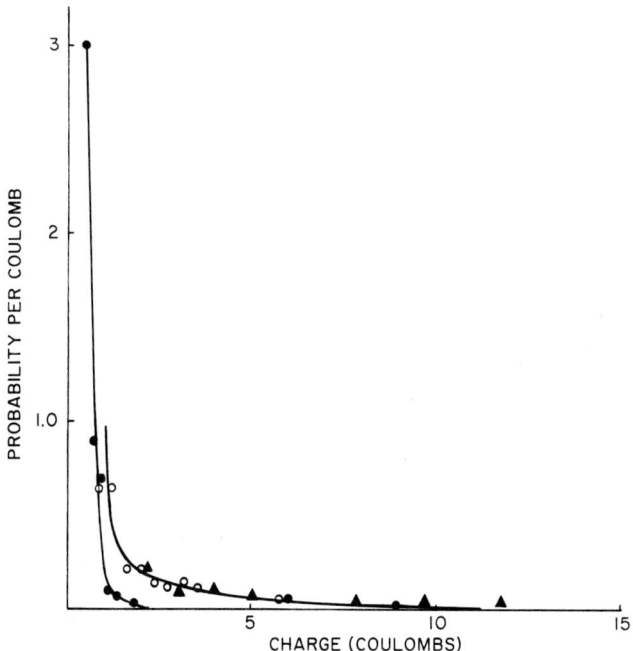

FIG. 7. Probability of charge transported, by single return strokes of greater than 10 kA, in the range Q, dQ. Cloud negative. ▲, first down; ○, down; ●, up. [Data by Berger (14).]

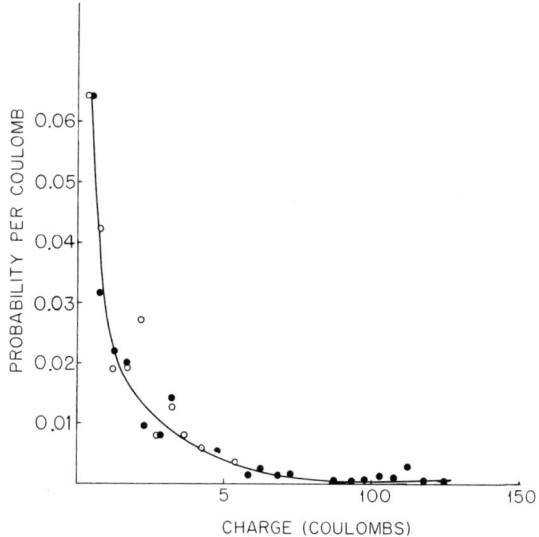

FIG. 8. Probability of charge delivered by continuing current in the range Q, dQ. Cloud negative. ○, down; ●, up. [Data by Berger (14).]

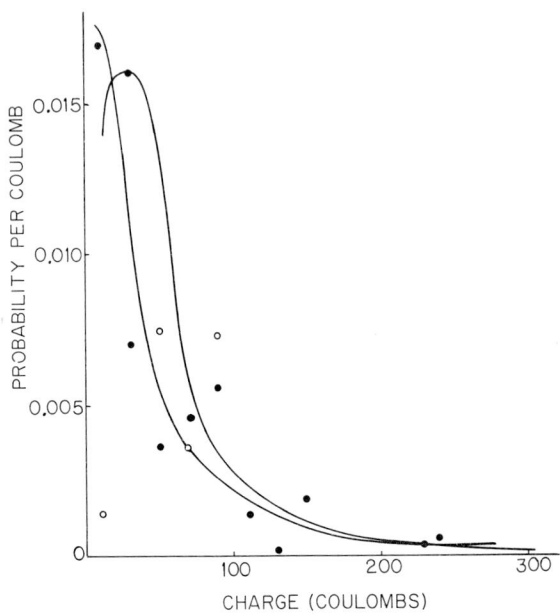

FIG. 9. Probability of charge in a single stroke in the range Q, dQ. Cloud positive. Strokes are followed by continuing currents. ○, down; ●, up. [Data by Berger *(14)*.]

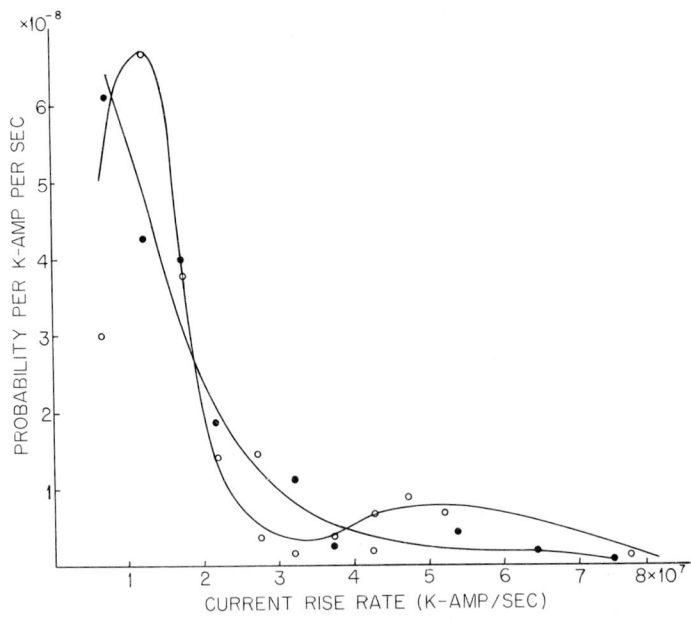

FIG. 10. Probability of a given maximum current rise rate. Cloud negative. ○, down; ●, up. [Data by Berger *(14)*.]

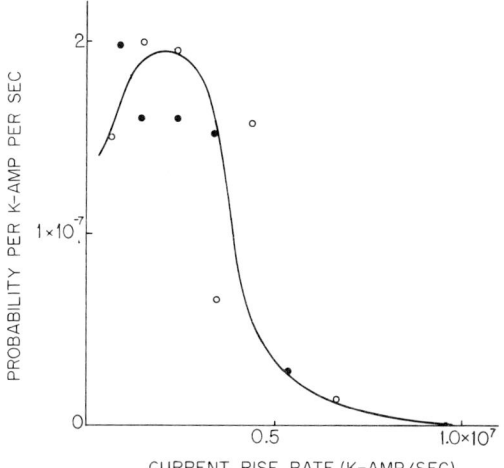

Fig. 11. Probability of a given maximum current rise rate. Cloud positive. ○, down; ●, up. [Data by Berger (14).]

from 0.004 to 0.025 C for the four cases. The cause seems likely to reside in the electrode processes by which the return wave is launched.

Finally, data (Fig. 12) taken on the number of strokes per flash show that the preponderance of flashes is single. Not shown by this graph, which distinguishes strokes only by leader direction, is the fact that strokes with positive clouds are essentially 100% single, so that Berger's data are slightly weighted toward singles, as is shown by comparison with Schonland's results, which are purely negative cloud-to-ground discharges. The data of Fig. 12 may have a slight bearing on our topic of concern, indicating possibly the relative efficiency of clouds as anode and cathode structures, but even more likely are of interest chiefly to students of the electrification process.

5. Dart Leaders

Evidently because of the clarity with which it stands out, largely free from the complexity of branching and relatively intense, the wave which rekindles the dying arc channel for a subsequent stroke has received more precise study than the stepped leader. The distribution in velocity observed by Schonland is given in Fig. 13. There is evidently a sharp lower limit on the wave velocity, no waves having been observed by either Schonland or Brook et al. (28) below 1.9×10^6 m/sec which were not artificially slowed by being partially stepped at their lower ends.

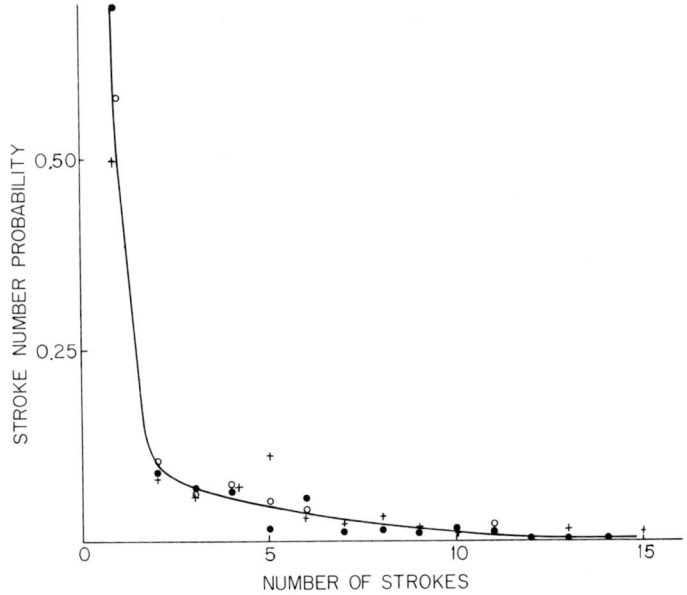

FIG. 12. Probability of a flash with n strokes. ○, down; ●, up [data by Berger (14)]; +, negative cloud, down [data by Schonland (4)].

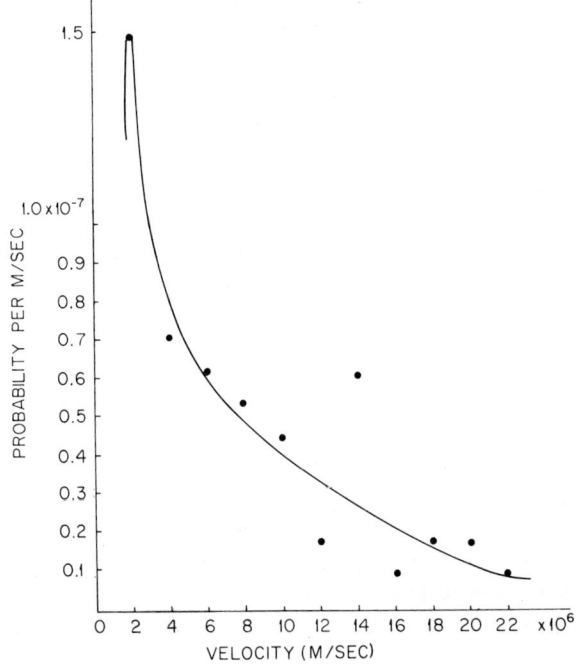

FIG. 13. Probability of a dart leader velocity in the range v, dv. Cloud negative. [Data by Schonland (4).]

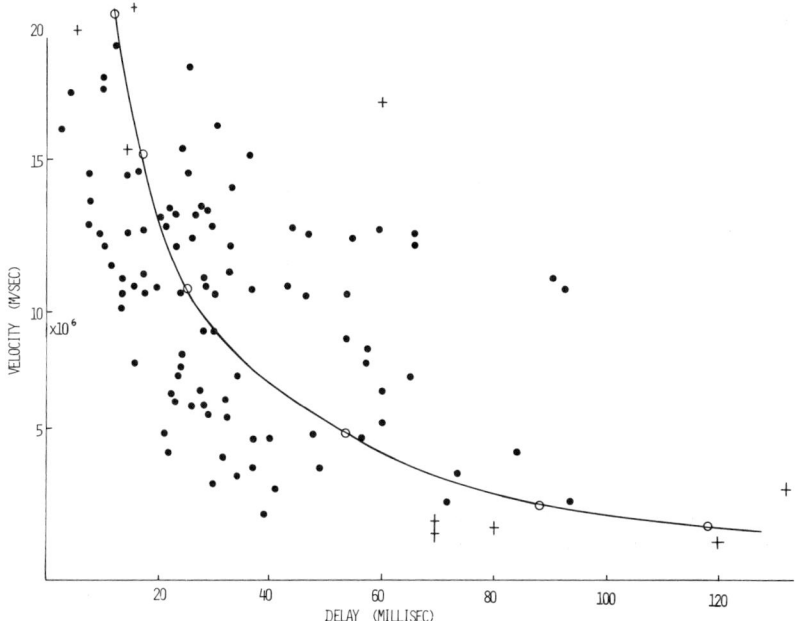

FIG. 14. Velocity of dart leaders as a function of channel age. ●, Brook et al.; +, Schonland; ○, recombination curve. [After Brook et al. (28), with permission of Journal of Geophysical Research; data by Schonland (4).]

The velocity of the dart leader is a function of the age of the channel at the moment of initiation. Data on this by Brook et al. and by Schonland are given in Fig. 14. As would be expected, other factors also enter in this effect, but except for one aberrant point by Schonland a strong trend is shown. The cause of the relation lies in either the electron concentration or temperature in the channel, or both. The former is the more probable, and if one assumes that velocity is proportional to electron concentration and electron loss is by recombination, the time dependence could follow the curve shown on the figure which has been fitted to the data centroid.

The brightest return stroke in a flash is usually, but not always, the first stroke, but in three of the four cases noted by Schonland the dart leader which preceded the inordinately bright stroke was partially stepped and followed an exceptional delay between strokes. A statistical distribution of time intervals between strokes is given in Fig. 15. Whether the delay is caused by preparation of the channel for renewed wave propagation or relates to problems of charge collection in the cloud is not immediately clear. The absence of multiple strokes with positive clouds tends to indicate the latter. Another bit of information which bears on this relates to the length of

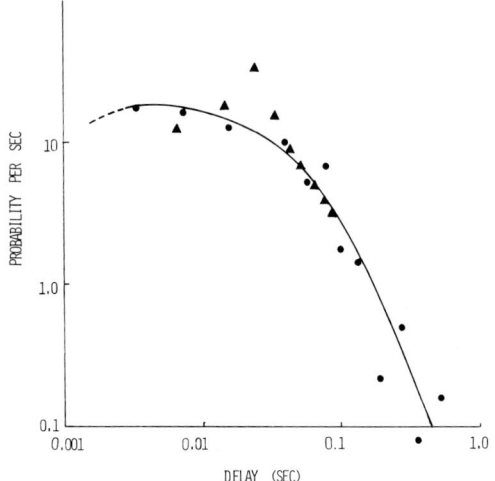

Fig. 15. Probability of time delay between strokes in the range t, dt. ▲, Kittagawa et al.; ●, Schonland. [Data by Schonland (4), and Kittagawa et al. (28).]

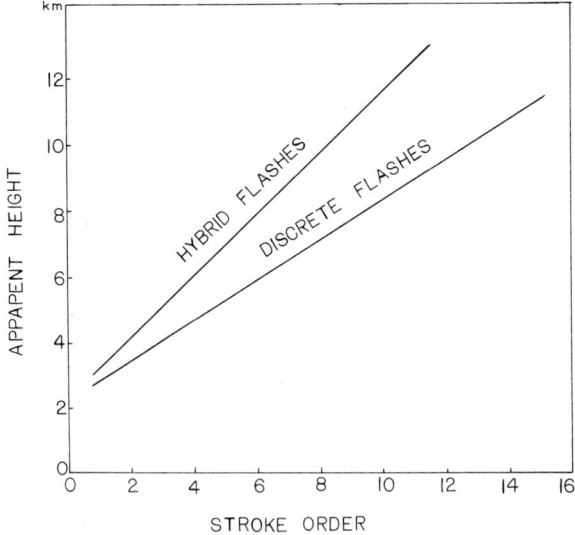

Fig. 16. Apparent height vs. stroke order for strokes in discrete and hybrid (continuing current) flashes.

each stroke. In general the height from which dart leaders come increases steadily with stroke number. Data obtained on this effect by Brook *et al.* are used in Fig. 16. These data indicate that the charge center from which successive strokes come moves up an average of 0.5 km in an average interval of 0.02 sec if there is no current between strokes, and 1.0 km if there is.

Electrical measurements further show that the charge delivered to the channel by the leader is almost exactly drained off subsequently by the return stroke process.

The term dart leader was chosen because the wave front which progresses down the column has a finite propagating thickness of about 50 m. Schonland notes that this length, like that of the return stroke, is consistent with the relaxation time of excited states and the wave velocity.

No definite observations on the diameter of dart leaders seem to exist. Uman and Voshall (29) have made some calculations of the growth in size and the cooling rate of the return stroke column which show that the time interval between multiple strokes may indeed be determined by the need to cool the column. In the course of this he estimates that the column has expanded by a factor of 10 at the time a dart leader is launched.

6. *M Components*

In the middle of the continuing luminosity following a return stroke Schonland observed that the channel is occasionally relit once or more by waves moving from negative cloud to ground. These were accompanied by an identifiable change in the electric field which was termed a "hook process" from the shape of the oscillogram. This pattern indicated that a small amount of negative charge moved down and distributed itself along the channel, and afterward drained on down to the ground. It is apparently an amount of charge insufficient to cause a return stroke, being about 0.01 to 0.1 of that in return strokes. The longer the interval between the return stroke and the M component, the longer the duration of the whole M process, presumably showing that more charge comes down when more time is allowed for it to build up. Velocities of M components do not seem to have been measured, but if we assume that most of the time spent in the hook process is wave transit time, the velocity is bounded by 1.5×10^6 and 5×10^7 m/sec, with an average value of 8×10^6 m/sec. Since this value lies well in the middle of the dart leader range, it seems reasonable to consider the M component as a wave of similar initial mechanisms at the cloud end, but one whose normal return stroke is strongly modified by the high channel conductivity prevailing.

7. Conclusion

The active research program which is currently going on will certainly add new and more accurate information to these many blank areas now existing. Much of the difficulty in our understanding of lightning centers around a popular thirst for knowledge on the subject, which has resulted in widespread premature distribution in vehicles such as encyclopedias and thence a diffusion into elementary books of information that was based on hasty conjecture. Typical of this is some early and excellent observational work of Simpson (30), which was accompanied by the unfortunate guess, derived from contemporary inexperience with electron plasma properties, that the normal cloud charge was positive, and the normal direction of charge motion was from ground to cloud. The confusion this guess created has still not been wholly cleared away.

B. Solar Phenomena

1. General Remarks

Conventional theory in astrophysics historically has rejected electric fields arising from charge separation as having no importance on the sun owing to its great conductivity (31). The success of the local thermodynamic equilibrium theory in describing the low resolution gross detail behavior of the sun gave credence to this. Higher resolution, space exploration, and the need to take seriously such things as prominences, sunspots, flares, faculae, granules, etc. has led to abandonment of a rigid application of local thermodynamic equilibrium and has brought plasmas and magnetic fields into the realm of orthodoxy under the rubric of magnetohydrodynamics (MHD). Much ingenuity has often been resorted to in explaining (without invoking charge separation) effects which seem morphologically parallel to terrestrial ones where charge separation fields are known to be responsible. Bruce (32) has enumerated and discussed many of these situations, pointing out that with a slightly different model of the subphotospheric structure even the mechanisms of triboelectrification might be operative on the sun to produce breakdown analogous to lightning. Although the cool subsurface region needed to permit Bruce's electrification process is certainly unacceptable to most astrophysicists today, there are at least two strata of relatively low electrical conductivity in the sun's atmosphere (Fig. 17), one just above and one just below the photosphere. It has been suggested (33,34) that turbulent motions of partially ionized gas in the sun's magnetic field acting as homopolar generators could generate large potential differences which can break through these layers to provide electrical arcs and even pinches.

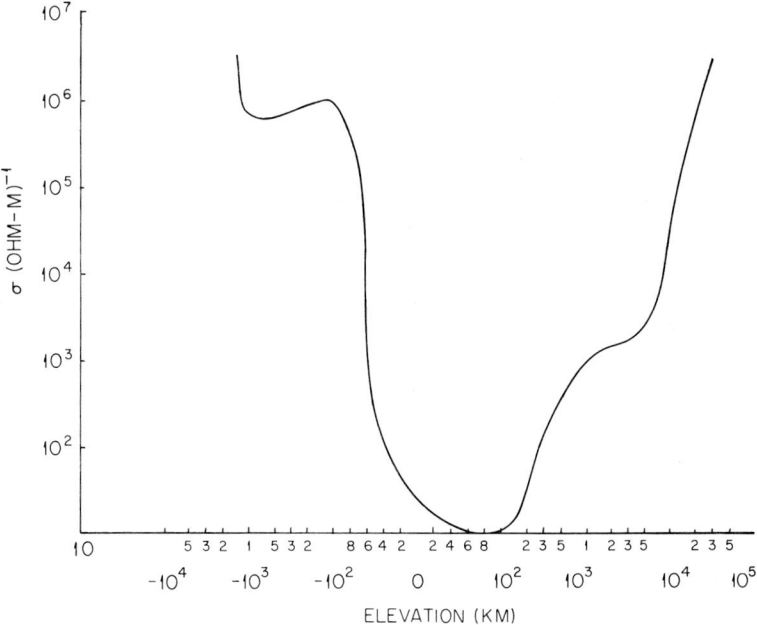

FIG. 17. Conductivity of the solar atmosphere, above and below top of photosphere.

Whether or not the phenomena in question are indeed electric in origin, it will be proper to review them here as phenomena which could be of this nature, and for which this hypothesis should therefore be kept in mind.

2. *Prominences*

Great luminous arches extending sometimes as much as a solar radius into space span the sun frequently, forming in times of the order of an hour and disappearing as quickly. At another extreme, vertical curtains of luminosity may seem to hang over the sun's surface for periods of months. Superficially the arches resemble large toruses of constant minor diameter which emerge from the surface and expand in major diameter as their center rises simultaneously. The largest one on record appeared sometime between May 10 and 23, 1946 on the far side of the sun. As it rotated into view it was a mass of interlacing streamers in an arch 280,000 km long and 112,000 km high. It rose slowly if at all in altitude until about 15:30 GMT June 4, when it suddenly began ascending, and vanished from the top at the end of its expansion shortly after 17:00 GMT. Its radial growth speed reached 3×10^5 m/sec. During the growth the outer circle of the torus remained essentially rooted on the surface at two points 1.1 solar radii apart. Toward the middle of its expansion the coherence of the torus clearly broke down into a hollow wickerwork of helical filaments, each about 1.5×10^7 m in diameter. Since the outwardly moving structure was not a spherical shell, it cannot have been composed of radially moving matter, but rather must have involved motion along the torus. A popular explanation is that plasma explosively injected into a tube of magnetic lines carries those lines outward as it expands. One major objection to a plasma injection explanation is that Doppler shifts of the order of 10 Å might be expected and none even as great as 1 Å have been observed, although small shifts do exist (~ 0.1 Å). To anyone experienced in electric technology, however, the observation of Bruce, that the prominence resembles nothing so much as an enormous arc discharge expanding away from its return conductor and extinguishing itself when its expansion becomes too great, will be appealing, and De (35) has discussed such a model. The anchoring of the outer roots of the arch and the dissimilarity between the two ends, one of which nestles on the solar surface, while the other stands almost cut off at a respectable distance, are strongly suggestive of anode and cathode behavior. In this view, the arc forms following breakdown across the solar surface between points at high potential difference, and lifts off the surface. Magnetic fields of 10^{-2} tesla have been observed, and if these are caused by the current filaments, they indicate currents of 10^{12} A. The circular section of the torus is caused by mutual repulsion of the helical current filaments of which it is composed. As far, however, as the electron fluid dynamic waves are concerned, only the transitional accelerated phase, when the 7×10^8 meter distance from anode to cathode must be bridged by a current augmenting wave in a matter of minutes, seems to have any bearing. Since the medium is already partially ionized, a dart leader would be the appropriate pattern of initial conduction, and a dart leader would have adequate velocity. At such a high initial level of ionization, times as short as 5 minutes could see the gap bridged. It is also possible that wave processes are present in the less well-defined initial period of formation of the prominences.

Further evidence for the existence of electric currents in prominences is shown by the forces which pairs of them exert upon each other from remote distances, and by the radial acceleration of 100 m/sec^2 which the great prominence of 1946 displayed up until the moment of its optical demise. An alternative to the breakdown and return stroke hypothesis which preserves the apparent electric arc structure of the prominence is an induction loop of current which somehow stands on its side and rises out of the solar surface. This would not need an aspect of electric charge separation and breakdown. Perhaps the chief objection to this proposal is the definite collector–emitter type asymmetry of the two ends of the prominence.

3. Surges

Surges are outward growing luminous columns associated with flares. Giovanelli and McCabe (36) have observed velocities which average at $1 \times 10^5 \pm 0.5 \times 10^5$ m/sec. One exceptional value which could have been as large as 1×10^6 was observed, and is not included above. Because of the problem of steric projection, these values should probably be increased by 30%. If they are themselves an electrohydrodynamic (EHD) phenomenon, and their indicated velocities barely permit it, then they must be a single-ended breakdown to a generalized ground conductor. They may equally well be plasma ejected from the pressure mechanism of the flare, which might be a current pinch. Pinches are well known to eject matter from their ends. Since the surges are seen to decelerate as they move outwards, an injection mechanism seems quite likely as a cause, but even that may be initially breakdown-related.

4. Lateral Waves

Moreton (37) and Athay and Moreton (38) have observed that luminosity patterns and luminosity-producing disturbances move across the face of the sun. These have been recognized at velocities of between 3×10^5 and 3.0×10^6 m/sec. They are associated with flares and move off in well-defined directions, which suggests that they involve moving charged matter guided by magnetic fields. The speeds are very large for plasma motions, however. If the plasma were, for example, ejected from an electric shock tube, the driver temperature would need to be $10^{9}\,^\circ$K. They are within the range of electron fluid dynamic waves and would not require excessively strong fields to initiate ($E/p \sim 4 \times 10^4$ V/m/torr) at the low pressures prevailing in the solar chromosphere.

5. Bursts

Bursts are solar radio noise manifestations involving electrons with sharply defined fronts moving outward from the photosphere at speeds ranging from 0.2 to 0.8 of the speed of light (39). Weiss and Wilde (40) propose that they result from sudden cancellation of opposing magnetic fields driven together by fluid motion (Sweet's mechanism). Beyond a demonstration of feasibility with respect to considerations of energy and magnetic escape windows, the acceleration mechanism is somewhat uncertain, and might again involve either the generation of an electron fluid wave or a simple particle acceleration by a moving field.

6. *Solar Wind Shocks*

In Part I of this review (*1*) it was noted that the shocks in the solar wind observed by Burlaga and Ness (*41*) might be of EHD character because of their speed (400 km/sec) and the temperature difference observed across the shock. Recently Burlaga and Scudder (*42*) have shown that the structure as seen in space is well described by the MHD theory advanced by Parker (*43*) for Sweet's mechanism, and hence the need for EHD theory can be eliminated.

C. *Ionospheric Effects*

1. *General*

The ionosphere is highly conducting, but only partially ionized, and stratified in conductivity. It grades above into the rarefied but still more highly conducting solar wind, and is bounded below by the nonconducting atmosphere. It is strongly controlled by the earth's magnetic field. Above 90 km the electrons undergo many cyclotron orbits between collisions. Above 200 km all charged particles are in long-term cyclotron orbits. The ionosphere is affected from above by shock waves from the sun, and from below by the tops of strong thunderstorms, which induce image charges into the ionospheric layer that undergo rapid changes as lightning discharges change the charge pattern in the cloud.

2. *Electrostatic Waves*

Because of its impact on the electromagnetic signal transmission problem, the electrical liveliness of the ionosphere has long been a subject of study. Most of the disturbances which propagate through it are of electromagnetic, magnetoacoustic, or Alfvén wave nature, but observations with satellites by MacPherson and Koons (*44*) have detected strong electrostatic waves in the VLF spectrum recently also. Their observations are summarized in Fig. 18. These waves are always and only observed near the South pole (invariant latitudes 70°–90°), except during periods of geomagnetic activity, when they are also present at between 65° and 72.5°. Vampola *et al.* (*45*) have shown that they exchange energy with energetic electrons (300 keV) in the process of scattering and trapping these electrons. It is believed that they are permanent features of the southern polar regions because it is here that the thermal plasma escapes into interplanetary space along the tail of the magnetosphere. Koons *et al.* (*46*) have shown that the relative velocities of H^+ and O^+ ions will make the thermal plasma unstable

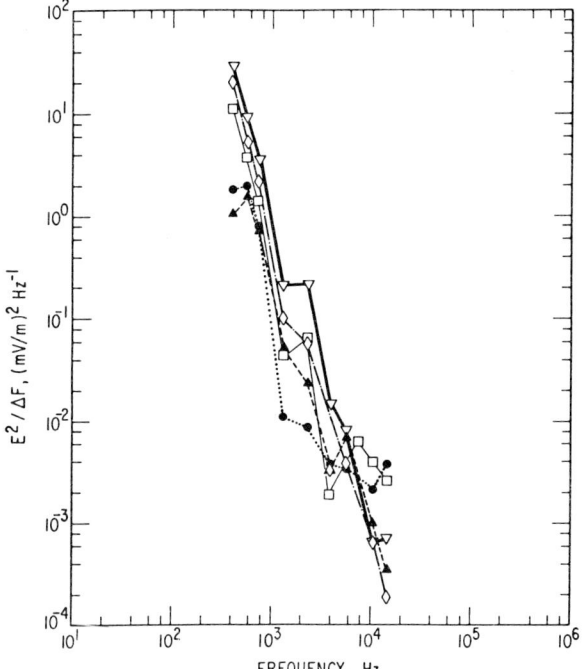

FIG. 18. Frequency dependence of the spectral density of electrostatic waves for five satellite polar passes. [After McPherson and Koons (44), with permission of *Journal of Geophysical Research*.]

to ion acoustic waves. Electron temperature in the magnetosphere is not firmly known, but that of the solar wind is given as 2×10^{5}°K. On this basis the frequency-determining wavelength of ion acoustic waves would range from 5 to 5×10^2 m, while that for electron acoustic waves would be 2×10^2 to 2×10^4 m. The characteristic lengths involved would seem too short to be physical in the ion acoustic case.

3. *Starfish Phase B Signals*

Following the Johnston Island "Starfish" high altitude hydrogen bomb explosion July 9, 1962, observing stations around the world reported very large amplitude signals, observed in a variety of ways, which were received with an elapsed time of the order of seconds after detonation. In many if not most cases this time was within the error in the local fiducial calibration, and often also within the error of reading of the records. This situation came about because almost no one had anticipated disturbances that would propagate on a time scale below tens of minutes except for possible prompt electromagnetic signals of no great interest. (An electromagnetic signal was indeed received, as were the expected slow signals accompanying pressure waves, Rayleigh waves, bomb debris outflow, Alfvén waves, etc.) Many observers therefore estimated an instant of first detection for this unexpected signal, which they usually gave as two or three seconds, without intending that it should be taken as a measurement. Thus, for example, on an ordinary quick-run magnetogram the fiducial marks are 3 sec wide, while the recording trace varies from 0.6 sec to 6 sec width depending on the rate of change of the signal. While some convention, such as reading the trace center, might improve the precision of data beyond these limits, it is unlikely that a retrospective calibration accuracy of better than 1–2 sec can be expected. Other observers did indeed have adequate fiducial accuracy, but seem to leave the implication that they were observing the arrival of the event at their station.

The recurrence of the value $2(\pm 2)$ sec as the reported arrival time of the signal at many stations around the world prompted Roquet *et al.* (47) to suggest that the whole phenomenon somehow took place at a worldwide instant. This idea seems at variance with the total information available, however, some of which has not been considered in previous analyses such as that of Caner (48), who termed this signal the Phase B signal (Phase A was the electromagnetic wave). In his extended report, he deals carefully with the strengths and weaknesses of the numerous theories offered in explanation, finding none wholly satisfactory, but finally favoring some kind of MHD wave coupled with electromagnetic conversion.

Caner sites the basic evidence which a theory must fulfill thus: (a) global synchronism (± 0.1–0.2 sec); (b) 2-sec delay after explosion (± 0.1–0.2 sec); (c) extremely sharp rise (over 20–30 gamma/sec); (d) initial period 3.5–4 sec (i.e., frequency 0.28–0.25 Hz); (e) amplitude heavily damped; (f) large initial amplitude (about 30–20 gamma); (g) altitude dependence (occurs only when a source is above the E layer); (h) period decreasing to about 2–2.5 sec; (i) horizontal polarization of the magnetic vector. Of these one can feel confident of the interpretations which led to conclusions (c), (f), (g), and (i). Conclusion (a), global synchronism, relates only to the instant of initial upturn of the magnetic field observed at four stations using very sensitive equipment, so sensitive in fact that all may well have been detecting the instant at which the disturbance formed in the vicinity of Johnston Island, which naturally was a worldwide instant, electromagnetically speaking. The interpretation given to item (b) is variable with author, but the effect seems most easily traced to a switching time as the cloud of bomb debris expanded and developed its charge separation and then made electrical connection between ionospheric layers. Facts (d), (e), and (h) are probably real, but they might also arise from the normal nonlinear response of an overdamped, overdriven electronic observing system, which when subjected to an intense brief pulse can be driven off its base line and will shift its apparent response frequency downward at the same time. The signal shown by the fluxgate magnetometer used by Caner could almost be a textbook example of this effect. This, however, seems unlikely to have been the case since nearly every station in the world, whatever its characteristics, saw rather similar effects. Any process proposed to explain the observed phenomena should therefore take account of all the observations as real.

It seems plausible that an electron fluid dynamical wave propagating in the ionosphere could explain most of these effects successfully (49). We will therefore present here the additional reported evidence that can be adduced for a progressive wave, as opposed to the concept of a worldwide instant.

Thus, at Brisbane, Bowman and Mainstone (50) found that ionosonde reflection from the F_2 layer was cut off abruptly at 9 : 09 : 12.8 sec $\pm$ 0.3 sec. This seems definite indication of the overhead arrival of some ionization. At nearly the same instant they also observed a solitary micropulsation. Even more striking is the result that can be obtained from study of the records of Sheridan and Joisce (51), taken with a radiospectrograph. This equipment monitored continuously all the transmitters (locations unknown) detectable at Sydney, from 5 to 40 MHz. Examination of this record shows that these dropped out of detection *at different instants over a time interval* of 2.0 $\pm$ 0.5 sec. Some strong local stations were properly unaffected at any time. This differential fading can most easily be interpreted as the development of reflection below the F_2 layer outward from Johnston Island to a distance at which all paths from the major sources of nighttime transmission were blanketed. While it is difficult to estimate this distance without knowing the transmitter locations, the geography of the night side of the earth at the moment in question suggests that the distance from Johnston Island to a point halfway between Sydney and Perth, Australia is the most likely critical distance (8000 km) spanned in this 2-sec interval.

Finally, one can attempt to make use of the data from the records of quick-run magnetometers, rubidium magnetometers, and micropulsation recorders. As previously remarked, it is not the point of initial onset of the magnetic signal which indicates the point of nearest approach of the moving charge zone (wave), but rather some point near the maximum. The point of onset depends purely on the amplification used, and is propagated at light speed. When, additionally, one considers that in most observations the maximum was badly off scale, no deductions better than ± 2 sec can be made of arrival time from the best magnetometer data. Even then, there is no certainty that the point of nearest approach is directly above the station, especially when one considers that in general the only records on which measurements could be made (because they did not go off scale) were

Z records. The discrepancies between excellently timed records such as those of Christchurch (England), State College (Pennsylvania), and Victoria (British Columbia), suggest strongly that one or more breakdown channels circled the world, passing these stations at different distances. All of the data have been publicly reported previously (52), but generally without analysis. In Fig. 19 the data chosen have been restricted for a presumed relevance and reliability. Some of the seemingly most reliable of the data are the Honolulu

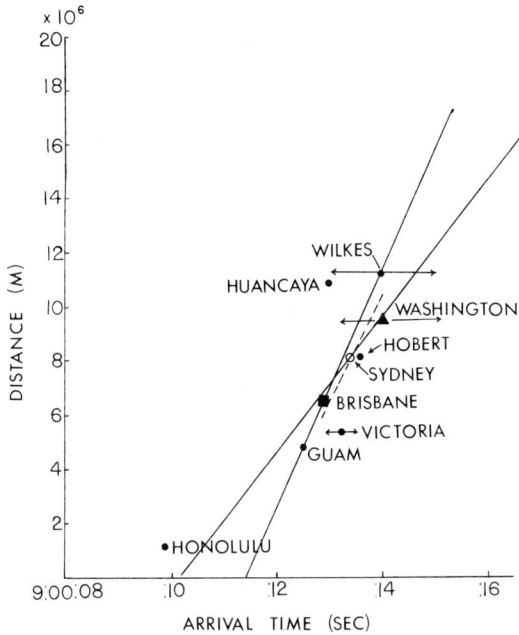

FIG. 19. Observations of possible prompt waves following Starfish high altitude explosion. Errors are as shown except for quick run magnetometer data, solid dots, which are unknown, and may be as large as ±2 sec.

and Guam quick-run magnetogram Z maxima, the Wilkes (Antarctica) quick-run H maximum, the Victoria H maximum run with a broad-band detector, the Hobart (53) micropulsation saturation onset, and the Huancaya (54) low channel B measurement. The results of Sheridan indicate a velocity, but where the line corresponding to it is to be placed on the time–distance chart depends on the assumed starting time for the event, since Sheridan's fiducial accuracy is not sufficient to determine it. We have assumed 9 : 00 : 11.4 as the starting time. We see that a line through the Brisbane point and 9 : 00 : 11.4 fulfills the slope requirements imposed by Sheridan's result, and is a reasonable fit to the other data. This line indicates a wave velocity of 4.5×10^6 m sec. Such a choice makes the Honolulu observation too early, and it is necessary to assume either that it is insufficiently calibrated, or that the most remote station received by Sheridan was not near Johnston Island. A compromise velocity might then be 2.6×10^6 m/sec.

Davis and Headrick (55) also detected a radial motion in the generally overhead E layer with good timing at $9:00:14 \pm 1$ at Washington, D.C., which was repeated at $9:00:28$ and $9:00:42$.

Electric fields may, in theory and practice, support breakdown waves with small currents in the region behind the wave. When present they will be accompanied by an increase of electron temperature. At Jicamarca, Peru, an increase in antenna temperature to $4.5 \times 10^{4\circ}$K was observed for 500 sec "immediately" following the explosion (56). At Armidale, Australia, a similar increase in thermal noise at 9.8 MHz was observed from $19:00:45$ until $19:05:52$ and indicated an antenna temperature of about $8 \times 10^{4\circ}$K (57). The onset of this signal at $19:00:45$ coincided with the restoration of 2.28 MHz reflection from the F_2 layer of the ionosphere at Brisbane, so that the source may have been overhead for some time, and merely invisible to the receiver. Temperatures such as these are produced by E/p ratios of about 10^4 V/m/torr.

III. Laboratory Experimentation

A. Primary Breakdown Waves

1. General Remarks

The investigation of propagating breakdown of a neutral gas is best conducted at low pressure where the scale of the active layers is large enough to offer a chance of resolution. This advantage is somewhat offset by the role that the necessary tube wall then plays in bounding the discharge, making the low pressure situation nonsimilar to high pressure breakdown. The work of Scott (58) which follows shows that this limitation may, however, be used to reveal new information about wave controlling factors.

No further experimental work has yet been done on secondary breakdown waves.

2. New Results

The basic relation between velocity and electric field at the wave front is the easiest point of contact between experiment and theory, and even here the agreement has not been outstanding. Extension of the range-limited work of Haberstich (59) by Blais and Fowler (60) suffered from assumptions concerning the relation between the potential applied to the launching electrode and the electric field at the front. They also assumed that they could extrapolate the decreasing velocities observed along the wave advance back to a value presumed to be prevailing at the electrode at the time the potential was known there.

Using the same apparatus as Blais, Scott undertook direct simultaneous measurements of the fields at the front of the wave as a function of the wave velocity. His effort involved calibration of a probe extending in 0.015 m from the sidewall of the ground cage using an open-ended full-size simulation of the Blais–Haberstich apparatus against the readings of field made with an axial probe thrust in from the open end. The wave and supporting plasma were simulated by metal cylinders of various diameters with ends of various shapes which were inserted halfway into the simulated wave tube. Known pulsed voltages were then applied to the cylinder. The result was a total calibration of the wall probe and its accompanying impedance matching local amplifier, which was uniquely developed to handle either positive or negative potentials at the nanosecond risetimes present in the experiments without the serious distortion otherwise present owing to impedance mismatch between probe and coaxial cable.

Scott also increased the voltage capabilities of the power supply to 100 kV by changing to the Marx circuit originally used by Beams. This had the added advantage of reducing the inductance of the spark channel in the spark-gap switches. The general intent of the research was to verify the residual pressure dependences found by Blais in the relation between wave velocity and reduced field (E/p), and to extend the results to other gases. Scott found that both the small residual pressure dependence of proforce waves and the exaggerated residual dependence of antiforce waves were again observed, in argon and in nitrogen as well as helium, and that the theoretical calibration used by Blais for his helium data was not seriously in error. Scott's data are given in Figs. 20–23.

Three features previously reported by Blais were observed. First, the pressure dependence of the wave speeds is analogous to the Paschen curve, which does not, however, imply that the two phenomena are related, but only that for a given field strength there is a pressure at which maximum speed occurs. The peak pressure for both argon and nitrogen was around 1.0 torr, in contrast to a pressure of 3.0 torr for helium as reported by Blais. Second, at very low pressures in the antiforce case the wave speed approaches a constant value independent of the electric field. Finally, a stepped leader phenomenon was observed in the time of flight measurements which resulted in some uncertainty in the wave speeds. This phenomenon was most pronounced at pressures below a few torr. It was not specifically investigated.

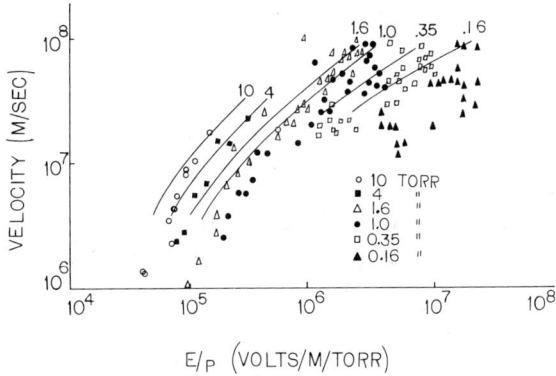

FIG. 20. Proforce wave velocity vs. E/p for nitrogen. Curves are theoretical; see Section IV. [After Scott and Fowler (58) with permission of *Physics of Fluids*.]

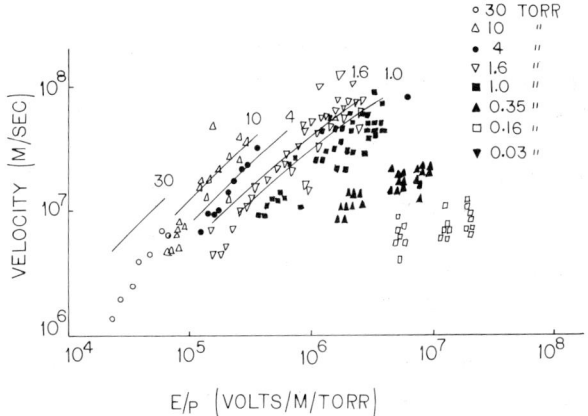

FIG. 21. Antiforce wave velocity vs. E/p for nitrogen. Curves are theoretical; see Section IV. [After Scott and Fowler (58) with permission of *Physics of Fluids*.]

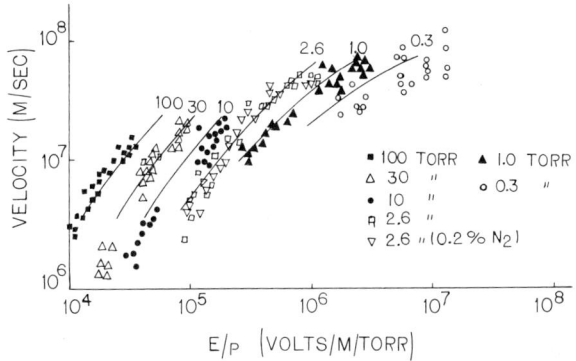

FIG. 22. Proforce wave velocity vs. E/p for argon. Curves are theoretical; see Section IV. [After Scott and Fowler (58) with permission of *Physics of Fluids*.]

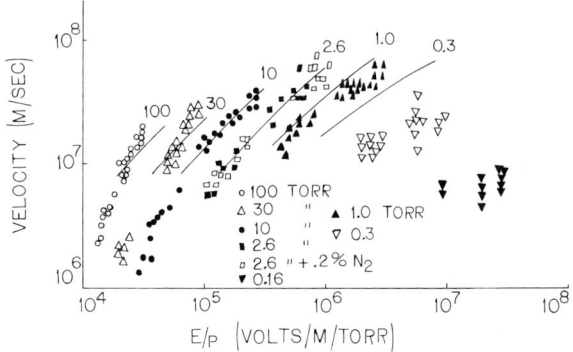

FIG. 23. Antiforce wave velocity vs. E/p for argon. Curves are theoretical; see Section IV. [After Scott and Fowler (58) with permission of *Physics of Fluids*.]

Three new effects were observed in these experiments. First, there is an upper limit for the wave speed on the order of 10^{10} cm/sec. The exact value was not ascertained because of the statistical scatter of the data. Second, at low pressures, for negative applied voltages, two progressive phenomena were clearly distinguished by both the electrostatic probe and the photomultipliers. The second phenomenon was easily identified by its sharp front as the strong proforce wave described by Shelton. This was preceded by a phenomenon characterized by a gradual buildup in potential and light intensity. This may be the weak proforce wave postulated by Fowler and Shelton (61) or a Townsend avalanche of electrons. It is quite possible that above the Shelton minimum velocity this phenomenon always undergoes transition to the strong proforce wave at some point in time which may occur too early or too late to be observed in the 7-m-long apparatus. However, for positive applied voltages only one phenomenon was observed which is consistent with the Shelton theory that denies the existence of strong antiforce waves. Finally, there was a distinct difference in the range of pressures for which wave initiation occurred in argon and in nitrogen. The pressure range for argon was from 0.1 torr to 100.0 torr in the antiforce case and from 0.3 torr to 100.0 torr in the proforce case. The range for nitrogen was from 0.03 torr to 30.0 torr in the antiforce case and 0.16 torr to 10.0 torr in the proforce case. In the proforce case, the minimum pressure was the lowest pressure at which strong waves could be recognized in this apparatus.

B. Microwave and Laser Breakdown

1. General Remarks

Microwave breakdown of gases, similar in nature to breakdown processes at any high frequency, has been known for many years (62). To this laser breakdown has recently been added. The directiveness and pervasiveness of microwave and laser energy give the illusion of controlled and simultaneous breakdown at all points, so that any propagating mechanisms have generally been obscured or ignored. Koopman (see Part I) has detected propagation in the laser breakdown, an effect which is at present generally restricted to focused laser energy, but which should become increasingly noticeable as longer and longer laser beams of higher and higher power are projected through the atmosphere, and must already exist in the internal breakdown of lasers which occasionally occurs.

2. *Microwave Breakdown*

Bethke and Reuss (63) discovered a situation in which propagating breakdown in a microwave field is clearly delineated, and Dawson and Lederman (64) have investigated its properties. The experiment consisted of reflecting microwaves off the advancing ionized shock wave in a conventional pressure-driven 0.025 m diam. shock tube used as a microwave guide. It was an outgrowth of the use of microwaves to probe the advancing shock front. At low energies a mere enhancement of the diffusion of electrons in

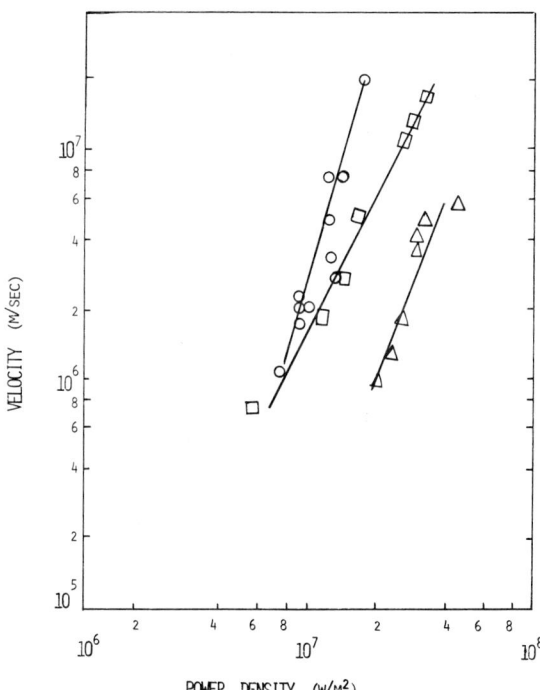

FIG. 24. Ionization front velocity as a function of input power flux. ○, 3 torr; □, 1 torr; △, 0.5 torr. [After Dawson and Lederman (64) with permission of *Physics of Fluids*.]

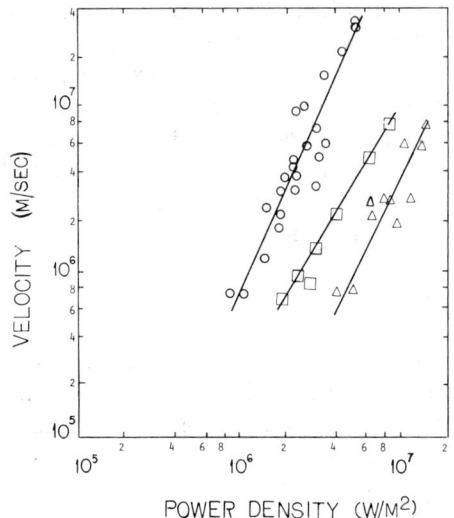

FIG. 25. Ionization front velocity as a function of power flux. ○, 3 torr; △, 0.5 torr; □, 1 torr. [After Dawson and Lederman (*64*) with permission of *Physics of Fluids*.]

front of the shock wave is observed, but above a critical energy a propagating ionization front takes over. The experimenters observed the wave with both photomultiplier stations and the microwave reflection probe itself. Their results for argon and air are given in Figs. 24 and 25. Because the microwave guide was operated in the TE mode, the forced diffusion of electrons to the walls at decreasing density results in greatly increased power for breakdown. This is similar to the situation found by Scott in steady field breakdown at pressures low enough for free diffusion.

3. *Laser Breakdown*

Breakdown with formation of a perceptible spark is commonly exhibited by focusing high power lasers in a gas, usually against a target. Pan *et al.* (*65*) have given an energy criterion for the threshold of breakdown. No one seems as yet to have investigated quantitatively the progress of breakdown itself. Nielsen and Canavan (*66*) have calculated the time required for electron cascade breakdown as a function of power flux.

C. *Charge Injection Waves*

Charge can be forcibly injected into a gas inside a Faraday cage from outside. If the energies are low the advance of the ionizing front can be quite slow (10^4–10^5 m/sec), and although a front is well defined, it does not seem to be describable by the same model and equations as the other waves in this class. Russell and Holzberlein (67) called these onset waves and investigated them for electron injection in a cylindrical geometry. They found that a simple compilation of electron trajectories into and out of the field free space described the advance of the luminous front. A sample of the result is given in Fig. 26.

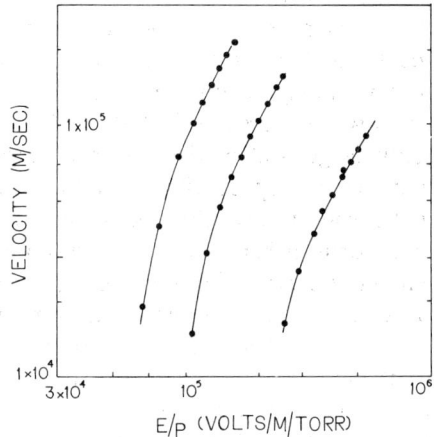

FIG. 26. Onset wave velocities as a function of E/p. [After Russell and Holzberlein (67).]

When the energies are high, breakdown of the gas in the avalanche mode can be added to the beam collisions that govern the low energy case. Thus, for example, when a relativistic electron beam at 6×10^6 V and 10^3 A is injected into a gas ranging from 0.2 to 100 torr in density, the beam is surrounded with a radially growing avalanche process (68). Very little information has been reported on the temporal behavior of this growth, and whether it ever intensifies enough to go into the self-sustained propagation mode is unknown, although the reported radial fields of 10^6 V/m should sustain waves. Some of the theoretical aspects of these beams, which might present interesting bow wave phenomena, have been discussed by Benford (69).

D. Avalanche Processes

1. General Remarks

J. S. Townsend formulated the concept that in a sufficiently strong uniform electric field, one electron begets an exponentially growing family of electrons in a spatially distributed array of ionizing events. In the early stages of this growth process the electrons are largely bunched into a near spherical swarm that moves along at electron drift velocity, and the positive ions are left behind in a conical volume whose base, coincident with the swarm, is where the growth is taking place. Such is the structure of a Townsend avalanche as delineated by the cloud chamber pictures of Raether (70). Raether has shown that this description of the avalanche growth process holds up to a critical number of electrons in the avalanche and then a transition sets in to a new process in which the front of the growth progresses at a significantly higher velocity than drift velocity in the uniform field, and becomes increasingly abrupt and well-defined in onset. The explanation which is now widely accepted is that the development of a plasma condition at sufficiently large electron densities results from a self-consistent redistribution of the electron space charge in the avalanche to provide locally the extra field needed to give the augmented velocity.

If the field is not uniform, as in a point-to-plane geometry, the weakening field as the avalanche moves away (from the point) results in a suppression of the growth process when the field nears the threshold value, so that avalanches may move outward and then dissipate. The phenomenon is termed a *corona*. As the process is repeated, a mass of ionization builds up around the point which can eventually organize itself into a self-propagating space charge structure, and breakdown can then proceed from the point to the plane. If the original field is sufficiently above threshold, the Raether transition to a self-propagating cloud will be possible before a field too weak for growth is reached, and then the self-propagating cloud may by itself connect point to plane in a breakdown. Such a geometry, while of great interest in problems of practical breakdown, where it usually occurs because points and projections provide locally strong fields, is not a useful one for inquiry into the fundamental problems of avalanche growth.

A detailed inquiry into avalanche processes is not necessary for this review, primarily restricted as it is to the self-consistent field propagation phase of ionization growth, but since every such wave must be traceable somewhere to an avalanche inception, this starting phase deserves limited attention.

2. *Normal Avalanches*

A normal Townsend avalanche grows from a single electron at a constant drift velocity governed by E/p. As the avalanche advances, it grows laterally as well, so that the resulting array has a conical shape. It is customary to treat this growth as governed by the electron diffusion equation with an initial point singularity. The number of generations is not large, however, and it is more appropriate to recognize that in a frame at rest with respect to the electron drift, the electron swarm must enlarge spherically as a consequence of random walks. The slowly moving ions then lie within an elongated cone whose axis runs from the location of the first ionizing impact to the instantaneous swarm center. The visualization of this cone, as accomplished by Raether in his cloud chamber photographs, depends on a sufficiency of drop density. The number of charges needed to form a drop is a controllable parameter, so that the cone is actually visualized in terms of a contour of *constant charge density*.

The cone angle θ is a function of E/p and the nature of the gas, and sample data are given in Fig. 27. To measure it is by no means easy, as the

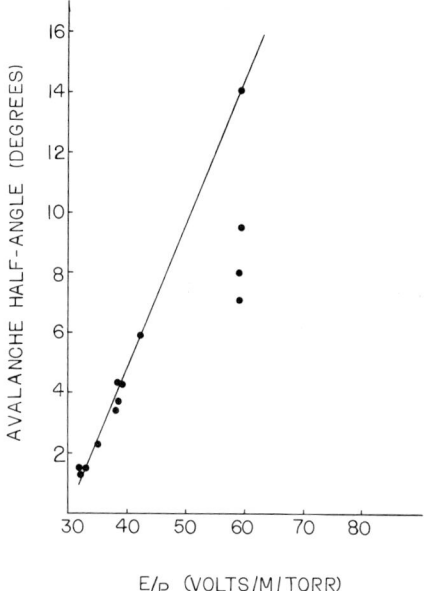

FIG. 27. Half-angle of normal avalanche cone; data from Raether (70); Wagner (72); and Allen and Phillips (71). Nitrogen.

pictures given by Allen and Phillips (71) show. To calculate it on the basis of the diffusion theory, Raether solved the free electron diffusion equation in cylindrical geometry

$$D\nabla^2 n + \beta n = \partial n/\partial t,$$

obtaining, with v representing the drift velocity,

$$n = \frac{e^{\beta t}}{(4\pi Dt)^{3/2}} \exp\left[-\frac{r^2 + (z - vt)^2}{4Dt}\right].$$

Then an investigation of the contours of constant n yielded

$$\tan\frac{\theta}{2} = \left(\frac{4\alpha kT}{eE}\right)^{1/2},$$

where $\alpha = \beta/v$. This result is not quite correct because positive ions rather than electrons form the observed cone, and their numbers depend on their exposure to the spherically diffusing and translating electron swarm. In this case

$$N = \alpha(2v/\pi)^{3/2} \exp(zv/2D) \int (4D\xi)^{-3/2} \exp[(\alpha - V/4D)\xi]$$
$$\times \exp[-(\rho^2 + z^2)v/4D\xi] \, d\xi$$

and

$$\tan\frac{\theta}{2} = \left(1 - \frac{1}{\alpha vt}\right)^{1/2} \bigg/ \left(\frac{eE}{4\alpha kT} - 1\right)^{1/2}.$$

The normal avalanche has been the source of some of the most widely accepted data on electron drift velocities and ionization rates. Both are peculiar to this process, however, because of uncertainties as to the degree of thermalization of the electron cloud, and so are not accurately related in a simple fashion to the corresponding cross section. It is customary, therefore, to leave the ionization data in the form of Townsend's first coefficient, α, the spatial exponential growth factor of the avalanche.

3. Abnormal Avalanches

If allowed sufficient running space, in a sufficiently strong field, avalanches show normal growth only until the electrons are numerous enough to form a polarized surface charge around the periphery of the avalanche which can "protect" the interior from penetration by the external field, as in a conductor. By Gauss's theorem, as this interior field-free state develops, the exterior field around the avalanche is enhanced, particularly in the forward and backward directions. At the same time the site of the ionizing action is transferred to the thin layer at the head of the avalanche where the electrons are most numerous, and the entire head structure transforms into a self-propagating ionizing wave of the proforce class. A short time later, the number and distribution of ions toward the tail of the avalanche becomes sufficient to develop a total field in its vicinity capable of extending a second locus of ionizing action in the backward direction as an antiforce wave. The entire structure now takes the form of a growing cylinder with rounded and distended ends, termed a plasma channel or streamer. The onset of these processes takes place characteristically when between 10^6 and 10^8 total free electrons are available in the avalanche.

Wagner (72) has examined the transition in several gases, and his data on the proforce wave speeds, antiforce wave speeds, normal drift velocities, and transition electron numbers are given in Table II. Based on the model just described, the transition electron total N can be estimated from the contribution of the separated ion and electron charges to the internal fields as

$$N/\ln^2 N = (\pi\varepsilon_0 E/e\alpha^2) \tan^2 \frac{\theta}{2}.$$

Most of the contribution to the field alteration comes from the relatively immobile ion cone.

TABLE II

AVALANCHE TRANSITION PROPERTIES[a]

Gas, P	E/p (10^3 V/m-torr)	V_- (10^5 m/sec)	V_+ (10^5 m/sec)	V_0 (10^5 m/sec)	$N \times 10^8$	V_-/V_0
$N_2 + CH_4$	6.43	4.5	2.0	1.7	3.25	2.65
(1:1)	6.70	5.7	2.7	1.8	5.36	3.17
50 torr	6.94	6.2	2.9	1.9	1.78	3.26
	6.94	5.8	2.4	1.9	1.78	3.05
	7.35	7.0	3.0	2.0	——	3.50
	7.45	8.3	3.1	2.1	6.55	4.29
	7.63	8.4	3.3	2.2	3.59	3.81
$N_2(20\%CH_4)$	5.70	3.9	0.8	1.7	——	2.29
88.5 torr	5.85	4.9	1.7	1.7	——	2.88
	6.17	4.1	1.1	1.7	——	2.41
	6.33	6.0	2.3	1.8	——	3.33
	6.44	8.0	4.3	1.8	——	4.44
N_2(pure)	7.60	5.7	1.3	2.2	6.54	2.59
50.3 torr		4.3	0.5			1.95
	7.23	3.2	−1.0	2.1	8.84	1.52
		4.2	0.0			2.00

[a] From Wagner (72) with permission of *Zeitschrift für Physik*.

When the transition is full blown in weak fields, the field at the front of the wave will be approximately that at the front of a conducting sphere in a uniform electric field, where it is well known that the field is three times the remote field. Shelton's theory of the proforce wave tells us that it advances at nearly electron drift velocity in the local field, so we would expect Wagner's proforce waves to move at three times the velocity of his normal avalanches.

Table II shows how well this expectation is confirmed. As the strength of the field increases, however, a shape change evidently sets in, because the factor is larger than 3. This may arise because in stronger fields the transition occurs earlier in the run of the wave, and more space is allowed for the electron cloud to stretch out of its initial sphericity. The extent to which these wave speeds themselves are predicted by the theory will be discussed later.

E. Long Sparks

1. General Remarks

Long laboratory sparks take place between electrodes at distances large with respect to the size of those electrodes, and hence in nonuniform fields. The largest potentials available for these studies are generated by Van de Graaff machines, which can provide more than 2×10^7 V and stored charges of 10^{-2} C, but have not generally been applied to this problem. Transformers have been constructed up to 1×10^6 V, but the usual method is the Marx (cascade or impulse) generator which can give as much as 6.5×10^6 V and 6×10^3 A (73) for the short duration needed. The longest sparks which have been produced so far have been 15 m long (74).

When the potential available is near the threshold for the gap in question, breakdown takes place via the accumulation of avalanches known as the corona, with streamers exploring the space around the cathode and anode structures until one finally closes the gap. When the potential is well above the threshold, the breakdown occurs via the leader and return stroke mechanism found with lightning. Whether the wave originates at cathode or anode, and whether therefore it is proforce or antiforce in nature, depends on the intensities of the fields around the two structures. In the opinion of Uman (2), the long spark is not a good simulation of lightning for a number of reasons, chief among which is that two conducting electrodes are present in the former and only one in the latter. As a consequence it does not seem to exhibit the stepped leader, but more nearly resembles a weak subsequent stroke. Moreover, the power levels differ by an order of magnitude at least.

One difficulty in investigating long sparks is the random walk of the avalanches which complete the original conduction path. Koopman and Saum (75) have applied laser preionization to guide the breakdown of long sparks along a straight line path. The guidance is excellent, and the velocity of propagation of breakdown is a function of the degree of preionization in the channel, as would be expected from theory. A velocity of 3×10^6 m/sec was observed for 500 kV applied and $\sim 3 \times 10^{16}$ electrons/m³ preionization.

Exactly what one observes in the early moments of the spark discharge depends very much on the electrode configuration, generator impedance, gas, and especially the overvoltage (excess of applied voltage over the breakdown value) on the gap. Much attention has been directed to the interval that ensues between voltage application and breakdown of the gap. This interval has been subdivided into a part which is entirely unpredictable, called the statistical time lag, and a part which depends on gap width and electron mobility, called the formative time lag. Both of these lags are maximized in a gap with small overvoltage. The statistical time lag is not germane to this article, but the formative time lag may or may not be. If breakdown occurs as the result of bridging the gap with normal avalanches, it is not. If the avalanches go over into the self-propagating wave called a leader, this governs at least a portion of the formative time lag, which then becomes pertinent. Language then becomes a problem in discriminating between situations, because in the early research with slow and insensitive methods of observation, usually involving inadequate spatial resolution, any luminous process crossing the gap may have been labeled a leader. The literature is replete with claims and counter claims about observation of processes which agreed or disagreed with Schonland's observations and deductions about lightning leaders. It also abounds in much colorful morphological language, intended to describe a situation that is only currently becoming capable of exact quantitative measurement and one which is deserving of very many more detailed studies. We begin with the work of Uman et al. (76) as one clearly defined point of departure.

Uman et al. examined sparks across a 4 m rod-to-plane gap in both polarities with an applied potential of 3.3×10^6 V. They used an image converter framing and streak camera, photodiodes, and oscilloscopic current and voltage detection. The gap was estimated to be overvolted by 5×10^5 V for rod negative and 1.4×10^6 V for rod positive. Although there was evidence of earlier current and luminosity in the gap, the image converter gave no response until between 1 and 2 μsec for rod positive and between 3 and 4 μsec for rod negative after voltage application, at which instant a luminous leader emerged from the rod and propagated downward toward the plane. The proforce (rod negative) leader advanced at about 2×10^6 m/sec. Before it had half crossed the gap, one or more upward antiforce leaders left the plane, all but one aborting, and that one advancing at about 3×10^6 m/sec to meet the downward bound leader very near midgap. After their junction, luminosity propagated up and down the established column simultaneously in waves that might either require interpretation as wholly new processes, or simply as modified continuations of the old ones, guided now by each other's channel. That there was a modification at least is shown by the fact that their speeds increased to 3×10^7 m/sec, as would be expected of secondary breakdown waves.

The antiforce (positive rod) leader, on the other hand, traversed the entire gap, or at least all but the last few centimeters, eventlessly, except that its speed increased from 1×10^6 m/sec to 2×10^6 m/sec. Upon arrival at the plane a return stroke moving at about 3×10^7 m/sec advanced back up the channel. Its intensity was much less than that of proforce return strokes.

In both polarities luminosity was observed at a slot stationed 2.3 m above the plane long before the appearance of the leader at the slot, but the observation was equivocal because it could have been light scattered into the observation cone. The currents measured in series with the gap could not be artifacts, however, and in the proforce experiment after a 1 μsec null period the current rose steadily at 10^8 A/sec until the instant when the leader appeared. The antiforce leader current differed in being preceded by a 0.25 μsec burst of 500 A, again after the first microsecond.

Since theory says that the propagation of leaders is governed by their self-generated field, knowledge about the leader diameter would be valuable, and is generally unavailable for long sparks. Lightning leaders are much larger than their corresponding return strokes, and the return strokes of sparks have been measured at from 0.01 to 0.03 m in diameter. The only trace of data on the leader diameter is the published image converter photographs themselves. The diameters indicated by those photographs seem likely to be distorted by overexposure because they suggest typical diameters of perhaps 6×10^{-2} m for antiforce leaders and 0.2 m for proforce leaders. Minor confirmation of the theoretical leader velocity relation is shown by a photograph (76) of the downward bound proforce leader encountering an upward bound antiforce leader. The velocity of the latter is nearly 60% larger than that of the former, although primary antiforce waves are usually the slower of the two with equal driving fields. When that factor is included, the ratio of column diameters taken from the photograph is in the right proportion for the electric intensities required, suggesting that the scale enlargement of the column image was linear, even if the column was in fact smaller than recorded. Stekol'nikov (77), Stekol'nikov and Shkilev (78), and Aleksandrov *et al.* (74) have worked under conditions much nearer threshold than Uman. [Udo (79) has given the curve in Fig. 28 for the breakdown of long rod–plane gaps in air.] Stekol'nikov observed many coronal streamers leaving a 0.125 m diam. negative spherical electrode to which 4×10^5 V potential had been applied with respect to a plane 1.05 m away. The streamers advanced with a velocity of about 5×10^7 m/sec near the sphere, which decreased rapidly to 2×10^6 m/sec at a distance of 0.40 m. The velocity ratio of 25 is in reasonable agreement with the inverse square law decrease of field, suggesting that in this case the avalanches were all normal low space charge avalanches with full field penetration. Stekol'nikov and Shkilev changed to a negative rod and plane configuration with a separation of 2.7 m. Depending

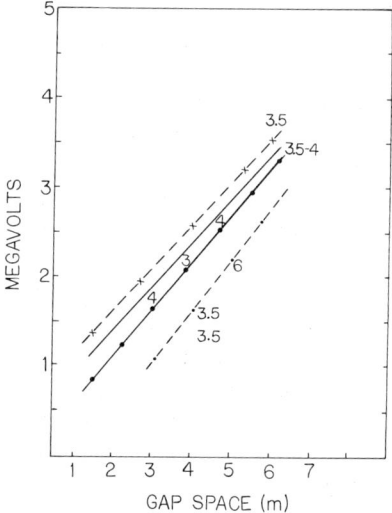

Fig. 28. Breakdown potential of long gaps in air. [After Udo (79) with permission of the Institute of Electrical and Electronics Engineers.]

on the rate of rise of the voltage applied to the gap the further developments around the rod electrode were seen to be simple or complex. For suddenly applied voltages (like those of Uman et al.), whether positive or negative, after an initial brief period of avalanches running away from the rod, a column formed out of the avalanche ionization, and many intermittent waves of ionization ran down and successively extended the ionized column at speeds of 10^6 m/sec, with a burst of avalanche branching occurring at the end of each run. At the instant that the column reached the plane, a bright wave left the rod end of the column. In the positive rod case it traversed the entire column at steadily increasing (but generally $\sim 10^6$ m/sec) speed. In the negative rod case a companion wave left the plane at the same moment and the two met somewhere midway in the column, both having speeds $\sim 10^6$ m/sec.

When the voltage across the gap was increased slowly, a few widely spaced column-extending waves with much avalanche branching crossed the gap with consequences similar to the fast potential rise case. As either kind of column grew, a portion of the column near the rod seemed to take on a special character and extend its dominance slowly (5×10^4 m/sec) down the column. Stekol'nikov believes that when the bright wave begins its run it originates at and is in fact an accelerated extension of this region of special character, but the actual records presented seem always to show a luminous process which begins back at the rod and runs all the way. At intermediate values of (negative) potential rise rate, this characteristic region has been seen to undergo a rapid and discontinuous step of growth midway through the breakdown from a point about twice its distance out from the rod, and growing back toward the original region.

The results of Stekol'nikov and his co-workers seem consistent with those of Uman *et al.* if we assume that Uman's image converter did in fact see only the waves traversing the column after the final avalanche-led connection was made. Reading again from photographic diameters, which can be quite unreliable, Stekol'nikov and Shkilev's pictures indicate typical column diameters of 0.06 m and bright wave diameters of 0.12 cm. Stekol'nikov and Shkilev point out that the propagation pattern of their leaders does not correspond to that of the lightning stepped leader. Nevertheless, these waves, which move down the same channel repeatedly and branch near their ends, do resemble the stepped leader in many ways. Using a photomultiplier, Kritzinger (80) saw the advance of the avalanching column as the motion of "globules" of light which generally behaved according to Stekol'nikov's description. Beginning with pure nitrogen gas for a breakdown medium, admixture with argon showed no effects, but the presence of oxygen produced a substantial change in the velocities of the "globules" (Fig. 29). Kritzinger suggests that this is evidence for photoprocesses in the wave, but that cannot be admitted to be conclusive unless it is shown that the column diameter is not altered by the presence of the electronegative gas, and that the more favorable cross section for ionization of oxygen does not affect the wave speed either.

Saxe and Meek (81) observed a positive rod to negative plane spark from a 10^6 V Marx generator in air with one pair of photoelectric timing stations, and interpretation of their results in terms of the foregoing morphology is thus somewhat uncertain. They find that there is an initial period which they call a corona, followed by a "leader" stroke, followed by a "main" stroke.

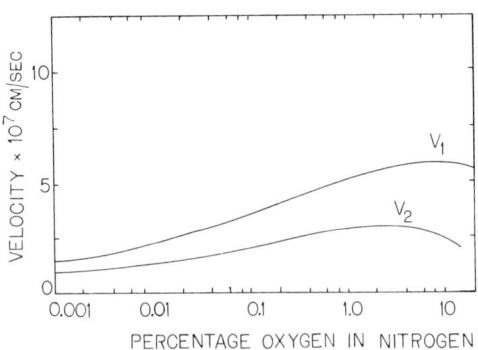

FIG. 29. Positive point to negative plane luminosity velocities. Voltage 100 kV; gap length 0.25 m; V_1 = average velocity over first 0.1 m. V_2 = average velocity over last 0.1 m. [After Kritzinger (80) with permission of *Nature*.]

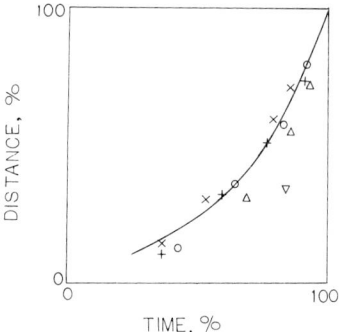

FIG. 30. Percentage-distance vs. percentage-time curves for the movement of the leader stroke across the gap. [After Saxe and Meek (81) with permission of the Institution of Electrical Engineers.]

Their data are principally for the leader. In a single measurement they estimated its general halo diameter at ~ 0.20 m, surrounding an intense core ~ 0.03 m in diameter. They found its terminal velocity (that pertaining at some distance from rod) to be the same for all separations (Fig. 30). They found that the resistance in the power supply affected leader velocity only to the extent that it determined the potential available for the leader at any given instant. They measured the current in the leader as a function of time under several conditions (Fig. 31). They also obtained data on velocity and on current at reduced pressure (Fig. 32). They also showed that the voltage drop along the channel is very small by plotting distance–time curves for three gap lengths with the minimum breakdown voltage for the longest gap used in each case, and the data plotted to coincide at the instant the leader stroke reached the plane. If there had been a strong difference in column resistance, this would not have shown the observed agreement (Fig. 33).

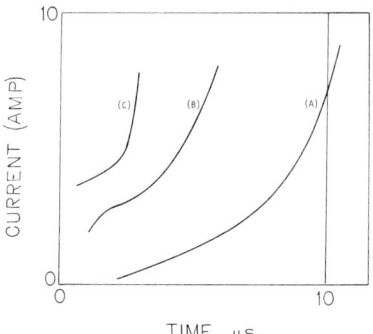

FIG. 31. Effect of reducing the gap length at fixed voltage on the current in the leader stroke. (A) 0.425 m gap; (B) 0.35 m gap; (C) 0.25 m gap. [After Saxe and Meek (81) with permission of the Institution of Electrical Engineers.]

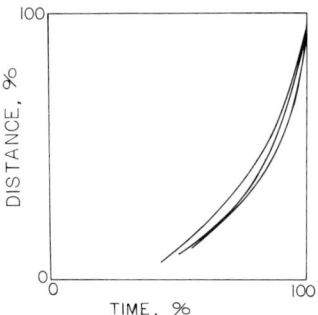

FIG. 32. Effect of the reduction of gas pressure on the percentage-distance vs. percentage-time curve. [After Saxe and Meek (81) with permission of the Institution of Electrical Engineers.]

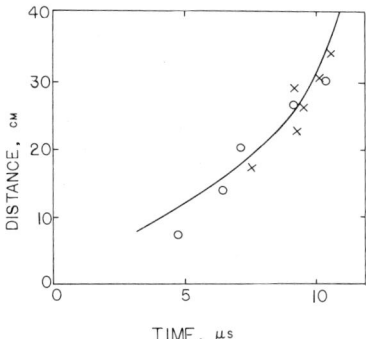

FIG. 33. Distance vs. time curves for gap lengths of 0.425, 0.35 and 0.25 m, using the minimum breakdown voltage for the 0.425 m gap, plotted so as to coincide at the instant that the leader stroke reaches the plane electrode. −, 0.425 m gap; ○, 0.35 m gap; ×, 0.25 m gap. [After Saxe and Meek (81) with permission of the Institution of Electrical Engineers.]

2. Return Stroke

The return stroke, being the most luminous and energetic portion of the discharge, received study. In 1943, using a drum camera, Flowers (81a) measured maximum currents and corresponding channel diameters of strong sparks, obtaining Fig. 34a, from which one concludes that the current density is something of an invariant at 1.1×10^7 A/m², although he also observed the interesting fact (Fig. 34b) that the current density is a function of how rapidly the channel is created. He observed that a faint glowing column was present before the return stroke, and that the channel he finally measured began with a smaller diameter and expanded. One point determined by Uman et al. (76) fits nicely on the same curve.

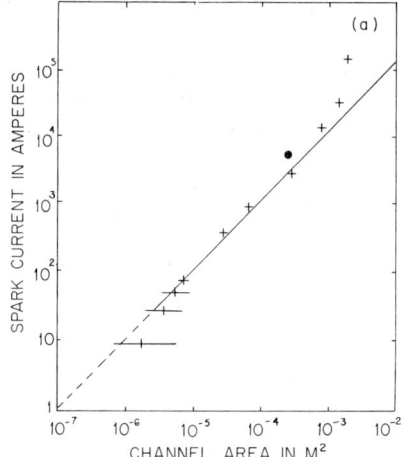

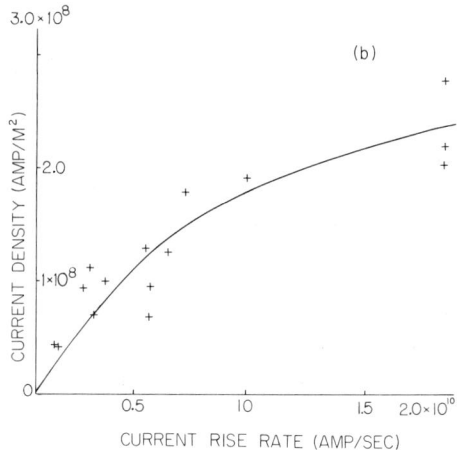

FIG. 34. (a) The relation of maximum current and channel cross section for the spark discharge in air. The slope of the curve corresponds to an average value of 1.1×10^7 A/m^2. (b) Current density at maximum current as a function of current rise rate. [After Flowers (81a) with permission of the *Physical Review*. ● is from Uman, *et al.* (76).]

To all practical purposes, no data exist on the velocity of the spark return stroke. Uman *et al.* quote 3×10^7 m/sec, and from the current records given one may also estimate that an interval of less than 0.1 μsec was involved in the rise of current to its maximum, and since this must have some relation to the transit time of the return stroke, a velocity of 4×10^7 m/sec is suggested by it. The proforce return stroke (from the negative plane) seems to have been slower than the antiforce return stroke (from the positive plane) by a factor of as much as 3.

Temperatures and electron densities in the return stroke channel are generally observed too late to be relevant to this article. The data at earliest time given by Orville (82) are at between 0 and 2 μsec for a negative rod and a 5-m spark using the 6.4×10^6 V generator. He finds a radiation (probably electron) temperature of $3.5 \times 10^{4\circ}$K and a density of 1.1×10^{24} electrons/m^3. The temperature drops to $2.8 \times 10^{4\circ}$K in the next 2 μsec, and the electron density falls to 3×10^{23} m^{-3}; thereafter it seems probable that the processes of interest have passed.

IV. Theories of the Electron Waves

A. Theories Based on Electron Impact Ionization

1. General Remarks

The Shelton theory of electron fluid waves ionizing by electron impact alone was originally carried to completion only for the proforce shock-fronted case with no preionization ahead of the wave. It fitted reasonably well to the data available, but the inability to carry through the solution for the antiforce case seemed a drawback to its acceptance. Sanmann and Fowler (83) have now offered a solution for the antiforce case so that this uncertainty has been removed. By the same technique a solution is made possible for the second kind of proforce wave discussed by Scott.

Since the preparation of Part I of this review (1), the author has had second thoughts on the terminology used there, in which the adjectives primary, secondary, and tertiary were used to designate classes of electron waves based on their initial condition. Such terminology creates discord and confusion because of the terms primary and secondary already applied by Loeb (83a) to streamers, structures for which these proposed theory classes are intended to furnish explanations. It is therefore desirable to revise the category naming as follows: instead of *primary*, those waves which move into a medium of substantially zero electron concentration will be called Class I waves. Instead of *secondary*, those waves which move into a medium of significant electron concentration will be called Class II waves. In Part I, tertiary waves were posited as being different from Class II by not fulfilling the zero-current condition, which was presumed to be possible with both Class I and Class II waves. At present it is not clear that such a class can be distinguished from Class II, because it is not clear that the zero current condition can be met physically if there is nonzero electron concentration ahead of the wave.

2. Shelton–Burgers Equations

The basic equations of the electron fluid rest on the general equations for interaction of three fluids: electrons, neutrals, and their ions. From these Shelton and Burgers independently derived a set of equations (having one major difference) in which only the electron properties appear. They did this by neglecting the changes in heavy particle velocities except where they were multiplied by the neutral particle mass density, and then finding expressions to eliminate these differences. The result was the following four equations, which have been restricted here to waves of steady profile.

$$dnv/dz = \beta n$$

$$\frac{d}{dz}[mn(v - V)v + nkT_e] = -enE - K_1 mn(v - V)$$

$$\frac{d}{dz}[mn(v^2 - V^2)v + nv(5kT_e + 2e\phi_i) + q] = -2envE - 2K_1 mn(v - V)V$$

$$\frac{dE}{dz} = \frac{e}{\varepsilon_0} n\left(\frac{v}{V_i} - 1\right).$$

The Class I solution of these equations rests on the conditions at the interface between the ionized and nonionized regions. These can be derived either from the above set, or more rigorously from the global equations of balance. They take the final form

$$n_1[v_1(v_1 - V_0) - k(T_e)_1/m] = 0$$
$$n_1 v_1[(v_1^2 - V_0^2) + (5k(T_e)_1/m) + (2e\phi_i/m)] = 0,$$

and from them Shelton obtained starting values for his solution in the discontinuous (shock proforce) case

$$v_1 = \frac{5V_0}{8} + \frac{9V_0^2 + 16(2e\phi_i/m)^{1/2}}{8},$$

which carries in it the implied lower limit on wave speed

$$V_0 > (2e\phi_i/m)^{1/2}.$$

3. Density Dependence of Wave Speeds

The Shelton and Fowler proforce wave theory described in Part I of this chapter, and the Sanmann and Fowler antiforce wave theory described in the next section do not account for any gas density dependence of the wave speed beyond the normal Townsend dependence E/p. The experimental data, although they are rather well centered around a simple theoretical curve calculated from first principles without any adjustable constants, still display a residual density dependence which leaves them fanned out over quite a region, as reference to Figs. 20–23 herein and to Figs. 17 and 18 of Part I will show.

Scott and Fowler have shown that the residual density dependence can be largely accounted for by two apparatus effects which could not be corrected out in their calibration. The first is a low pressure effect. As the electron mean free path increases, it is limited by the mean diffusion distance to the wall rather than the interparticle distance. This can be treated, at least in its incipience, by replacing the electron atom collision frequency K_1 by a pseudo collision frequency K_1', where

$$K_1' = K_1 + 2.405\bar{v}/a_0 .$$

Here $\bar{v}$ is the mean electron speed, a_0 is the tube radius, and 2.405 is the first root of the zero-order Bessel function.

The second effect sets in at high pressures and is caused by the constriction of the wave first noted by Blais and Fowler. Because the plasma column radius is not a_0, but is some $a < a_0$, the field is actually stronger by a factor a_0/a than was estimated from the calibration used. Since so little is known as yet about even the steady current constriction of a glow column, only an empirical approach was possible and it was established by these means that

$$a/a_0 = (1 + p/p_0)^{1/2}.$$

Then, with a single constriction constant for each gas, it was possible to fit the majority of proforce and antiforce data. The complete expression used was

$$E/p = \frac{(m/ek)V_0[K_1/p + (2.4\sqrt{3b}/a)V_0]}{(1 + p/p_0)^{1/2}} .$$

The quantity b is given by theory as 0.2 for proforce waves and 0.6 for antiforce waves. p_0 took the empirical values of 10 for helium, 0.42 for argon, and 0.39 for nitrogen.

Whenever wave velocities below 7×10^6 m/sec occur, the experimental curves again depart from the Shelton theory drastically, but this is to be expected, because the analysis was carried out under an assumption that the electron temperature remains constant throughout the wave front, and this assumption is in error by factors which rapidly exceed 2 for velocities lower than 7×10^6 m/sec.

4. Class I Antiforce Waves

Antiforce waves cannot fulfill discontinuous (shock) conditions at the ionizing front except for a very small range of velocities just above the ionization speed $(2e\phi_i/m)^{1/2}$. Since they are known to exist over a much wider range of velocities both above and below this value, it is evident that they must fulfill the initial conditions some other way. There remains the choice that the interface between ionized and non-ionized regions is a weak discontinuity rather than a strong one, i.e.,

$$n_0 = n_1 = 0; \quad (dn/dz)_0 \neq (dn/dz)_1, \quad (dn/dz)_1 \neq 0.$$

Sanmann and Fowler investigated this choice. There are immediate consequences with regard to the initial value of the electron velocity. Expanding the continuity equation and setting $n = 0$ at $z = 0$,

$$v_0(dn/dz)_0 = 0.$$

The assumption that $(dn/dz)_0 \neq 0$ then demands the additional condition that $v_0 = 0$, and this, introduced into Poisson's equation, shows that the initial conditions for the field are $E = E_0$, $(dE/dz)_0 = 0$. But if dE/dz has an initial value of zero, then as soon as n and v assume nonzero values, (dE/dz) becomes negative and continues so up to the point where $v/V_i = 1$. Since passing through the wave corresponds to proceeding in the negative z direction, this implies that the magnitude of the field increases in traversing the interior of the wave up until the electrons and the ions reach the same velocity. Beyond this point the electrons acquire velocity in excess of the ions and the field decreases in magnitude. Ultimately, the field must go to zero since a conductor or ion–electron plasma in equilibrium cannot support a field. To this extent the infinite plane wave model is unphysical, because true columnar waves need small currents in support of their growing column capacitance. These currents have a zero divergence, however, and distribute axial charge flow radially up onto the wave front, where the radial component vanishes, and so must the axial one.

For a large part of the wave structure, the electric field has a value which is larger than the initial value E_0. This has consequences regarding the validity of the approximations under which the momentum and energy conservation equations are presented. Shelton eliminated $(V_1 - V_0)$ in his study of the proforce wave to obtain an energy equation, but this equation requires a negative temperature when E is slightly greater than E_0, and well before the field reaches its maximum value. Sanmann (84) demonstrated that the difficulty lies in the heat conduction term which should be included in the energy equation but which Shelton chose to ignore because of the intractability it gave to the equations. Arguments for its negligibility hold force for Shelton's proforce case, but not for the antiforce waves. One must therefore find a replacement for the energy equation as given by Shelton. Sanmann and Fowler therefore employed the global momentum equation in its place, choosing to evaluate the difference $(V_i - V_0)$ as a function of position z in the wave profile by use of the Langevin equation, and obtaining

$$nkT_e = \frac{\varepsilon_0}{2}(E^2 - E_0^2) - \frac{nV_0 eE}{K_i}\left[1 - \exp\left(-\frac{K_i z}{2V_0}\right)\right].$$

The symbol K_i is the collision frequency for the positive ions.

Expressed in the dimensionless variables of Part I, with the addition of $\Omega = 3E_0 e/mV_0 K_i$ the electron fluid equations become

$$\frac{dj}{d\xi} = \mu v$$

$$\frac{d}{d\xi}[j(\psi - 1) + v\theta] = -v\eta - \kappa v(\psi - 1)$$

$$j(\psi - 1) + v\theta = \frac{\alpha}{2}(\eta^2 - 1) - \frac{\Omega}{3}j\eta\left[1 - \exp\left(-\frac{3\xi}{2\Omega}\right)\right]$$

$$\frac{d\eta}{d\xi} = \frac{v(\psi - 1)}{\alpha},$$

where j is a contraction for $v\psi$, the electron drift current.

Shelton showed that the dependent variables paired themselves in their dependence on ξ into the rapidly and slowly varying pairs (η, ψ) and (v, θ). A fifth physical dependent variable is present as a parameter in these equations. It is the wave speed, and is present in the form of κ. For complete determination of all these quantities a fifth equation is needed, and the spirit of the Shelton solution is to assume a relation between η and ψ which fulfills the boundary conditions on these quantities, but ignores their true microscopic relationship, which may be quite complex. For the antiforce wave, Sanmann was unable to find a single function which would connect the point $\eta = 1$, $\psi = 0$, where $(d\eta/d\psi) = 0$, and the point $\eta = 0$, $\psi = 1$ where again $(d\eta/d\psi) = 0$, while at the same time it has a maximum $(d\eta/d\psi$ also $= 0)$ on the line $\psi = 1$. He therefore chose the piecewise continuous function shown in Fig. 35, described by

$$\eta = 1 - (\eta_1 - 1)(2\psi^3 - 3\psi^2)$$

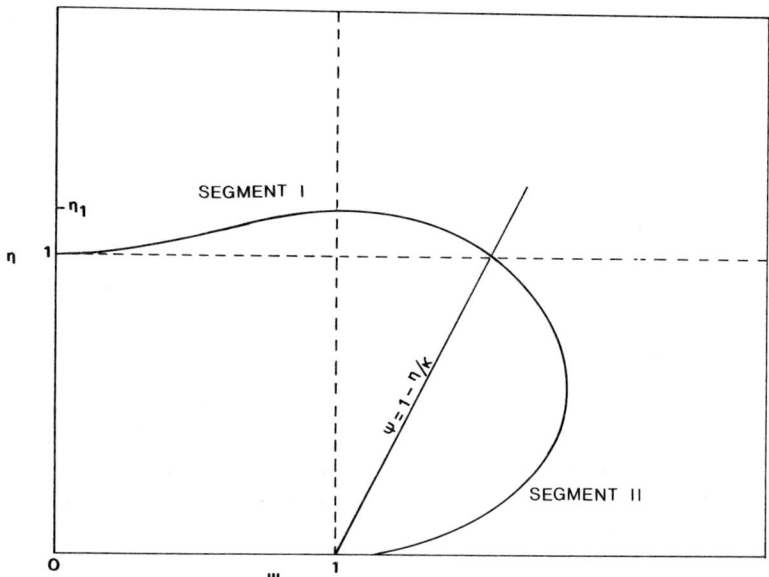

FIG. 35. Integration path for antiforce waves in the drift velocity (ψ) electric field (η) plane. [After Sanmann and Fowler (83) with permission of *Physics of Fluids*.]

on its first segment, and

$$\left(\frac{\eta - \eta_1/2}{\eta_1/2}\right)^2 + \frac{4\kappa^2}{\eta_1^2}(\eta_1 - 1)(\psi - 1)^2 = 1$$

on its second segment. η_1 is the value of η at its maximum.

By integrating the basic equations along this path he was able to obtain the wave velocity dependence given in Fig. 36 and the structural profile for a particular wave speed, i.e., $\alpha = 0.086$, given in Figs. 37 and 38. To obtain tangible results he found it necessary to abandon the global momentum equation on the second segment for reasons of integrability. Analysis based on the solution that resulted showed that heat conduction was no longer important on this segment, and that the Shelton energy equation was valid again. He therefore shifted to this equation to complete the set in solving this segment. The results of this investigation were plotted in Figs. 21 and 23, taking into account the high and low density apparatus corrections described in the previous section. The essential prediction, which experiment seems to bear out quite well, is that the wave velocity is primarily a function of E/p, and is a nearly linear function. Ionization processes are only of major importance in changing the thickness and structure of the wave.

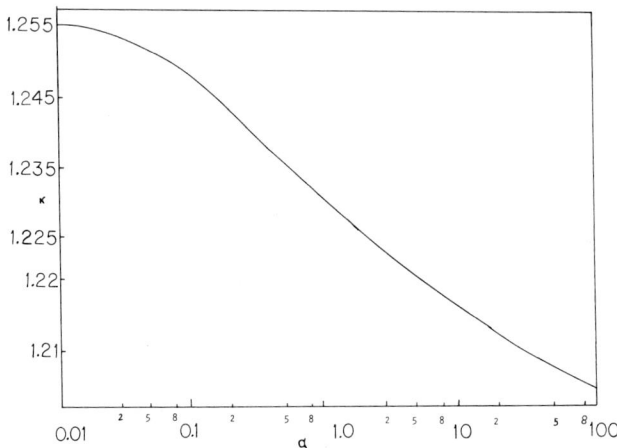

FIG. 36. Wave velocity parameter (κ) as a function of wave energy parameter (α) for antiforce waves. [After Sanmann and Fowler (83) with permission of *Physics of Fluids*.]

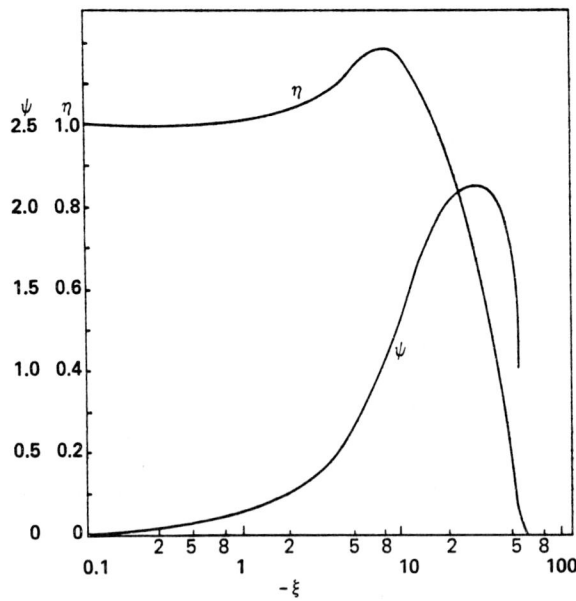

FIG. 37. Drift velocity (ψ) and electric field (η) as functions of position (ξ) in the antiforce wave. [After Sanmann and Fowler (83) with permission of *Physics of Fluids*.]

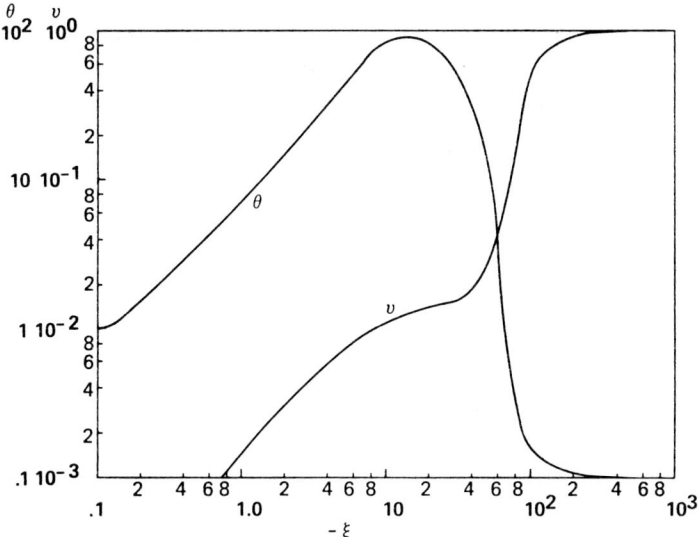

FIG. 38. Temperature (θ) and electron density (v) as functions of position (ξ) in the antiforce wave. [After Sanmann and Fowler (83) with permission of *Physics of Fluids*.]

5. Weak Proforce Waves

The relation suggested by Sanmann for handling the first segment of the antiforce wave can be used immediately as an attack on the second family of proforce waves observed by Scott, which run at speeds around and below the ionization velocity threshold. Since the Shelton wave, with its strong

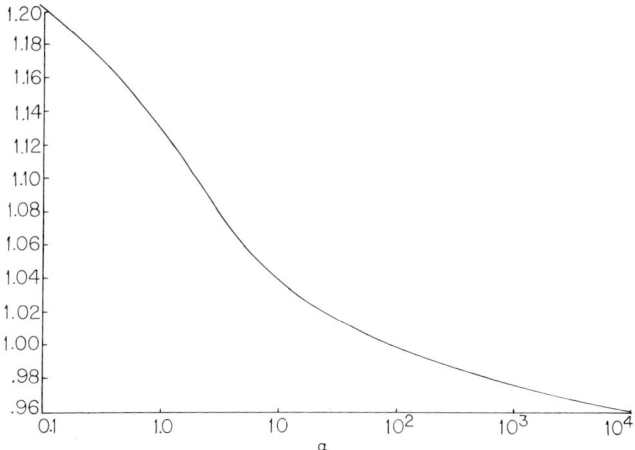

FIG. 39. Weak proforce wave constant (κ).

discontinuity, cannot exist below this threshold, the wave that does must obey the same initial conditions as the antiforce wave, except that η must decrease as ψ increases. This can be accomplished by reflecting Sanmann's (η, ψ) path about the $\eta = 1$ line, yielding

$$\eta = 1 - 3\psi^2 + 2\psi^3.$$

When this path assumption is used to integrate the basic equations, the resulting solutions are

$$v = 3\alpha\mu\psi$$

$$\xi = \frac{2}{\mu}\psi$$

$$\kappa = \frac{37/30}{1 + \alpha\beta/K_1}$$

$$\theta_2 = \frac{10K_1}{111\beta} - \frac{9\alpha}{37}.$$

The most important of these, the wave speed relation, is given in Fig. 39. It is important because it seems probable that all primary lightning leaders fall into this class rather than fitting Shelton's shock solution.

6. *Class II Waves*

The solution for Class II waves also differs from Class I in the shock condition at the front. For this class the shock conditions are no longer restrictive on the wave velocity because of the initial ionization present. The shock equations are

$$mn_1v_1^2 + n_1(kT_e)_1 = mn_0v_0^2 + n_0(kT_e)_0$$

$$mn_1v_1^3 + 5n_1(kT_e)_1 + 2n_1e\phi_iv_i = mn_0v_0^3 + 5n_0(kT_e)_0 + 2n_0e\phi_iv_0$$

$$n_1v_1 = n_0v_0.$$

The third equation of this set is not exact, but should be a very good approximation because of ion inertia and the relaxation time for new ionization. In the solution, $e\phi_i$ drops out and the equations become (in dimensionless notation)

$$\psi_1^2 + 5\theta_1 = \psi_0^2 + 5\theta_0 \qquad \psi_1^2 + \theta_1 = \psi_1\psi_0 + \psi_1\theta_0/\psi_0,$$

for which the solutions are $\psi_1 = \psi_0$, $\theta_1 = \theta_0$, and

$$\psi_1 = \tfrac{1}{4}(\psi_0 + 5\theta_0/\psi_0)$$

$$\theta_1 = \tfrac{1}{16}(3\psi_0^2 + 14\theta_0 - 5\theta_0^2/\psi_0^2).$$

Whether there is a weak discontinuity solution corresponding to the first case has not been investigated, but the strong discontinuity solution exists with $\theta_1 > \theta_0$ only if

$$\psi_0 > (5\theta_0/3)^{1/2}.$$

But ψ_1 must also be $< \psi_0$ or the wave is unstable, so

$$2 - (4 - 5\theta_0)^{1/2} < \psi_0 < 2 + (4 - 5\theta_0)^{1/2},$$

which also requires $\theta_0 < 0.8$.

An upper limit on θ_0 is precisely the behavior described by Uman for the dart leader in the lightning stroke, which seems to propagate only after the channel has cooled to a certain degree.

If the momentum equation is now integrated using the parabolic (ψ, η) relation $\eta = [(1 - \psi)/(1 - \psi_1)]^2$, the wave constant κ is

$$\kappa = \frac{[1/(1 - \psi_1)] + [(1 - j_0)/2]}{\frac{3}{2} + [\alpha\beta/K_1(1 - \psi_1)^2]}.$$

Since $j_0 = v_0 \psi_0$ and $\psi_1 = \frac{1}{4}(\psi_0 + 5\theta_0/\psi_0)$ it is apparent that the wave speed parameter κ is a strong function of the degree of initial ionization.

7. Class III Waves

If there is a third class of waves, it is not a matter of the relaxation of the zero current condition (which must already be relaxed for Class II), but probably that the divergence of the current density behind the front, in the plasma, is so large that the integration over wave thickness proceeds through a region which must now be treated as an axial current density discontinuity. This means that instead of ending the integration in the (ψ, η) plane at $\psi = 1$, $\eta = 0$, it must be terminated at some point along the electron drift velocity line

$$\psi = 1 - \eta/\kappa.$$

Thus it is not apparent that this is anything more than a special kind of Class II wave. The problem has not yet been fully investigated, but it seems likely that the return stroke falls into this category, and that since the approach to the drift line is more difficult to achieve mathematically when $\kappa > 0$ than when $\kappa < 0$, the existence of purely or at least preponderantly antiforce return strokes may be accounted for.

B. Diffusion Equation Theories

Turcotte and Ong (85) attacked the problem of ionizing fronts by means of the conventional diffusion equations:

$$j_e = -D_e(\nabla n_e + eEn_e/kT_e)$$

$$j_i = -D_i(\nabla n_i - eEn_i/kT_i)$$

$$\partial n_e/\partial t + \nabla \cdot j_e = \beta n_e - \rho n_e^2 n_i$$

$$\partial n_i/\partial t + \nabla \cdot j_i = \beta n_e - \rho n_e^2 n_i,$$

where ρ is the recombination coefficient. They met and overcame the formidable nonlinear difficulties which had generally held back the use of these equations before. They included in the particle balances implied by these equations both production of electrons by electron impact and loss by recombination. They specialized the basic equations to one-dimensional steady profile waves moving toward $+z$ at speed V_0, and chose to ignore all ion motions, obtaining the equations

$$-V_0 \frac{dn_e}{dz} = \frac{d}{dz}\left(D_e \frac{dn_e}{dz} + \frac{eD_e}{kT_e} neE\right) + \beta n_e - \rho n_e^2 n_i$$

$$-V_0 \frac{dn_i}{dz} = \beta n_e - \rho n_e^2 n_i,$$

and included Poisson's equation to complete the set

$$dE_z/dz = e/\varepsilon_0(n_i - n_e).$$

They then searched for a Class I wave solution where $n_e = n_i = 0$ as $z \to +\infty$, and $n_e = n_i = n$ as $z \to -\infty$. Because of the zero current condition on Class I waves, they also required that $E_z \to 0$ as $z \to -\infty$. They introduced dimensionless variables

$$\xi = V_0 z/D_e, \qquad M = n_e/n, \qquad N = n_i/n, \qquad F = \varepsilon_0 E_z V_0/neD_e$$

$$\gamma = e^2 D_e/\varepsilon_0 nkT_e, \qquad \lambda = D_e \rho n^2/V_0^2$$

and reduced the three equations to two:

$$\frac{dN}{dM} = \frac{M(1-MN)}{M-N+MF}; \qquad \frac{dF}{dM} = \frac{M-N}{M-N+MF}$$

subject to the boundary conditions $N = 0$ when $M = 0$; $N = 1$, $F = 0$ when $M = 1$.

The dimensionless parameter γ is a measure of the electrical interaction.

It is the ratio of the mean free path for electrons to the shortest Debye length, divided by the average number of electron–neutral collisions required for an ionizing collision. If $\beta < 1$, the Debye length is large and the ion and electron number densities may be appreciably different. If $\beta > 1$, space charge effects dominate, and the number densities of ions and of electrons are nearly equal throughout most of the ionization front. For values of n of interest, β is a large number.

The dimensionless parameter λ is an eigenvalue. The values of λ for which a solution can be obtained give the allowed propagation speeds of the ionization front. Turcotte and Ong obtained analytic solutions for extremes of γ and λ, and discussed the possibilities of a general solution. For $\gamma < 1$, $\lambda < 1$ they found

$$F = M - 1 \qquad N = [\tfrac{1}{2} + (\tfrac{1}{4} - \lambda)^{1/2}]M$$
$$M = \{1 + \exp[2\lambda(\xi - \xi_0)]\}^{-1/2},$$

while for $\gamma > 1$

$$F = (1 - M^2)/\beta \qquad N = M + (2\lambda/\beta)M^2(1 - M^2).$$

In this second extreme case M is again given by the same expression as for small γ. The first case describes proforce waves ($F < 0$), and the second describes antiforce waves ($F > 0$).

In their investigation of the general solution, Turcotte and Ong found that the parameters fall into regions of allowed solution, as shown in Fig. 40.

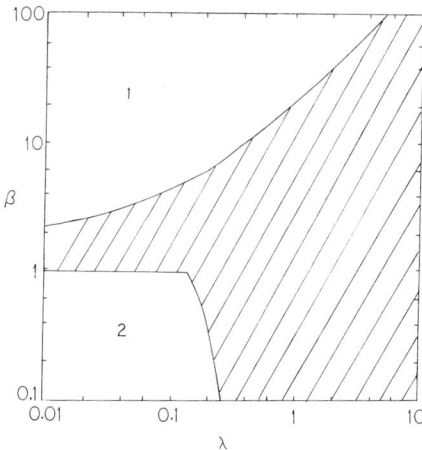

FIG. 40. Values of λ and β for which solutions can be obtained. Solutions exist in regions 2 and 3. [After Turcotte and Ong (85) with permission of *Journal of Plasma Physics*.]

These regions coincided reasonably well with the analytically possible extreme parameter solutions.

The method presented is interesting, and may be a more valid approach to the problem of wave structure than that afforded by the fluid equations. One great defect is the inability of the theory to determine a value of the wave speed V_0. Another is that recombination is invoked chiefly because it would not be possible to obtain convergence at $z = -\infty$ without some electron loss mechanism. But the time scale of recombination is profoundly pressure- and electron temperature-dependent, and notoriously long, so that its use should result in extraordinarily "thick" waves at low pressures where the waves are known to be very thin. A possible alternative, which has not been explored, would be to substitute radial ambipolar diffusion as the electron loss mechanism, replacing the term $-\rho n_e^2 n_i$ with a term $D_a(1/r) \times (\partial/\partial r)(r\, \partial n_e/\partial r)$, and to a first approximation replacing it in turn by $-D_a n_e/\Lambda^2$, where Λ is the so-called "diffusion radius" ($\Lambda = 2.405/a$ for a cylinder of radius a). The solution of this problem might well be simpler than that employing a recombination term.

C. Time-Dependent Solutions

Albright and Tidman (86) conducted an investigation of the time dependent electron fluid equations, with a particular view to discussing the stability of steady profile solutions. Their initial equations were not exact because they did not include the momentum and energy of nascent electrons, and they therefore overlooked a kinetic term which is first order in the electron velocity rather than quadratic. They employed Kirchhoff's equation in place of Poisson's, and in so doing neglected the ion current, an action which is possible in the laboratory frame, but makes transformation to the wave frame inaccurate. They considered electron production by electron impact and loss by recombination, but eliminated the production term between the continuity and energy equations. They neglected time changes and electron kinetic energy, assumed constant temperature in the energy equation, and integrated it into the form

$$\left(\frac{3kT_0}{2} + e\phi_i\right) n_e + \frac{\varepsilon_0}{2}(E^2 - E_0^2) + \frac{\varepsilon_0}{e}\left(\frac{5}{2}kT_0 + e\phi_i\right)\frac{dE}{dz} = 0.$$

Dropping similar terms from the momentum equation, they obtained

$$\partial E/\partial t + (e/mK_1\varepsilon_0)(en_e E + kT_0 \nabla n_e) = 0.$$

Introducing as dimensionless variables $\mathscr{E} = E/E_0$,

$$\tau = (eE_0)^2 t/mK_1(3kT_0 + e\phi_i),$$

$\xi = -eE_0 z/(3kT_0 + 2e\phi_i)$, and $\rho = kT_0/2e\phi_i$, they eliminated n_e between the equations to obtain a single equation in $\mathscr{E}$ for proforce waves

$$\frac{\partial \mathscr{E}}{\partial \tau} + \left(\mathscr{E}^2 + \frac{\partial \mathscr{E}}{\alpha \xi} - 1\right)\mathscr{E} + \rho\left(\frac{\partial \mathscr{E}^2}{\partial \xi} + \frac{\partial^2 \mathscr{E}}{\partial \xi^2}\right) = 0.$$

For $\rho = 0$, a Lagrange transformation can provide an exact solution of the residual equation, and if the profile is assumed to be linear in over a finite step, the result

$$V_0 = \left(\frac{5/2kT_0 + e\phi_i}{3/2kT_0 + e\phi_i}\right)\frac{eE_0}{mK_1}$$

is obtained. Since $e\phi_i$ is generally greater than kT_0 (the electron temperature somewhere in the wave), the factor in parentheses is of order unity. This result then compares almost identically with the results of the Shelton theory, where the parameter $\kappa = mK_1 V_0/eE_0$ is always found to be very nearly unity. Use of the Lagrange transformation, however, may introduce a hidden violation of the assumption that $N_i V_i = 0$, which was used in deriving the equation for $\mathscr{E}$.

Albright and Tidman then investigated the stability of this solution, and found that it was stable only for κ equal to unity. Presumably this limitation to such a narrow value reflects the large number of restrictions already placed on the solution, although κ is always found to be limited to a narrow range in the Shelton theory.

As Shelton had also found, a solution was not possible for antiforce waves with the equations in this form.

D. Photoionization Models

Klingbeil et al. (87) attacked the problem of a model for ionizing waves (such as those in streamers), which would include photoionization in the electron fluid equations. They began with the same equation set as Albright and Tidman, except that they elaborated the electron production function to include photoionization. They transformed their equations to the wave (stationary) frame without correcting the Kirchhoff equation for the ion current so introduced, and linearized the equations. Integrating the linearized continuity equation

$$\frac{dn_e}{dz} + \frac{\beta}{V_0 - v}n_e = -\frac{\beta_v}{V_0 - v}\exp(-z/\lambda_v)$$

(where β_v is the photoelectron density produced at z in the transit time of the wave, assuming that all photons originate at the back of the wave, and λ_v is the absorption length for these photons) one obtains

$$n_e = A \exp[-\beta z/(V_0 - v)] + \frac{\beta_v \exp(-z/\lambda_v)}{V_0 - v - \beta\lambda_v}.$$

It is now assumed that all original electrons are photon produced, so that $A = 0$, and that nevertheless the original photoelectron flux is very small ($\beta_v \sim 0$), so that $V_0 - v - \beta\lambda_v$ must also be very small. This then becomes deterministic of the wave velocity

$$V_0 = v + \beta\lambda_v.$$

By evaluating V from the momentum equation and β from the energy equation, Klingbeil et al. obtain an expression for V_0:

$$V_0 = \lambda_v n\bar{v}\left(\frac{\sigma/3kT_0[eE_0\lambda - (\lambda/\lambda_v)kT_0]^2 - \sum e\phi_j^*\sigma_j^*}{3/2kT_0 + e\phi_i + (6kT_0)^{-1}[eE_0\lambda - (\lambda/\lambda_v)kT_0]^2}\right)$$
$$- \frac{eE_0}{mK_1} + \frac{kT_0}{mK_1\lambda_v}$$

In this equation $\bar{v}$ is the mean electron speed, λ is the elastic mean free path, ϕ_j^* is the excitation energy of state j, and σ_j^* is its cross section. Klingbeil et al. have computed this equation for air at 760 torr and their result is given in Fig. 41.

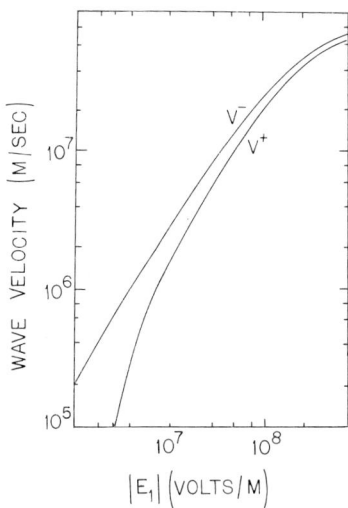

FIG. 41. Negative and positive-wave velocities as a function of E_0 at atmospheric pressure. [After Klingbeil et al. (87) with permission of Physics of Fluids.]

If there were significant numbers of original electrons which were not photon produced, i.e., if $A \neq 0$, the entire argument on which V_0 rests would collapse. Therefore one must conclude that agreement here either proves the hypothesis that photoionization *is* the *major* process, or that the agreement is fortuitous. One of the most persuasive arguments against photoionization is still the small effect that impurities of almost any description have on wave speed. Wagner, for example, found that the addition of arbitrary amounts of CH_4 to N_2 had only a minor effect on the forward and backward velocities of the streamers he studied. It should be noted that he added CH_4 for the specific purpose of intercepting and converting the N_2 vacuum UV radiation into visible quanta to aid in illuminating the avalanche, so that now there was presumably a vast reduction in available photoionizing radiations.

The small effect of impurities can be understood in the electron impact model. Whether it can be equally well explained by means of the above equation is not discussed by the authors.

E. Computer Avalanche Simulations

Ward (*88*), Davies *et al.* (*89*), and Kline and Siambis (*90*) have studied avalanche and streamer growth by means of computer simulations. The attempt is to account for the observed values of α/p as a function of E/p, and to describe the abnormal avalanche velocities observed by Wagner. All treated the problem as one-dimensional. Ward and Davies *et al.* used the continuity equations and Poisson's equation and injected secondary discharge-generated photoelectrons from the cathode. They assumed that the solution of the momentum equation is $v = eE/mK_1$. Kline and Siambis introduced a gas photoionization term based on the photoionization efficiency ψ of an avalanche as measured by Penney and Hummert (*91*). (A sample of their data on ψ is given in Fig. 42.) Kline and Siambis followed the trajectories of a group of individual particles in a self-consistent field determined by Poisson's equation. They decided on collisions by calculating the probability of survival and comparing it with a pseudo random number generated in the range (0, 1). This decision process was also extended to decide on the type of collision.

All found that they could account equally well for the abnormal anode- and cathode-directed avalanche growth observed by Wagner, although in the cases of Ward and Davies *et al.* the effect must be regarded as an optical illusion based on luminosity profiles of avalanche overlays, while in the case of Kline and Siambis it would be a real wave progression. A major objection to the work based on γ processes supplying later electrons into the avalanche as a source of explanation of Wagner's effects is that these effects are ob-

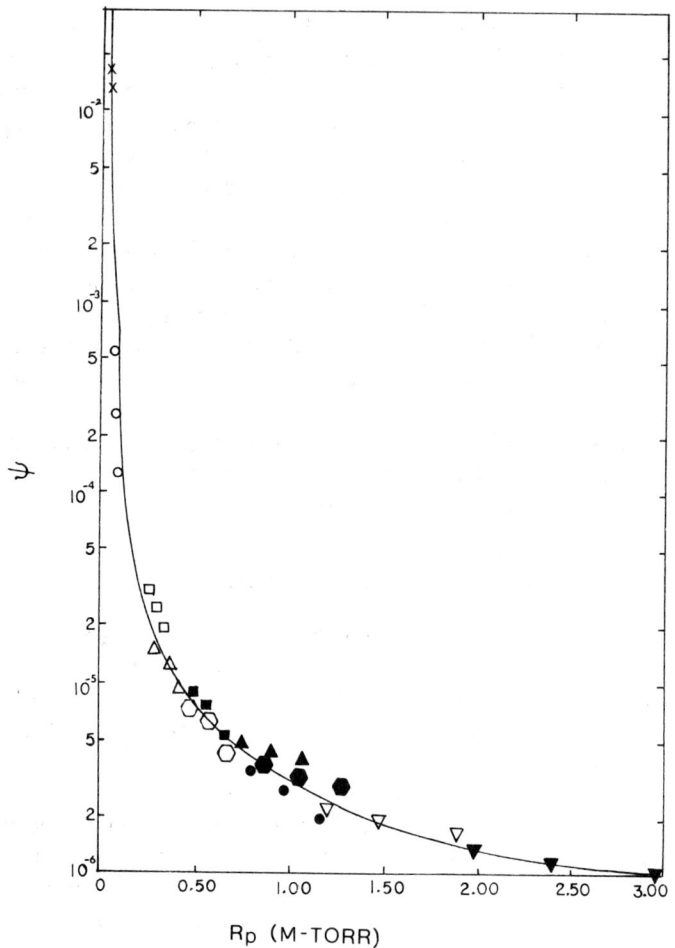

FIG. 42. Photon emitting efficiency of a corona (ψ) vs. distance times pressure. ψ = photoions generated per discharge ion-sr-m-torr in nitrogen. [After Penney and Hummert (91) with permission of *Journal of Applied Physics*.]

served with *single* avalanches, and that he knows by timing when secondary electrons appear from his cathode to augment the avalanche.

Kline and Siambis obtain excellent agreement between their calculations and observations of α/p over two orders of magnitude (Fig. 43). It would be expected that their method should describe normal avalanches well. The agreement found with the behavior of abnormal avalanches (Fig. 44) is less convincing, since it really amounts to having chosen a successful value of ψ somewhere in the observed range. The functional distance R given by Penney and Hummert does not refer precisely to the length of free photon flight, but is a rather vague "distance from the corona." When one considers that the Kline and Siambis method was one-dimensional rather than three-dimensional, and that it ignores electron–electron and electron–ion collisions (i.e., electron pressure), and that it in fact does show a rather strong

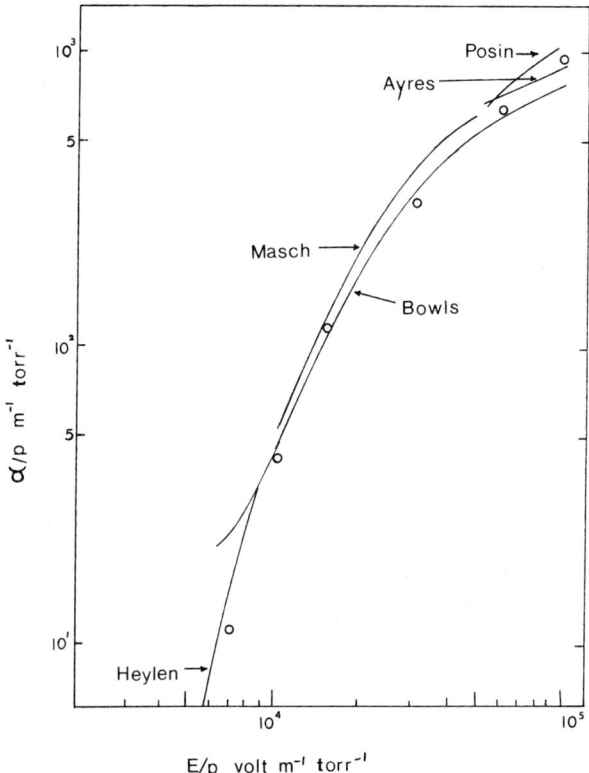

FIG. 43. Calculated and experimental values of α/p, the first Townsend ionization coefficient divided by pressures vs. E/p. Experimental results are shown as smooth curves. [After Kline and Siambis (90) with permission of the *Physical Review*.]

effect of space charge in the absence of photoionization even with these omissions (Fig. 45), their conclusion that "photoionization is an essential mechanism in streamer formation" does not seem fully justified. A more realistic electron impact model might succeed just as well.

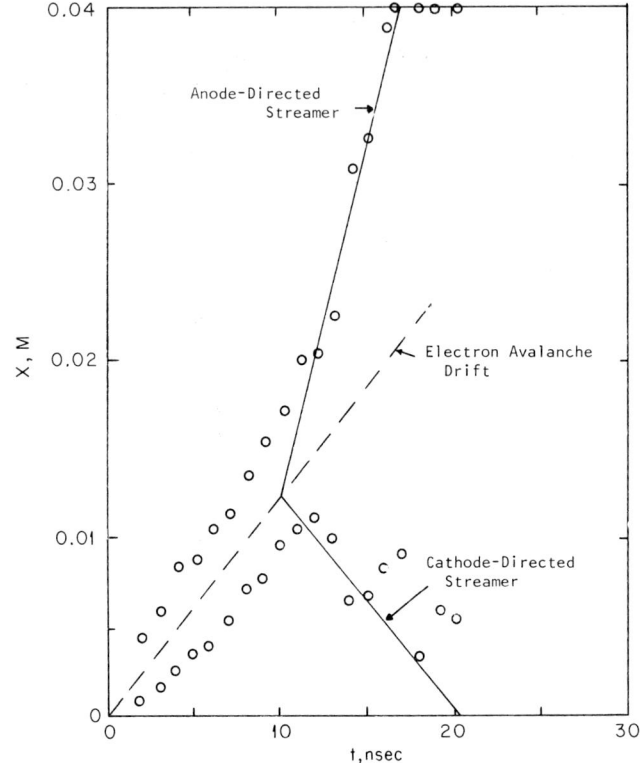

FIG. 44. Calculated luminosity vs. position and time. The points are the leading and trailing edges (where the luminosity falls to two orders of magnitude below the peak luminosity) of the luminous region at each time. Streamer development begins at $t \approx 12.5$ nsec when $\ln N(t = 0) + \alpha x = 21.3$ and the total number of electrons $N = 1.42 \times 10^9$. [After Kline and Siambis (90) with permission of the *Physical Review*.]

V. Conclusion

The foregoing review has been an attempt to present the new knowledge in the still-controversial field of electron fluid behavior. Much remains to be done, but the advances in both theory and experiment show much promise of future usefulness to electrical engineers and students of natural phenomena in treating some of their persistent problems.

The question of the importance of photoprocesses remains unresolved.

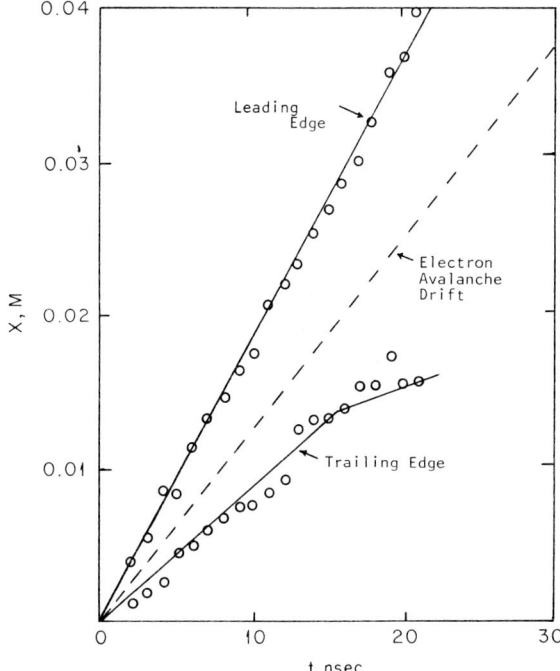

FIG. 45. Calculated luminosity vs. position and time. The points are the leading and trailing edges of the luminous region at each time. Space-charge distortion of the applied electric field begins at $t \approx 7$ nsec. [After Kline and Siambis (90) with permission of *Physical Review*.]

In Fig. 46, the totality of available data for nitrogen and air is plotted together with the theories of Klingbeil *et al.*, Shelton and Fowler, and Sanmann and Fowler. All work excellently in the mid-range of fields and velocities, probably because both of these variables depend principally on the electron mobility in this range. At the top of the range the Shelton–Sanmann theories seem slightly more successful, but at the low end the KTF theory is in striking agreement, while the SS theory fails completely in its present form. Before a final verdict is returned, however, it should be noted that not only do the abnormal avalanche velocities of Wagner fit on the KTF curve (a result the theory was intended to achieve), but so do the normal velocities. We must then perforce conclude that if this curve describes photon controlled processes, then all avalanche processes are photon controlled, a conclusion which seems unacceptable.

Much additional work is badly needed.

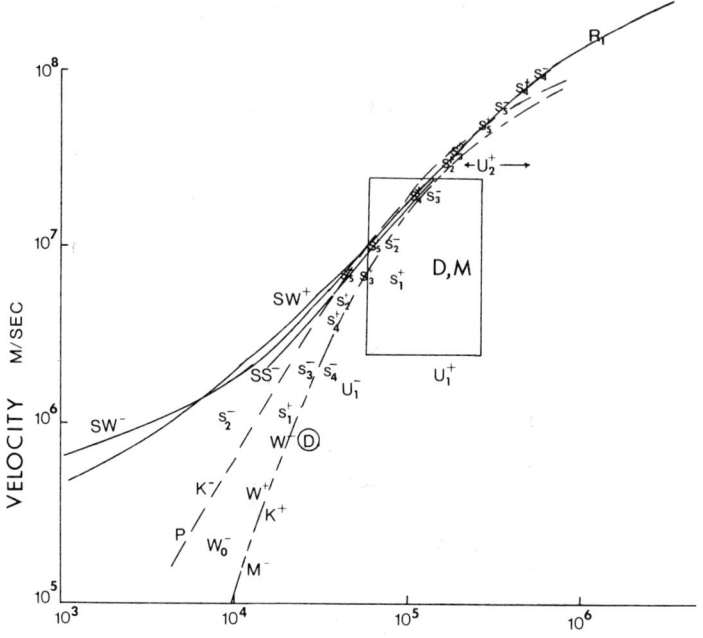

NITROGEN E/p volts/m/torr

FIG. 46. Compilation of available data on lightning, low pressure laboratory waves, avalanches, and long sparks, shown with curves of current theories. In nitrogen and air. [After Scott and Fowler (58) with permission of *Physics of Fluids*.]

S_i^+ = Scott i = 1 : 30 torr
  (anti) 2 : 10 torr
      3 : 4 torr
S_i^- = Scott 4 : 1.6 torr
  (pro) 5 : 1 torr
P = pilot leader, lightning
D = dart leader, lightning
M = M component, lightning
R_1 = first return, lightning
U_1^+ = Uman, spark leader (anti)
U_1^- = Uman, spark leader (pro)
U_2^+ = Uman, return stroke (anti)

M^- = Saxe and Meek spark leader (pro)
W_0^- = Wagner normal avalanche
W^+ = Wagner abnormal avalanche (anti)
W^- = Wagner abnormal avalanche (pro)
Ⓓ = Dawson and Lederman, microwave
K^- = Klingbeil *et al.* theory (pro)
K^+ = Klingbeil *et al.* theory (anti)
SS^- = Shelton and Fowler (strong pro)
SW^- = Shelton and Fowler (weak pro)
SW^+ = Sanmann and Fowler (weak anti)

REFERENCES

1. R. G. Fowler, *Advan. Electron. Electron Phys.* **35**, 1 (1974).
2. M. A. Uman, *Proc. IEEE* **59**, 457 (1971).
3. B. F. J. Schonland and H. Collens, *Proc. Roy. Soc., Ser. A* **143**, 654 (1934).
4. B. F. J. Schonland, D. J. Malan, and H. Collens, *Proc. Roy. Soc., Ser. A* **152**, 595 (1935).

5. D. J. Malon and H. Collens, *Proc. Roy. Soc., Ser. A* **162**, 175 (1937).
6. B. F. J. Schonland, *Proc. Roy. Soc., Ser. A* **166**, 56 (1938).
7. B. F. J. Schonland and H. Collens, *Proc. Roy. Soc., Ser. A* **166**, 56 (1938).
8. B. F. J. Schonland, D. J. Malan, and H. Collens, *Proc. Roy. Soc., Ser. A* **168**, 455 (1938).
9. D. J. Malan and B. F. J. Schonland, *Proc. Roy. Soc., Ser. A* **191**, 485 (1947).
10. D. J. Malan and B. F. J. Schonland, *Proc. Roy. Soc., Ser. A* **209**, 158 (1951).
11. B. F. J. Schonland, *Proc. Roy. Soc., Ser. A* **220**, 25 (1953).
12. M. A. Uman, "Lightning." McGraw-Hill, New York, 1969.
13. K. B. McEachron, *J. Franklin Inst.* **227**, 149 (1939).
14. K. Berger, *J. Franklin Inst.* **283**, 478 (1967).
15. G. A. Shelton and R. G. Fowler, *Phys. Fluids* **11**, 740 (1968).
16. F. E. Allibone and J. M. Meek, *Proc. Roy. Soc., Ser. A* **166**, 97 (1938).
17. G. N. Aleksandrov, *Zh. Tekh. Fiz.* **37**, 288 (1967).
18. D. B. Hodges, *Proc. Roy. Soc., Ser. B* **67**, 582 (1954).
19. M. A. Uman, *J. Geophys. Res.* **78**, 3530 (1973); M. A. Uman, D. K. McLain, R. J. Fisher, and E. P. Krider, *J. Geophys. Res.* **78**, 7911 (1973).
20. M. L. Pruett, *J. Geophys. Res.* **68**, 803 (1963).
21. M. A. Uman, *J. Atmos. Terr. Phys.* **26**, 1215 (1964); *J. Geophys. Res.* **74**, 949 (1969).
22. M. A. Uman and R. E. Orville, *J. Geophys. Res.* **70**, 549 (1965); **69**, 5151 (1964).
23. R. E. Orville, *J. Geophys. Res.* **73**, 6999 (1968).
24. W. H. Evans and R. L. Walker, *J. Geophys. Res.* **68**, 4455 (1963).
25. M. A. Uman, *J. Geophys. Res.* **69**, 583 (1964).
26. R. D. Hill, *J. Geophys. Res.* **68**, 1365 (1963).
27. R. C. Jones, *J. Geophys. Res.* **73**, 809 (1968).
28. M. Brook, N. Kittagawa, and E. J. Workman, *J. Geophys. Res.* **67**, 649 (1962); N. Kittagawa, M. Brook, and E. J. Workman, *J. Geophys. Res.* **67**, 637 (1962); M. Brook and N. Kittagawa, *J. Geophys. Res.* **65**, 1189 (1960).
29. M. A. Uman and R. E. Voshall, *J. Geophys. Res.* **73**, 497 (1968).
30. G. C. Simpson, *Proc. Roy. Soc., Ser. A* **111**, 56 (1926).
31. V. A. Ambartsumyan, "Theoretical Astrophysics." p. 265. Pergamon, New York, 1958.
32. C. E. R. Bruce, *Engineer* **224**, 676 (1967).
33. H. Alfvén, "Cosmical Electrodynamics." p. 7. Oxford Univ. Press, 1950.
34. R. G. Fowler and J. Hashemi, *Nature* **230**, 518 (1973).
35. B. R. De, *Solar Phys.* **31**, 437 (1973).
36. R. G. Giovanelli and M. K. McCabe, *Aust. J. Phys.* **11**, 191 (1958).
37. G. E. Moreton, *Sky & Telescope* **22**, 145 (1961).
38. R. G. Athay and G. E. Moreton, *Astrophys. J.* **133**, 935 (1961).
39. J. P. Wilde, K. V. Sheridan, and A. A. Meylan, *Aust. J. Phys.* **12**, 369 (1959).
40. A. A. Weiss and J. P. Wilde, *Aust. J. Phys.* **17**, 282 (1964).
41. L. F. Burlaga and N. I. Ness, *Solar Phys.* **9**, 467 (1969).
42. L. F. Burlaga and J. D. Scudder, *Astrophys. J.* **191**, L149 (1974).
43. E. N. Parker, *Astrophys. J. Suppl. Ser.* **177**, 77 (1963).
44. D. A. MacPherson and H. C. Koons, *J. Geophys. Res.* **75**, 5559 (1970).
45. A. L. Vampola, H. C. Koons, and D. A. MacPherson, *J. Geophys. Res.* **76**, 7009 (1971).
46. H. C. Koons, D. A. MacPherson, and M. Schulz, *J. Geophys. Res.* **75**, 6122 (1970).
47. J. Roquet, R. Schlich, and B. Selzer, *J. Geophys. Res.* **68**, 3731 (1963).
48. B. Caner, *Can. J. Phys.* **40**, 1846 (1962).
49. R. G. Fowler *Phys. Today* **26**, 23 (1973).
50. G. G. Bowman and J. S. Mainstone, *Aust. J. Phys.* **17**, 409 (1964).
51. K. Sheridan and J. Joisce, *Aust. J. Phys.* **16**, 584 (1963).

52. Y. Kato and S. Takei, *Sci. Rep., Tohoku Univ.* **15**, 6 (1963).
53. P. J. Edwards and J. S. Keid, *J. Geophys. Res.* **69**, 3607 (1964).
54. M. Cassaverde and A. Giesecke, *J. Geophys. Res.* **68**, 2603 (1963).
55. J. R. Davis and J. M. Headrick, *J. Geophys. Res.* **69**, 911 (1964).
56. G. R. Ochs, D. T. Farley, Jr., and K. L. Bowles, *J. Geophys. Res.* **68**, 701 (1963).
57. R. Smith, private communication.
58. R. P. Scott, Ph.D. Dissertation, University of Oklahoma, 1975; R. P. Scott and R. G. Fowler, *Phys. Fluids* (in press).
59. A. Haberstich, Ph.D. Dissertation, University of Maryland, 1964.
60. R. N. Blais and R. G. Fowler, *Phys. Fluids* **16**, 2149 (1973).
61. R. G. Fowler and G. A. Shelton, Jr., *Phys. Fluids* **17**, 334 (1974).
62. S. C. Brown "Basic Data of Plasma Physics," p. 251. MIT Press, 1964.
63. G. W. Bethke and A. D. Reuss, *Phys. Fluids* **7**, 1446 (1964); **12**, 822 (1969).
64. E. F. Dawson and S. Lederman, *Phys. Fluids* **16**, 205 (1973).
65. Y. L. Pan, J. R. Simpson, and A. F. Bernhardt, *Appl. Phys. Lett.* **24**, 871 (1974).
66. R. E. Nielsen and G. H. Canavan, *J. Appl. Phys. Lett.* **44**, 4224 (1973).
67. G. Russell and T. M. Holzberlein, *J. Appl. Phys.* **40**, 3071 (1969).
68. E. J. Lauer, T. C. Owen, and R. W. Bauer, *Bull. Amer. Phys. Soc.* **16**, 1228 (1971).
69. G. Benford, *J. Plasma Phys.* **10**, 203 (1973).
70. H. Raether, "Electron Avalanches and Breakdown in Gases." p. 6. Butterworths, Washington, 1964.
71. K. R. Allen and K. Phillips, *Proc. Roy. Soc., Ser. A* **274**, 163 (1963).
72. K. Wagner, *Z. Phys.* **189**, 465 (1966).
73. G. A. Dawson, C. N. Richards, E. P. Krider, and M. A. Uman, *J. Geophys. Res.* **73**, 815 (1968).
74. G. N. Aleksandrov, B. N. Gorin, V. P. Redkov, I. S. Stekol'nikov, and A. V. Shkilev, *Dokl. Akad. Nauk SSSR* **183**, 1048 (1968); [*Sov. Phys.-Dokl.* **13**, 1246].
75. D. W. Koopman and K. A. Saum, *J. Appl. Phys.* **44**, 5328 (1973).
76. M. A. Uman, R. E. Orville, A. M. Sletten, and E. P. Krider, *J. Appl. Phys.* **39**, 5162 (1968).
77. I. S. Stekol'nikov, *Dokl. Akad. Nauk SSSR* **141**, 1076 (1962).
78. I. S. Stekol'nikov and A. V. Shkilev, *Dokl. Akad. Nauk SSSR* **145**, 712 (1962) [*Sov. Phys.-Dokl.* **7**, 712]; *Dokl. Akad. Nauk SSSR* **151**, 837 (1963) [*Sov. Phys.-Dokl.* **8**, 825]; *Dokl. Akad. Nauk SSSR* **151**, 1085 (1963) [*Sov. Phys.-Dokl.* **8**, 829].
79. T. Udo, *IEEE Trans.* **PAS83**, 471 (1964).
80. J. J. Kritzinger, *Nature* **197**, 1165 (1963).
81. R. F. Saxe and J. M. Meek, *Proc. Inst. Elec. Eng.* **102C**, 221 (1955).
81a. J. W. Flowers, *Phys. Rev.* **64**, 225 (1943).
82. R. E. Orville, *J. Atmos. Sci.* **25**, 827 (1968).
83. E. E. Sanmann and R. G. Fowler, *Phys. Fluids* **18**, 1433 (1975).
83a. L. B. Loeb, *Science* **148**, 1417 (1965).
84. E. E. Sanmann, Ph.D. dissertation, University of Oklahoma, 1974.
85. D. L. Turcotte and R. S. B. Ong, *J. Plasma Phys.* **2**, 145 (1968).
86. N. W. Albright and D. A. Tidman, *Phys. Fluids* **15**, 86 (1972).
87. R. Klingbeil, D. A. Tidman, and R. F. Fernsler, *Phys. Fluids* **15**, 1969 (1972).
88. A. L. Ward, *Phys. Rev. A* **138**, 1357 (1965).
89. A. J. Davies, C. S. Davies, and C. J. Evans, *Proc. IEEE* **118**, 816 (1971).
90. L. E. Kline and J. G. Siambis, *Phys. Rev. A* **5**, 794 (1972).
91. G. W. Penney and G. T. Hummert, *J. Appl. Phys.* **41**, 572 (1970).

Atomic Photoelectron Spectroscopy, Part I

STEVEN T. MANSON

Department of Physics
Georgia State University
Atlanta, Georgia

I. Introduction 73
II. Theoretical Description of the Photoionization Process 75
   A. Photoionization Cross Sections 76
   B. Photoelectron Angular Distributions 97
   C. Relationship of Ionization by Fast Charged Particles to Photoionization 105
III. Experimental Techniques 106
   References 108

I. Introduction

Photoionization is the process of a photon of energy $h\nu$ colliding with and ionizing an atom or molecule in state i with the resulting positive ion being left in state j,

$$h\nu + X(i) \to X(j)^+ + e^-. \tag{1}$$

In most instances i and j refer to the ground state of X and X^+, but they can, in principle, be excited states as well. The fundamental relation governing the photoionization process (Einstein, 1905) is

$$\varepsilon = h\nu - I_{ij}, \tag{2}$$

where ε is the kinetic energy of the ejected photoelectron and I_{ij} is the energy required to just remove an electron from $X(i)$, leaving $X(j)^+$. If i and j refer to the ground states of X and X^+, respectively, then I_{ij} is the binding energy of the electron.

Until recently, photoionization was investigated experimentally almost entirely via photoabsorption measurements, but in the last 15 years direct observations of the ejected photoelectrons (photoelectron spectroscopy) have been made in ever-increasing numbers.

These measurements have been stimulated by a number of factors. First, there was the pioneering experimental work of Vilesov *et al.* (1961, 1962) in the ultraviolet region and Siegbahn *et al.* (1967, 1969) in the X-ray range.

Second is the fact that the photoelectron spectrum includes, in addition to *all* of the information which can be obtained from photoabsorption measurements, a great deal of additional information which photoabsorption measurements do not provide. Photoelectron spectroscopy provides probably the most accurate method for the determination of atomic and molecular subshell binding energies. This can rarely be done with photoabsorption spectroscopy because the precise energy for the absorption edge is often obliterated by the Rydberg states leading up to the edge and for open shell systems by multiplet structure. In addition, the size of the photoionization cross section at the absorption edge for the opening channel may be small compared to the cross section background of the other open channels and, thus, experimentally difficult to pick out above background and scatter. Intensity measurements in photoelectron spectroscopy also give the contribution of each atomic or molecular subshell to the total photoionization cross section since, as seen from Eq. (1), the photoelectrons from each subshell will have a different energy. In addition, intensity measurements will give directly the contribution of channels in which the ion is left in an excited state (so-called "shake-up" process) simply by energy analysis of the photoelectrons. A continuous distribution of photoelectrons from multiple photoionization processes (so-called "shake-off" processes) is also detectable in principle; in practice, the cross section is often so small that it cannot be observed above the background.

The angular distribution of the photoelectrons can also be measured using photoelectron spectroscopy. Such measurements, as will be discussed in the next section, give information on the phases of the matrix elements, rather than just their absolute squares. The angular distribution also contains a dependence on the phase shifts of the various possible continuum waves which describes the outgoing photoelectron. Thus, the manifest advantages of the photoelectron spectroscopy are clear.

In addition to the information provided by photoelectron spectroscopy about the photoionization process and many details of the structure of the atomic or molecular target, one can also learn something of the

$$c^\pm + A \rightarrow A^+ + e^- + c^\pm \tag{3}$$

process for any charged particle, $c^\pm$, moving sufficiently swiftly. In other words, any charged particle moving fast enough behaves like a photon and, for such particles, the ionization process is closely related to the photoionization process. The relationship will be discussed in detail in the following section.

From an experimental point of view, photoelectron spectroscopy is generally divided into two general categories corresponding to ultraviolet radiation (UPS) or X-rays (XPS) producing the photoelectrons. In this review, however, we take the theoretical point of view and attempt to unite, rather than separate, the two categories by looking at phenomena as a function of hv. In addition, the subject matter will be confined to atoms where the theoretical situation is fairly well understood and a fair amount of experimental results have been reported. The experimental situation is much better for molecules, there being many more molecules than atoms which are gaseous at room temperature, but *ab initio* theoretical work is in a decidedly fledgling state. This will be discussed further in the final section.

The primary purpose of this review is, thus, to synthesize what is known experimentally *and* theoretically about atomic photoelectron spectroscopy into a single body of knowledge. Experimental and theoretical results are compared, where both are available, in an effort to assess the accuracy of the results, particularly the theory. Further theoretical results are also presented, once their accuracy is ascertained, to give an overview of atomic photoelectron spectra over a broad range of atoms, subshells, and photon energies. This review therefore complements recent reviews by Bahr (1973), Shirley (1973), Price (1974), Hercules and Carver (1974), and Krause (1975), as well as the books by Turner *et al.* (1970), Zaidel and Shreider (1970), Sevier (1972), Damany *et al.* (1974), Marr (1967), and Eland (1974).

In Section II of this chapter a theoretical description of the photoionization process including discussions of cross sections and angular distributions is presented. Section III gives a brief guide to the recent literature on experimental techniques employed in photoelectron spectroscopy. Section IV presents a comparison of experimental and theoretical cross sections and an overview of the theoretical cross-section results. In Section V, the same is done for photoelectron angular distributions. Finally, Section VI sums up the review and presents concluding remarks as well as a prospectus for future efforts in the field of atomic photoelectron spectroscopy. (Sections IV, V, and VI will be published as Part II of this review.)

II. Theoretical Description of the Photoionization Process

The intensity of a photoelectron line is proportional to the photoionization cross section of that energy. In this section, the basic theory and the various approaches to the calculation of photoionization cross sections are discussed. In addition, the theory of the variation of photoelectron line intensity with angle of observation, i.e., photoelectron angular distributions, is presented and methods of calculation are reviewed. The final portion of this section is devoted to the relationship of atomic photoionization cross sections and angular distributions to the energy and angular distribution of electrons ejected from atoms by fast charged particle impact ionization.

A. Photoionization Cross Sections

1. General Theory

The cross section for photoionization of a system in state i by an unpolarized photon beam of energy $h\nu$ leaving the system in a final state f consisting of photoelectrons of energy ε plus ions in state j is given by (Bates, 1946)

$$\sigma_{ij}(\varepsilon) = (4\pi^2 \alpha a_0^2/3g_i)(\varepsilon + I_{ij})|M_{if}|^2, \qquad (4)$$

where α is the fine structure constant (1/137), a_0 is the Bohr radius (5.29 × 10^{-9} cm), g_i is the statistical weight of the initial discrete state,* and the ionization energy I_{ij} and the photoelectron energy ε are expressed in Rydbergs (13.60 eV). The matrix element, expressed in Rydberg atomic units,† is given by (Bethe and Salpeter, 1957)

$$|M_{if}|^2 = \frac{4}{(I_{ij} + \varepsilon)^2} \sum_{i,f} \left| \langle f | \sum_\mu \exp(i\mathbf{k}_v \cdot \mathbf{r}_\mu) \nabla_\mu | i \rangle \right|^2, \qquad (5)$$

with the summation over i, f being the sum over the degenerate initial and final states respectively, $\mathbf{r}_\mu$ is the position coordinate of the μth electron, $\mathbf{k}_v$ is the propagation vector of the photon ($|\mathbf{k}_v| = 2\pi\nu/c$), and the wave functions are normalized such that for the initial discrete state $|i\rangle$

$$\langle i | i \rangle = 1, \qquad (6)$$

and for the final continuum state $|f\rangle (= |j, \varepsilon\rangle)$

$$\langle j, \varepsilon | j', \varepsilon' \rangle = \delta_{jj'} \delta(\varepsilon - \varepsilon'). \qquad (7)$$

Up to this point the theory is essentially exact. By "essentially" is meant that Eq. (4) is really a first-order perturbation theory result (Heitler, 1954). It is to be noted, however, that the second-order perturbation result is a factor of α (1/137) smaller than the first-order. Thus, it is an excellent approximation to neglect it and all higher order contributions.

In addition, for incident photon energies below several keV, the $\exp(i\mathbf{k}_v \cdot \mathbf{r}_\mu)$ term in the matrix element Eq. (5) can be approximated. This is done by noting that the major concentration of wave function amplitude is around a distance from the nucleus, $\mathbf{r}_\mu$, of the order of the Bohr radius. Thus, for photon energies below several keV, $\mathbf{k}_v \cdot \mathbf{r}_\mu$ is small enough so that

---

* Statistical weight is another way of saying the number of degenerate sublevels at the energy of the initial state, e.g., the number of magnetic sublevels (in the absence of a magnetic field).

† By Rydberg atomic units we mean that length is in units of $a_0 = \hbar^2/(me^2)$, energy in units of the Rydberg $R = e^2/(2a_0)$ and wave number in units of $1/a_0$.

$\exp(i\mathbf{k}_v \cdot \mathbf{r}_\mu)$ can be well approximated by unity.* This approximation simplifies the matrix element considerably and is known as the "dipole approximation" or "neglect of retardation." The matrix element M_{if} can then be written

$$|M_{if}|^2 = \frac{4}{(I_{ij} + \varepsilon)^2} \sum_{i,f} \left|\langle f | \sum_\mu \mathbf{V}_\mu | i \rangle\right|^2 = \sum_{i,f} \left|\langle f | \sum_\mu \mathbf{r}_\mu | i \rangle\right|^2, \quad (8)$$

and the problem of calculation of photoionization cross sections reduces to one of finding wave functions for initial and final states. The details of the transformation of the matrix element in Eq. (8) are given below.

2. Alternative Forms of the Dipole Matrix Element

Before we proceed to a discussion of wave functions used in photoionization calculations, it is of interest to examine the dipole matrix element in further detail. Consider the exact nonrelativistic Hamiltonian of an atomic system

$$H = \sum_\mu \frac{P_\mu^2}{2m} + V(\mathbf{r}_1, \mathbf{r}_2, \ldots), \quad (9)$$

where the momentum of the μth electron $\mathbf{P}_\mu$ and the $\mathbf{r}_\mu$ satisfy the basic commutation relations

$$[x_\mu, P_{\mu'y}] = 0, \quad [x_\mu, P_{\mu'x}] = i\hbar\delta_{\mu\mu'}, \text{ etc.} \quad (10)$$

We thus find that

$$[\mathbf{r}_\mu, H] = i\hbar P_\mu/m, \quad (11)$$

noting that

$$[a, b^2] = [a, b]b + b[a, b]. \quad (12)$$

Then, taking the initial and final atomic states of the photoionization process to be eigenstates of the exact nonrelativistic Hamiltonian [Eq. (9)], i.e.,

$$H|i\rangle = E_i|i\rangle, \quad H|f\rangle = E_f|f\rangle, \quad (13)$$

* Actually, the approximation for cross sections is much better than is implied by the discussion. This is because expanding the exponential out gives $1 + i\mathbf{k}_v \cdot \mathbf{r}_\mu$ and taking the absolute square yields $1 + (\mathbf{k}_v \cdot \mathbf{r}_\mu)^2$ so we really need have only $(\mathbf{k}_v \cdot \mathbf{r}_\mu)^2$ very small compared to unity for the approximation to be valid.

the matrix element of Eq. (11) is

$$\langle i|[\mathbf{r}_\mu, H]|f\rangle = \frac{i\hbar}{m}\langle i|\mathbf{P}_\mu|f\rangle$$
$$= (E_f - E_i)\langle i|\mathbf{r}_\mu|f\rangle \quad (14)$$

by virtue of the fact that i and f are eigenstates of H. Thus, using Eq. (14), the dipole matrix element [Eq. (8)] can be written

$$|M_{if}|^2 = \frac{\hbar^2}{m^2(E_f - E_i)^2}\sum_{i,f}\left|\langle i|\sum_\mu \mathbf{P}_\mu|f\rangle\right|^2$$
$$= \frac{4}{(E_f - E_i)^2}\sum_{i,f}\left|\langle i|\sum_\mu \mathbf{V}_\mu|f\rangle\right|^2. \quad (15)$$

Note that

$$E_f - E_i = I_{ij} + \varepsilon \quad (16)$$

in our previous equation. Equation (15) is known as the dipole-velocity (or just "velocity") form of the matrix element, while Eq. (8) is called the dipole-length form.

A third, formally equal, form of the dipole matrix element can be derived from the relation

$$(E_f - E_i)\langle i|\mathbf{P}_\mu|f\rangle = \langle i|[\mathbf{P}_\mu, H]|f\rangle = \langle i|[\mathbf{P}_\mu, V]|f\rangle. \quad (17)$$

Then, since the potential V of a general atomic system is

$$V = -Z\sum_\mu \frac{e^2}{r_\mu} + \sum_{\mu<\mu'}\frac{e^2}{|\mathbf{r}_\mu - \mathbf{r}_{\mu'}|}, \quad (18)$$

Eq. (17) can be written

$$(E_f - E_i)\langle i|\mathbf{P}_\mu|f\rangle = \langle i|[-i\hbar\mathbf{V}_\mu, V]|f\rangle$$
$$= -i\hbar\langle i|\mathbf{V}_\mu V|f\rangle = -iZe^2\hbar\langle i|\frac{\mathbf{r}_\mu}{r_\mu^3}|f\rangle. \quad (19)$$

Thus, the dipole matrix element can be written

$$|M_{if}|^2 = \frac{16Z^2}{(E_f - E_i)^4}\sum_{i,f}\left|\langle i|\sum_\mu \frac{\mathbf{r}_\mu}{r_\mu^3}|f\rangle\right|^2, \quad (20)$$

the dipole-acceleration form of the matrix element.

Thus we have three alternative formally equal forms of the dipole matrix element. We emphasize that what has been shown is that using exact wave functions, i.e., eigenfunctions of the exact nonrelativistic Hamiltonian,

Eqs. (8), (15), and (20) are equal. Of course, for atomic systems other than hydrogen, exact wave functions are not available. In that case (approximate wave functions) the results of using the various expressions for the dipole matrix element can differ considerably from each other and from the correct expression. On the other hand, it is possible that all three expressions might give the same result with approximate wave functions, and that this result might still be incorrect. This point will be discussed further in connection with central field wave functions below. We thus see that equality among the results of the alternative forms of the dipole matrix element is a necessary but not sufficient condition for the correctness of that result. To conclude this discussion we note that, as a practical matter, it is only the dipole-length and dipole-velocity expressions that are usually computed and compared. This is because approximate wave functions are usually generated by a variational principle on the energy which is not very sensitive to the details of the wave function near the nucleus, while the dipole-acceleration form of the matrix element [Eq. (20)] emphasizes this region quite strongly by the r_μ^{-3} in the matrix element.

3. Central Field Calculations

The simplest type of wave functions which are useful in calculating photoionization cross sections are those based on a central-field approximation to the exact Hamiltonian, i.e., one considers the solution to the approximate Hamiltonian

$$H_0 = \sum_\mu [(P_\mu^2/2m) + U(r_\mu)] \qquad (21)$$

for the initial and final states of the atom. Note that $U(r_\mu)$, the central potential seen by each electron, is a function of *scalar* r_μ only. The solutions to H_0 are antisymmetric products of one-electron wave functions, $r^{-1}P_{nl}(r)Y_l^m(\theta, \Phi)$ [$r^{-1}P_{\varepsilon l}(r)Y^m(\theta, \Phi)$ for continuum electrons]. The radial parts of the one-electron functions are solutions to the one-body Schrödinger equation

$$\left[\frac{d^2}{dr^2} - \frac{l(l+1)}{r^2} - U(r) + E\right]P(r) = 0. \qquad (22)$$

For both discrete *and* continuum functions, Eq. (22) has r in atomic units and energies in rydbergs. In using central-field wave functions, then, only the one electron is permitted to change quantum numbers in the photoionization process or the matrix element vanishes. Thus, multiple transitions are specifically excluded. Further, since the initial and final states are solutions to the Schrödinger equation in the *same* central potential $U(r)$, the orbitals not directly involved in the photoionizing transition remain unchanged. The rearrangement of the remaining electrons after a transition is known as core relaxation, i.e., core relaxation effects are excluded in the central-field model. Therefore, the nonparticipating orbitals integrate out to unity in the dipole matrix element and the photoionization cross section for an nl electron can be written in dipole-length form as (Cooper, 1962)

$$\sigma_{nl}(\varepsilon) = \frac{4}{3}\pi^2\alpha a_0^2 \frac{N_{nl}(\varepsilon - \varepsilon_{nl})}{2l + 1}[lR_{l-1}(\varepsilon)^2 + (l + 1)R_{l+1}(\varepsilon)^2], \quad (23)$$

with ε_{nl} the binding energy of an nl electron (intrinsically negative), N_{nl} the occupation number of the nl subshell, and the matrix element

$$R_{l\pm 1}(\varepsilon) = \int_0^\infty P_{nl}(r) r P_{\varepsilon, l\pm 1}(r)\, dr, \quad (24)$$

where the continuum normalization, from Eq. (7), takes the form

$$P_{\varepsilon l}(r) \xrightarrow[r \to \infty]{} \pi^{-1/2}\varepsilon^{-1/4} \sin[\varepsilon^{1/2}r - \tfrac{1}{2}l\pi$$

$$- \varepsilon^{-1/2} \ln 2\varepsilon^{1/2}r + \sigma_l(\varepsilon) + \delta_l(\varepsilon)], \quad (25)$$

where $\sigma_l(\varepsilon) = \text{Arg } \Gamma(l + 1 - i\varepsilon^{-1/2})$ and $\delta_l(\varepsilon)$ is the phase shift. It is thus seen that the central-field calculation reduces the problem to a one-electron model of the photoionization process involving only the wave function of the photoelectron before and after the photoionization, and a single-electron Hamiltonian

$$h_0 = (P^2/2m) + U(r) \quad (26)$$

to which each is a solution.

An interesting property of the one-electron model is that

$$[\mathbf{r}, h_0] = i\hbar\mathbf{p}/m, \quad (27)$$

which is formally equivalent to Eq. (11), the basis for the transformation from length to velocity matrix elements. Thus, an *exact* transformation of the dipole matrix element within the framework of the central-field calculation is

$$R_{l\pm 1}(\varepsilon) = \frac{2}{\varepsilon - \varepsilon_{nl}} \int_0^\infty P_{nl}(r)\left[\frac{d}{dr} \pm \frac{2l + 1 \pm 1}{2r}\right] P_{\varepsilon, l\pm 1}(r)\, dr, \quad (28)$$

the dipole-velocity form. Hence it is seen that length and velocity forms are necessarily equal in the central-field calculation. This shows that equality of the alternative forms of the dipole matrix element is no guarantee of agreement with experiment. The derivation of the acceleration matrix element goes through similarly, but since it is used infrequently, we omit the details.

The alternative forms of the dipole matrix element, then, give no information as to how close to experiment the results of a central-field calculation are. They do, however, provide a powerful check on the numerical methods used in the computation. It is almost impossible, for example, to have an error in the calculation of the dipole matrix element and still retain the equality of length and velocity.

Up to this point, no mention has been made of the detailed form of the central potential. The simplest form is, of course, a hydrogenic potential with a fixed effective charge Z_{eff} (Slater, 1930), i.e., a potential (in rydbergs)

$$U(r) = -2Z_{\text{eff}}/r. \tag{29}$$

Photoionization results using the hydrogenic potential* were worked out some time ago (Hall, 1936). The major defect in this potential is that it is far too small for small r (near the nucleus) where an electron will not be screened by the other electrons and will "see" a full nuclear charge Z and a potential $-2Z/r$. In addition, for large r, atomic electrons should "see" the charge of the nucleus screened by the $Z-1$ other atomic electrons for a net charge of unity, i.e., a potential $-2/r$. Thus, the hydrogenic result will be grossly in error when the major contribution to the dipole matrix element comes from small r or large r which occurs for low photoelectron energy and high photoelectron energy, respectively. This is borne out in direct comparison of the hydrogenic vs. more sophisticated results (Manson and Cooper, 1968). Thus, for reasonable results, one needs the central potential to have boundary conditions

$$U(r) \xrightarrow[r \to 0]{} -2Z/r, \qquad U(r) \xrightarrow[r \to \infty]{} -2/r. \tag{30}$$

The Thomas–Fermi potential (Thomas, 1927; Fermi, 1928) has such boundary conditions, but being a statistical model, it does not include shell effects which can be extremely important (Fano and Cooper, 1968). A Hartree self-consistent-field potential (Hartree, 1928) can be used and has been used in the work of Cooper (1962). This includes shell effects but is devoid of exchange, which is very important in atomic systems. To include exchange fully is not possible within the context of a central-potential calculation because exchange is explicitly a noncentral nonlocal interaction (Hartree,

---

* The actual potential used in hydrogenic calculations is not quite so simple, but includes a constant term as well known as outer screening to give a realistic ionization energy.

1957; Slater, 1960). One can approximate the effects of exchange using a central potential (Slater, 1951). The Slater approximation to exchange amounts to assuming the form of the exchange potential to be that of a free-electron gas. If the total charge density (both spins) is ρ, this exchange potential is given by (Herman and Skillman, 1963)

$$V^{\text{ex}}(r) = -6\left[\frac{3}{8\pi}|\rho(r)|\right]^{1/3} \qquad (31)$$

which is clearly a central potential. If this approximate exchange potential is introduced into a Hartree self-consistent-field calculation and the resulting equations are solved self-consistently, one obtains the Hartree–Slater (HS) central-field wave functions and potential.* These results are discussed in great detail and tabulated for neutral atoms for $Z = 2$ to $Z = 103$ by Herman and Skillman (1963). They are often referred to as Herman–Skillman potentials and wave functions.†

Quite a number of photoionization calculations employing HS wave functions have been reported to date. Among the many are (Combet Farnoux and Heno, 1967; Manson and Cooper, 1968; McGuire, 1968).

As shall be seen in Section IV, the results of HS calculations provide quite reasonable agreement with experiment (20% or better) in most energy ranges. The HS central-field potential is computationally convenient and the calculation is quite fast on present-day computers. Overall, the HS approximation provides the best central field for photoionization calculations.

Central-field calculations, however, suffer from a major difficulty: the inexact treatment of exchange is such that multiplet structure is completely ignored. This can be quite important for open-shell atoms, particularly near threshold. Further, the Slater approximation to exchange, Eq. (30), is always attractive. While "real" exchange forces are often attractive, it is well known that they can be repulsive as well (Kennedy and Manson, 1972). Before going to more exact and complicated treatments, however, we note that a significant portion of what is now known about atomic photoionization cross sections and photoelectron spectroscopy has been learned from HS central-field calculations (Fano and Cooper, 1968).

* It is important to note, however, that HS potential obtained from the self-consistent solution goes to zero at infinity. Thus, to have the correct asymptotic form, the Latter (1955) cutoff is introduced. This is that the potential $U(r) = U^{\text{HS}}(r)$ for $r < a$, $U(r) = -2/r$ for $r > a$ with a being defined as the value of r such that $U^{\text{HS}}(a) = -2/a$.

† Originally, this was referred as a Hartree–Fock–Slater calculation. This is felt to convey the false impression that it is better than Hartree–Fock, which it most assuredly is not. It is, however, better than Hartree alone, so that the Hartree–Slater terminology is entirely appropriate. In addition, the acronym for Hartree–Slater is HS, which is the same for Herman–Skillman, thus aiding in identification.

4. Hartree–Fock Calculations

The simplicity of wave functions consisting of single Slater (1960) determinants, i.e., antisymmetric products of one-electron functions, can be maintained while still treating exchange correctly. This is the Hartree–Fock (HF) method (Hartree, 1957). For calculations of discrete state wave functions, the method has been reviewed by Hartree (1946, 1957) and Slater (1960) among others. In addition, many recent tabulations of extensive sets of HF discrete state wave functions have been reported (Clementi, 1965; Mann, 1968; Fischer, 1968; Clementi and Roetti, 1974). Basically the method involves setting up a wave function ψ for the system in question, which is an antisymmetric product of one-electron functions, $r^{-1}P_{nl}(r)Y_l^m(\theta, \Phi)$, or, more generally, a linear combination of such products so as to correctly represent the angular momentum couplings of an open-shell many-electron system. The $P_{nl}(r)$ are treated as unknowns and the so-called energy functional $\langle \Psi | H | \Psi \rangle$ is constructed; H is the exact nonrelativistic Hamiltonian. The variation principle is then applied to this functional subject to the constraints of the orthonormality of the one-electron functions. This results in a set of self-consistent coupled integrodifferential equations for the $P_{nl}(r)$ which can then be solved, yielding the HF wave functions for the given state. The mathematical details can be obtained in the above works and references therein; a detailed discussion of HF theory is beyond the scope of this review. Note, however, that the HF wave function is the most accurate independent-particle wave possible since it is obtained via the variation principle.

Dealing with the final continuum state resulting from the photoionization process is more difficult since the HF problem is not defined for wave functions containing continuum orbitals. This is because the HF method solves for each orbital in the field generated by the charge distribution of the other orbitals. The charge distribution for a continuum orbital is not defined since continuum orbitals are non-normalizable. Thus, one proceeds as follows: first a HF calculation for the residual ion core minus photoelectron is performed. This can be done by ordinary discrete state HF procedures as described above, although some extra care must be taken when the photoelectron comes from an inner shell and the ion core is in an excited state well above the ionization threshold (Bagus, 1965). This done, the core orbitals are frozen and the above HF procedure can be carried out for the total ion core plus photoelectron final state with only the radial part of the continuum orbital, $P_{\varepsilon l}(r)$, unknown. This procedure yields a single integrodifferential equation for $P_{\varepsilon l}(r)$ which is known as the continuum HF equation. The details of this method for various cases are given by Seaton (1951), Dalgarno et al. (1964), Amusia et al. (1969), and Kennedy and Manson

(1972). As an example, however, the continuum HF equation for $P_{\varepsilon d}(r)$ resulting from the Xe 5p → εd photoionization is given by (Kennedy and Manson, 1972).

$$\left[\frac{d^2}{dr^2} - \frac{l(l+1)}{r^2} + \varepsilon + V(r)\right]P_{\varepsilon l}(r) + U(r) = 0 \tag{32}$$

where

$$V(r) = \frac{2}{r}\left[Z - 2\sum_{n=1}^{5} Y^0(ns, ns) - 6\sum_{n=2}^{4} Y^0(np, np)\right.$$

$$\left. - 10\sum_{n=3}^{4} Y^0(nd, nd) - 5Y^0(5p, 5p) + \frac{1}{5}Y^2(5p, 5p)\right] \tag{33}$$

with

$$Y^k(nl, n'l') = \frac{1}{r^{k+1}}\int_0^r P_{nl}(r')r'^k P_{n'l'}(r')\,dr' + r^k\int_r^\infty \frac{P_{nl}(r')P_{n'l'}(r')}{r'^{k+1}}\,dr' \tag{34}$$

and

$$U(r) = \frac{2}{r}\left[\frac{1}{5}\sum_{n=1}^{5} Y^2(ns, \varepsilon d)P_{ns}(r) + \sum_{n=2}^{4}\left\{\frac{2}{5}Y^1(np, \varepsilon d) + \frac{9}{35}Y^3(np, \varepsilon d)\right\}P_{np}(r)\right.$$

$$-\left\{\frac{14}{15}H^1(5p, \varepsilon d) - \frac{9}{35}Y^3(5p, \varepsilon d)\right\}P_{5p}(r) + \sum_{n=3}^{4} Y^0(nd, \varepsilon d)$$

$$\left. + \left\{\frac{2}{7}Y^2(nd, \varepsilon d) + \frac{2}{7}Y^4(nd, \varepsilon d)\right\}P_{nd}(r)\right] + \lambda_{3d} P_{3d}(r) + \lambda_{4d} P_{4d}(r) \tag{35}$$

with the λ's off-diagonal parameters to insure orthogonality between $P_{\varepsilon d}(r)$ and P_{3d}, P_{4d}.

Within the framework of the HF approximation for the wave function, the general expression for photoionization of an electron from an $(nl)^q \; {}^{2S+1}L$ to an $\{[(nl)^{q-1} \; {}^{2S_c+1}L_c], (\varepsilon l')\} \; {}^{2S+1}L$ final state is given by (Bates, 1946)

$$\sigma_{nl}(LS, L_cS_c, \varepsilon l'L') = \frac{4\pi^2\alpha a_0^2}{3g_i}\frac{I_{ij} + \varepsilon}{R}\frac{1}{4l_>^2 - 1}\zeta(LS, L_cS_c, l'L')\gamma|R_{l'}(\varepsilon)|^2. \tag{36}$$

Here l' ($= l \pm 1$) is the angular momentum of the photoelectron, $l_>$ is the greater of l and l', ζ is the relative multiplet strength, L and S are the total orbital and spin angular momenta, respectively, for the initial state, L_c and S_c for the residual ion core, and L' is the total orbital angular momentum of the total ion plus photoelectron final state. The overlap integral, γ, is given by

$$\gamma = \pi \prod_{\substack{\text{passive} \\ \text{electrons}}} \left| \int_0^\infty P_{nl}^i(r) P_{nl}^f(r) \, dr \right|^2 \tag{37}$$

where the superscripts i and f refer to orbitals from the initial and final states respectively, and the dipole matrix element is

$$R_{l'}(\varepsilon) = \int_0^\infty P_{nl}^i(r) r P_{\varepsilon, l'}^f(r) \, dr. \tag{38}$$

The overlap integral γ appears because the discrete orbitals of the initial and final states are solutions to different HF equations due to the removal (photoionization) of one of the electrons. Thus, the ion core "relaxes" and γ represents the effects of this core relaxation. Note that core relaxation was not included in the central-field calculation discussed previously. The cross section for a particular physical photoionization channel $i(LS) \to j(L_c S_c)$ is

$$\sigma_{ij}(\varepsilon) = \sum_{L'} \sum_{l'} \sigma_{nl}(LS, L_c S_c, \varepsilon l L'). \tag{39}$$

The dipole-velocity form of the matrix element is given by Eq. (27), just as for the central-field model, with the modification that $\varepsilon - \varepsilon_{nl}$ is to be replaced by $\varepsilon + I_{ij}$. In the HF calculation, however, there is no assurance that length and velocity will give the same results. Thus, the two forms give some idea of the accuracy of the calculated result, rather than being a check on the numerical procedure as in the central-field calculation.

One would like to have an *a priori* guide to which form of the dipole matrix element, length or velocity, will be the more accurate. It has been argued by Starace (1971, 1973) that the dipole-length is the appropriate form when dealing with approximate nonlocal potentials. On the other hand, Grant (1974) has made the case that dipole-velocity is the correct form. The connection between the two arguments has been summarized by Grant and Starace (1975). They also point out that much work needs to be done before the question of length vs. velocity can be answered definitively.

Before closing this section on the HF calculation, it must be pointed out that the above procedure is not the only one possible. One can, for example, use the same discrete orbitals for initial and final states. Among the common choices are initial state HF orbitals and HS orbitals. This procedure is certainly less accurate than the one described above, as it neglects, e.g., core relaxation. The advantage, however, of using a common set of discrete orbitals in initial and final states is that this makes systematic improvements over the HF calculation considerably easier to implement.

A final note on the HF calculation is that within the framework described, resonances in the photoionization cross section due to autoionizing and Auger states are entirely omitted. It is true that these appear only in a range of the order 10 eV below each ionization threshold beyond the first, but where they appear the cross section can be affected enormously. To include them, however, requires going beyond the HF approximation.

5. Beyond the Hartree–Fock Calculation: Continuum Configuration Interaction, Close-Coupling, and Many-Body Perturbation Theory

To go beyond the HF theory means that one no longer assumes a wave function which is composed of antisymmetric products of one-electron orbitals. Great progress has been made in developing wave functions including correlation (or configuration interaction) for discrete states, and methods exist (Hartree, 1957; Slater, 1960; Lefebvre and Moser, 1969; Weiss, 1973) for obtaining discrete wave functions to any desired degree of accuracy. Thus, we shall not deal with the discrete wave function problem here. The problem of improving the wave function for the unbound continuum states of the final ion plus photoelectron system, however, is more difficult. Several approaches are outlined below.

One can introduce configuration interaction into continuum wave functions via the approach outlined by Fano (1961) and extended by Mies (1968). A set of solutions to an approximate Hamiltonian H_0 is generated

$$H_0 \psi_{i\varepsilon} = \varepsilon \psi_{i\varepsilon}, \tag{40}$$

with i designating all the quantum numbers of the channel, discrete wave functions normalized to unity, and continuum to $\delta_{ij}\delta(\varepsilon - \varepsilon')$. The total Hamiltonian $H = H_0 + V$ has matrix elements

$$\langle i\varepsilon | H | j\varepsilon' \rangle = \varepsilon \delta_{ij} \delta(\varepsilon - \varepsilon') + V_{i\varepsilon, j\varepsilon'}, \tag{41}$$

with the δ function becoming a Kronecker δ for discrete functions. We expand the solution to H, Ψ_E, in terms of the ψ_i:

$$\Psi_E = \sum_i \int a_{i\varepsilon}(E)\psi_{i\varepsilon}\, d\varepsilon, \tag{42}$$

where the integral includes the sum over discrete $\Psi_{i\varepsilon}$. We wish Ψ_E to be a solution to

$$H\Psi_E = E\Psi_E \tag{43}$$

which leads to

$$a_{i\varepsilon}(E) = \frac{1}{E - \varepsilon'} \sum_j \int d\varepsilon'\, a_{j\varepsilon'}(E) V_{i\varepsilon, j\varepsilon'}, \tag{44}$$

which formally has a singularity at $\varepsilon = E$. To remove this difficulty, define

$$a_{i\varepsilon}(E) = \frac{P}{E - \varepsilon} A_{i\varepsilon}(E) + B_i(E)\delta(\varepsilon - E), \tag{45}$$

with P as the principal value of an integral, so that $A_{i\varepsilon}(E) = (E - \varepsilon)a_{i\varepsilon}(E)$ [since $x\delta(x) = 0$] and the $B_i(E)$, defined only for open channels, are integration constants to be determined. Thus,

$$A_{i\varepsilon}(E) = \sum_j P \int \frac{d\varepsilon' A_{j\varepsilon'}(E) V_{i\varepsilon,\, j\varepsilon'}}{E - \varepsilon'} + \sum_j{}' V_{i\varepsilon,\, jE} B_j(e), \tag{46}$$

where the first sum is over all channels and the second (primed) sum only over open channels. By open channels we mean final states which are energetically accessible; closed channels refer to those which are not. From Eq. (46) it is seen that the A's and B's are linearly related. Defining the K matrix (Fano and Prats, 1964)

$$A_{i\varepsilon}(E) = \sum_j K_{i\varepsilon,\, jE} B_j(E), \tag{47}$$

so that, combining Eqs. (46) and (47)

$$K_{i\varepsilon,\, jE} = V_{i\varepsilon,\, jE} + \sum_l P \int \frac{d\varepsilon' V_{i\varepsilon,\, l\varepsilon'} K_{l\varepsilon',\, jE}}{E - \varepsilon'}. \tag{48}$$

In terms of the K matrix, the Ψ_E is then given by

$$\Psi_E = \sum_i{}' B_i(E)\Psi_{iE} + \sum_i\sum_j{}' P \int \frac{K_{i\varepsilon,\, jE} B_j(E)\Psi_{i\varepsilon}\, d\varepsilon}{E - \varepsilon}. \tag{49}$$

The B's are determined by normalizing Ψ_E, yielding

$$\sum_{i,j}{}' B_i^*(E)[\delta_{ij} + \pi^2 \sum_l K^*_{iE,\, lE} K_{lE,\, jE}] B_j(E) = 1. \tag{50}$$

The K-matrix elements are obtained from Eq. (48) using a perturbative or iterative procedure, and the B's are obtained subsequently from Eq. (50). Then the operations indicated in Eq. (49) can be carried out to get the continuum wave function Ψ_E, and thus the dipole matrix element. The fact that complete sets of solutions, including discrete, are used means that the effect of autoionizing states on the photoionization cross section comes out directly (Fano, 1961; Fano and Cooper, 1965). A number of *ab initio* calculations using this method primarily with autoionizing states have been published (Altick and Moore, 1966; Starace, 1970).

The practical limitations of this method are that all possible channels cannot be included since they form an infinite set and the energy integrals cannot be carried out to infinity. The method does, however, offer the advantage of being able to spotlight easily the important contributions to the continuum wave function.

The choice of H_0 is also of great importance. If, for example, H_0 is chosen to be a central-field Hamiltonian, the problem is fairly complicated since both $V_{i\varepsilon, i\varepsilon'}$ connecting states of a given channel but different energy (*intra*channel coupling) *and* $V_{i\varepsilon, j\varepsilon'}$ connecting different channels (*inter*channel coupling) are, in general, nonzero. One can simplify the problem by taking H_0 to be the HF Hamiltonian within each channel. This diagonalizes the intrachannel coupling, i.e., $V_{i\varepsilon, i\varepsilon'} = 0$ in this case. As a matter of fact, diagonalizing the intrachannel coupling and solving the continuum HF differential equation are equivalent problems (Fano and Cooper, 1968). The HF integrodifferential equation method is generally numerically more accurate since the problems of summing to infinite energies are avoided (Starace, 1970; Kennedy and Manson, 1972), but the diagonalization procedure gives more insight into the physical nature of the corrections to a simple continuum wave function.

The interchannel problem can also be solved via coupled integrodifferential equations known as the close-coupling equations (Percival and Seaton, 1957). This is essentially a multichannel extension of the continuum HF calculation discussed previously, or, alternatively, the HF method is the single-channel case of close coupling. A great deal of literature exists on the close-coupling method (Burke, 1968; Burke and Seaton, 1971; Harris and Michels, 1971; Smith, 1971; Truhlar *et al.* 1974). The general principles of the method, employing the Kohn (1948) variation principle, are briefly outlined below.

In the close-coupling method, one considers wave functions of the form

$$\Psi_E = \sum_\alpha a_\alpha \psi_\alpha F_\alpha(r), \tag{51}$$

where the a_α are the expansion coefficients, the ψ_α are the (discrete) wave functions of the residual positive ion, and the F_α represent the continuum photoelectron, and proper antisymmetry including spin functions is assumed. If the sum is carried out over a complete (infinite) set and the ψ_α are exact, then Eq. (51) is an exact representation of Ψ_E. In practice, however, only for photoionization of He (or photodetachment of H^-) are the ψ_α known exactly. In addition, the sum over an infinite set must be truncated to be dealt with computationally. Then the finite sum is used in the Kohn (1948) variation

$$\int \psi_\alpha^*(H - E)\Psi_E \, d\tau' = 0 \tag{52}$$

with the $d\tau'$ integration over all coordinates except r, the radial coordinate of the F_α. Equation (52) yields a set of *coupled* integrodifferential equations for the F_α which, as shown by Kohn (1948), give the S matrix correct to first order. If only one term is included in Eq. (51), then the result is precisely the HF equation discussed previously. Close coupling, therefore, may be thought of as a multiconfigurational continuum HF. The boundary conditions on the F_α are standard scattering boundary conditions (Percival and Seaton, 1957; Burke, 1968) and the general form of the close-coupling equations can be written

$$\left[\frac{d^2}{dr^2} - \frac{l(l+1)}{r^2} + \varepsilon\right] F_\alpha(r) = \sum_\alpha (V_{\alpha\alpha'} F_{\alpha'}(r) + W_{\alpha\alpha'}) \tag{53}$$

with the $V_{\alpha\alpha'}$ the local direct interaction and the $W_{\alpha\alpha'}$ the nonlocal exchange terms (Percival and Seaton, 1957). The $W_{\alpha\alpha'}$ are extremely complicated, being sums of Slater (1960) integrals over all of the continuum F_α; Eqs. (33) and (35) give some idea of the complexity in just a single-channel case. The close-coupling equations have as many linearly independent solutions as there are open channels. These solutions can then be combined into a wave function having the correct boundary conditions for the problem of interest, i.e., a wave function which asymptotically represents an ion in a given state, α, times a (phase-shifted) Coulomb wave (Smith, 1971). This wave function is then used, along with the wave function of the initial discrete state, to calculate the dipole matrix elements and, thus, photoionization cross sections.

The close-coupling method is limited by the necessity for truncation of the expansion [Eq. (51)]. Recent advances in the theory, however, have ameliorated the problem somewhat. First, the method of *pseudostates* (Damburg and Karule, 1967; Burke and Webb, 1970) has been introduced. This method involves using a set of ψ_α in the expansion, Eq. (51), which are not eigenfunctions of the target atom for closed channels, i.e., pseudostates, while for open channels the ψ_α are eigenfunctions. This is done in the hope that the terms containing the pseudostates will be a closer approximation to the true wave function than similar terms using eigenfunctions of the target. One needs to retain the eigenfunctions for the open channels to insure that Ψ_E has the correct asymptotic form. Second, algebraic noniterative methods of solution of the close-coupling equations have been employed (Callaway *et al.*, 1970; Truhlar *et al.*, 1974), making it computationally feasible to include more terms in the close-coupling expansion. Third, there is the so-called R-matrix method (Burke and Seaton, 1971; Burke and Robb, 1972; Fano and Lee, 1973; Lee, 1974; Burke and Robb, 1975). This approach is actually an outgrowth of, and related to, close-coupling theory and was originally developed much earlier in connection with nuclear physics (Wigner and

Eisenbud, 1947). The essence of this method derives from the fact that when the outgoing electron is still in the vicinity of the atom [r is small in Eq. (51)] all the electrons should be treated in a similar manner, while when the photoelectron is at large distance, the close-coupling expansion is suitable. Thus, configuration space is separated into two regions; an inner region ($r < a$) and an outer region ($r > a$), where a is a suitably chosen constant radius. In the inner region one does essentially a discrete bound state problem, while in the outer region one solves the close-coupling equations and the solutions are matched at $r = a$. The inner region discrete-like solutions ψ_j with eigenvalues E_j are found and the connection between the wave functions of the inner and outer regions can be expressed in terms of the R matrix

$$R_{\alpha\alpha'} = \sum_j \frac{g_{\alpha j} g_{\alpha' j}}{E_j - E}, \qquad (54)$$

where $g_{\alpha j}$ are energy-independent. This method has the great advantage that it can also be applied to true bound states so that initial and final states can be simultaneously improved. In addition, one calculation applies to a range of energies. The results of this method, however, must be insensitive to the choice of a if they are to be meaningful.

As with the continuum configuration interaction technique, the close-coupling method includes the effects of autoionizing resonances on photoionization cross sections (Smith, 1971) which come about from the interaction of a closed channel (or channels) with the open channels. Quite a few results of close-coupling calculations dealing primarily with the positions and shapes of these resonances have been reported. Among them are Burke et al. (1963), Ormonde et al. (1967), Smith et al. (1967), Burke and Moores (1968), Dubau and Wells (1973), Ormonde et al. (1973), Burke et al. (1974), Rountree et al. (1974), and Burke and Robb (1975).

At this point, it is worthwhile to reemphasize that the HF and close-coupling methods are the single-channel and multichannel cases of the *same* integrodifferential equation formalism for obtaining the final state wave function. In addition, the same wave function could be obtained via a matrix diagonalization approach, the continuum configuration interaction method with the *intra*channel problem being formally equivalent to the single channel HF and the *inter*channel problem corresponding to the multichannel close-coupling method.

A somewhat different approach to improving on the HF method is taken by many-body perturbation theory (MBPT) introduced by Brueckner (1955, 1959) and Goldstone (1957). The basic philosophy of this method is to improve both initial and final state wave functions simultaneously via the matrix elements of the exact Hamiltonian itself. A crucial point, however, is that the emphasis is on perturbation theory. The correlation effects must be sufficiently weak for reasonably rapid convergence to be obtained for this method to be successful. To illustrate, we wish to solve the N-particle problem

$$H\Psi = E\Psi, \tag{55}$$

where E and H are the exact nonrelativistic energy and wave function respectively. The Hamiltonian, broken into one-particle and two-particle terms, can be written

$$H = \sum_{\mu=1}^{N} T_\mu + \sum_{\mu<\mu'} v_{\mu\mu'}. \tag{56}$$

For atoms the T_μ are the sum of the kinetic energy and nuclear attraction for the μth electron and $v_{\mu\mu'}$ is the interelectron repulsion, $e^2/r_{\mu\mu'}$. One can then consider an approximate one-body Hamiltonian H_0

$$H_0 = \sum_\mu (T_\mu + V_\mu), \tag{57}$$

where V_μ is a one-body potential, $V_\mu = V(\mathbf{r}_\mu)$. The solutions to the one-body problem are

$$(T + V)\psi_n = \varepsilon_n \psi_n \tag{58}$$

with the ψ_n forming an orthonormal set. In terms of this basis set, then,

$$H_0 = \sum_n \varepsilon_n \eta_n^+ \eta_n, \tag{59}$$

$$H' = H - H_0 = \tfrac{1}{2} \sum_{p,q,m,n} \langle pq|v|mn\rangle \eta_p^+ \eta_q^+ \eta_m \eta_n - \sum_{p,m} \langle p|V|m\rangle \eta_p^+ \eta_m. \tag{60}$$

The η^+ and η operators are creation and annihilation operators, respectively, which satisfy the usual Fermi–Dirac anticommutation relations. By the use of Wick's theorem (Wick, 1950), the terms in the perturbation expansion may be represented by diagrams (Goldstone, 1957) significantly the same as Feynman diagrams. Thus the name of "diagrammatic perturbation theory" is sometimes applied to this method.

To calculate photoionization cross sections, one must include in the exact Hamiltonian [Eq. (56)] the term coupling the radiation field with the electron motion, as well as in H' [Eq. (60)]. The summation over the various terms in the perturbation expansion relevant to photoionization can be accomplished in various ways. Direct summation of the diagrams can be performed, and the fact that certain diagrams cancel out certain others simplifies the task (Kelly, 1964; Kelly, 1971; Kelly and Ron, 1972; Ishihara and Poe, 1972a,b; Wendin, 1973; Kelly, 1975). The random phase approximation (RPA) can be applied to MBPT theory to effectively sum certain classes of terms in the perturbation expansion (Altick and Glassgold, 1964; Amusia, 1971; Amusia et al., 1971; Lin, 1974b; Chang and Fano, 1975). In addition, RPA can be formulated as a coupled equation problem (Chang and Fano, 1975) and solved that way. An advantage of RPA is that initial and final state wave functions are improved in such a way that length and velocity matrix elements are equal (Amusia et al., 1971).

A critical factor in MBPT is the choice of a basis set ψ_n, or, alternatively, the approximate one-body potential V. This is because the appropriate choice of V causes a number of terms in the perturbation expansion to cancel each other out (Amusia et al., 1971; Kelly, 1975). The basis set generated from a central potential is easily obtained, but the cancellation does not then occur. A HF basis set, on the other hand, retains the cancellation property, but has the undesirable feature that the HF basis for the initial state may be a very poor zeroth approximation to the final state orbitals. Employing different basis sets for initial and final states leads to many complicated extra terms in taking matrix elements. Methods for overcoming this difficulty have been proposed by Silverstone and Yin (1968) and Huzinaga and Arnau (1970) by using a potential, V, containing projection operators. This removes the nonorthogonality problem, but the matrix elements of this potential are still difficult because of the projection operators. There is, thus, no one best choice and the specific problem must dictate the most efficacious procedure to employ.

Despite the difficulties, however, the methods which improve both initial and final states simultaneously seem to be capable of giving the best results for photoionization cross sections. They have, in addition to the advantages of being continuously improvable, the advantage of allowing one to see clearly which interactions are the important ones for generating the dipole matrix element.

6. Satellite Lines

Among the most intriguing results of recent investigations in photoelectron spectroscopy are the occurrences of satellite lines in the spectra (Shirley, 1972; Siegbahn et al., 1967; Carlson, 1973). Satellite lines arise from the fact that the photoionization process can leave the positive ion not only in the ground state, but in excited states as well. Thus, from the Einstein (1905) relation [Eq. (2)], the differing ionization energies I_{ij} lead to different photoelectron energies and, therefore, more than one photoelectron line. The cross section for producing photoelectrons leaving the ion in the ground state is usually, but not always, considerably greater than the cross section leaving the ion in an excited state, so that the "ground state line" is usually much more intense than the "excited state lines." It is the latter that are called satellite lines. An example of the reverse situation has been found recently in barium (Hotop and Mahr, 1975), where the double ionization cross section can be more than a factor of 2 larger than the single ionization.

In other words, satellite lines result from ionization plus excitation, while the normal line involves only ionization. These, however, are in no sense distinct physical processes. The major confusion in the discussion of satellite lines comes from using wave functions for the initial and final state based on a one-electron model in which the cross section for satellite lines vanishes. It is important to realize, however, that *a photoelectron line will appear for every transition where the exact initial state $|i\rangle$ and the exact final state $|j, \varepsilon\rangle$ have a nonzero dipole matrix element between them*. This point cannot be overemphasized, and provides the basis for most of the following discussion of satellite lines. Of course, all of these lines will not be observed since various experimental limitations such as resolution, background, etc., will obscure some of them.

The methods described in Section II,A,5 are suitable for the calculation of satellite line as well as "normal" line intensities without modification. Due to the importance of satellite lines, however, it is of use to show the characteristics of the calculation and spotlight specifically the inadequacies of the one-electron model which require correction to do a proper calculation (Manson, 1976).

Before discussing this point, it is worthwhile to note that satellite lines, i.e., photoionization leaving the ion in an excited state, can occur in certain instances within the one-electron central field calculation. This occurs when the initial state of the atom has an open shell (of more than one electron) so that the various possible angular momentum couplings in the *ion* give rise to multiplet structure, i.e., several distinct multiplet energy levels arising from different couplings of the same configuration of one-electron states. For example, the ground configuration of O^+ is $(1s^2, 2s^2, 2p^3)$, which has multiplets ^4S, ^2D, ^2P at energies, I_{ij}, of 13.6 eV, 16.9 eV, and 18.6 eV, respectively (Moore, 1949), above the ground state of atomic oxygen. Thus, photoionization of 2p electrons of atomic oxygen will lead to three photoelectron lines, corresponding to leaving the ion in the ^4S, ^2D, ^2P states (Starace et al., 1974; Manson et al., 1974), within the framework of the one-electron approximation. This is not to say that the results of a one-electron calculation are necessarily accurate, but merely that they are nonzero.

To see clearly the inadequacies of the one-electron calculation, we expand the exact wave functions in terms of a basis set of antisymmetric products of one-electron functions. In practice, the one-electron functions are generally HS or HF radial orbitals multiplied by spherical harmonics and Pauli spinors, but in principle any complete set may be used. For an N-electron system, then, we denote this antisymmetrized product as $\psi_\alpha^{(N)}$, with α symbolizing all of the quantum numbers of the individual one-electron functions including spin as well as the coupling scheme, and the exact initial state wave function, $|i\rangle$, can be written

$$|i\rangle = \sum_\alpha a_\alpha \psi_\alpha^{(N)}, \qquad (61)$$

where the a_α are the configuration interaction expansion coefficients.

As an example, consider the ground $^1S_0$ state of the helium atom. If we characterize the basis functions for the two-electron system as $\psi_\alpha^{(2)} \equiv (nl, n'l')$, the exact wave function can be written

$$\begin{aligned}|i\rangle &= \sum a_{nl,\,n'l'} \psi_{nl,\,n'l'}^{(2)} \\ &= a_{1s,\,1s}(1s,\,1s) + a_{1s,\,2s}(1s,\,2s) + a_{2s,\,2s}(2s,\,2s) + \cdots \\ &\quad + a_{2p,\,2p}(2p,\,2p) + a_{2p,\,3p}(2p,\,3p) + a_{3p,\,3p}(3p,\,3p) + \cdots \\ &\quad + a_{3d,\,3d}(3d,\,3d) + a_{3d,\,4d}(3d,\,4d) + \cdots + a_{4f,\,4f}(4f,\,4f) + \cdots \\ &\quad + \cdots, \end{aligned} \qquad (62)$$

where the dependence on magnetic and spin quantum numbers is suppressed for simplicity. Note that this state is generally characterized (1s, 1s) [usually written $(1s)^2$], but from Eq. (62) it is seen that this is only an approximation. With a reasonable basis set, it is true that $a_{1s,\,1s}$ will be quite close to unity and all of the other $a_{nl,\,n'l'}$ will be small; they will not, however, vanish. This is a crucial point in connection with satellite lines, as will be seen.

Using the same one-electron spin-orbitals that were employed in the initial state expansion, the exact wave functions $|j\rangle$ of the ion core $N-1$ electron system can be constructed:

$$|j\rangle = \sum_\gamma b_\gamma \psi_\gamma^{(N-1)}. \qquad (63)$$

Here again, one of the b_γ will generally be quite close to unity and the others quite small, but the respective deviations from unity and zero are quite important for satellite lines.

The final state in a photoionization process is not, of course, an $N-1$ electron ion but an N-electron ion plus photoelectron system. In terms of the exact ionic wave function [Eq. (63)] the exact wave function for the final state of a photoionization process with the ion left in state $|j\rangle$ can be written as

$$|j,\varepsilon\rangle = \sum_k C_k^j(\varepsilon)|k\rangle F_k^j(\varepsilon), \qquad (64)$$

where the $|k\rangle$ are the ionic wave functions (including $|j\rangle$) given by Eq. (63), the C_k^j are the expansion coefficients, and the F_k^j represent the photoelectron. This is the close-coupling expansion given in Eq. (51); the only difference is that the various possible eigenfunctions (one for each open channel) have already been combined to give the correct boundary conditions for the given physical problem. The superscript j in Eq. (64) refers to the fact that the ion is left in state $|j\rangle$. In this expansion [Eq. (64)] C_j^j is generally quite close to unity and the C_k^j for $j \neq k$ are small but nonzero.

To illustrate how the various inadequacies of the one-electron approximation translate into satellite lines, consider the simple example of photoionization of He in the ground state leaving He$^+$ in the $\bar{n}\bar{l}$ state. Note that He$^+$ is hydrogenic, so its wave functions are known exactly. The exact final state wave function is given by

$$|\bar{n}\bar{l}, \varepsilon\rangle = \sum_{n,l} C_{nl}^{\bar{n}\bar{l}}(\varepsilon)|nl\rangle F_{nl}^{\bar{n}\bar{l}}(\varepsilon)$$

$$= C_{1s}^{\bar{n}\bar{l}}(\varepsilon)|1s\rangle F_{1s}^{\bar{n}\bar{l}}(\varepsilon) + C_{2s}^{\bar{n}\bar{l}}(\varepsilon)|2s\rangle F_{2s}^{\bar{n}\bar{l}}(\varepsilon) + \cdots \quad (65)$$

The $F_{nl}^{\bar{n}\bar{l}}$ have the correct spherical harmonics and spin functions so that each term has ^1P symmetry, the only allowed final state in a dipole transition from the $^1S_0$ ground state of He. If the He$^+$ ion is left in the 2s (excited) state and the expansion of the initial state wave function, Eq. (62), is made in terms of the He$^+$ (hydrogenic) eigenfunctions, Eqs. (62) and (65) give the dipole matrix element

$$\langle a_{1s,1s}(1s, 1s) + a_{1s,2s}(1s, 2s) + a_{2s,2s}(2s, 2s)$$
$$+ a_{2p,2p}(2p, 2p) + \cdots |\mathbf{r}_1 + \mathbf{r}_2|$$
$$\times C_{1s}^{2s}|1s\rangle F_{1s}^{2s} + C_{2s}^{2s}|2s\rangle F_{2s}^{2s} + C_{2p}^{2s} F_{2p}^{2s} + \cdots\rangle$$
$$= a_{1s,1s}^* C_{1s}^{2s} \langle 1s|\mathbf{r}|F_{1s}^{2s}\rangle$$
$$+ a_{1s,2s}^* C_{1s}^{2s}\langle 2s|\mathbf{r}|F_{1s}^{2s}\rangle + a_{1s,2s}^* C_{2s}^{2s}\langle 1s|\mathbf{r}|F_{2s}^{2s}\rangle$$
$$+ a_{1s,2s}^* C_{2s}^{2s}\langle 2s|\mathbf{r}|F_{2s}^{2s}\rangle + a_{2p,2p}^* C_{2s}^{2s}\langle 2p|\mathbf{r}|F_{2p}^{2s}\rangle + \cdots \quad (66)$$

Examining the terms in the matrix element, it is seen that the first, third, and fourth terms contain one of the coefficients that are close to unity, $a_{1s,1s}$ and C_{2s}^{2s}, so they are *a priori* the largest contributor to the matrix element. The situation is shown schematically in Table I, where it is seen that terms (3) and (4) arise from initial state configuration interaction, term (1) from the interchannel coupling, and terms (2) and (5) from a combination of the two. That is to say, contributions to the satellite line intensity arising from initial state configuration interaction are of approximately the same order of magnitude as those arising from interchannel coupling. Thus, to predict satellite intensities, both must be considered.

TABLE I

SCHEMATIC REPRESENTATION OF THE CONTRIBUTIONS TO THE DIPOLE MATRIX ELEMENT FOR THE PHOTOIONIZATION OF He IN THE GROUND STATE TO He$^+$ IN THE 2s STATE[a]

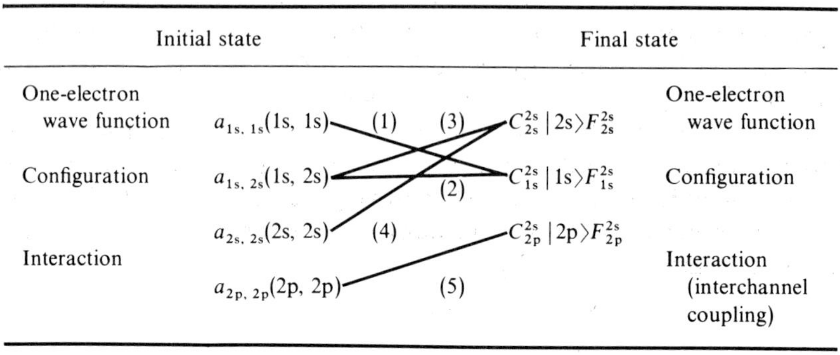

Initial state		Final state
One-electron wave function	$a_{1s,1s}(1s, 1s)$ —(1)— (3)— $C^{2s}_{2s}\|2s\rangle F^{2s}_{2s}$	One-electron wave function
Configuration Interaction	$a_{1s,2s}(1s, 2s)$ —(2)— $C^{2s}_{1s}\|1s\rangle F^{2s}_{1s}$ $a_{2s,2s}(2s, 2s)$ —(4)— $C^{2s}_{2p}\|2p\rangle F^{2s}_{2p}$ $a_{2p,2p}(2p, 2p)$ —(5)—	Configuration Interaction (interchannel coupling)

[a] The lines between terms of the initial and final state wave function refer to nonzero dipole matrix elements.

Simultaneous photoionization plus excitation of He has been investigated both experimentally (Samson, 1969a; Krause and Wuilleumier, 1972) and theoretically (Salpeter and Zaidi, 1962; Jacobs and Burke, 1972; Åberg, 1969, 1973). The ratio of cross sections for leaving He$^+$ in the 2s state plus the 2p state to leaving He$^+$ in the ground state, i.e., $(\sigma_{2s} + \sigma_{2p})/\sigma_{1s}$, was found, in the most sophisticated calculation including both initial state and interchannel coupling breakdowns of the one-electron model (Jacobs and Burke, 1972), to have a maximum value of about 14%; leaving out the interchannel effects, the maximum ratio was 9% (Salpeter and Zaidi, 1962), and using one-electron wave functions with core relaxation, the "sudden" approximation, the ratio came to about 4% (Åberg, 1973), thus confirming the results indicated in the previous paragraph.

In addition to the fact that a considerable body of data exists on satellite lines in He, this example was chosen for its simplicity—no final ionic state configuration interaction, since He$^+$ is a one-electron system. In a more general case, however, inadequacies in the one-electron approximation due to final ionic state configuration interaction will contribute significantly to the intensity. It is to be noted, however, that the effects on the dipole matrix element of initial and final ionic state configuration interaction are independent of photoelectron energy, while the strength of the interchannel coupling generally decreases (eventually) with increasing photoelectron energy (Fano, 1961). Thus, at the higher photoelectron energies, interchannel coupling is probably not too important, but further work is necessary to delineate the details.

B. Photoelectron Angular Distributions

1. Basic Theory

The differential cross section for the angular distribution of photoelectrons for an $|i\rangle \to |j, \varepsilon\rangle$ electric dipole photoionization process by linearly polarized light is given by (Bethe, 1933; Cooper and Zare, 1968, 1969)

$$d\sigma_{ij}(\varepsilon)/d\Omega = [\sigma_{ij}(\varepsilon)/4\pi][1 + \beta_{ij}(\varepsilon)P_2(\cos\phi)], \tag{66}$$

where ϕ is the angle between the photoelectron direction and the polarized photon electric vector, $\sigma_{ij}(\varepsilon)$ is the photoionization cross section, $\beta_{ij}(\varepsilon)$ is the asymmetry parameter, and $P_2(x) = (3x^2 - 1)/2$. Unpolarized light can be considered the incoherent superposition of two polarized beams with one polarization lying in the plane defined by the photon beam and photoelectron directions, and the other perpendicular to it. If ϕ is defined as above for the polarization in the photon–photoelectron plane, ϕ' as the angle with the perpendicular polarization, and θ as the angle between photon beam and photoelectron directions, then the expression for the photoelectron angular distribution becomes (Cooper and Manson, 1969)

$$\begin{aligned}\frac{d\sigma_{ij}(\varepsilon)}{d\Omega} &= \frac{1}{2}\frac{\sigma_{ij}(\varepsilon)}{4\pi}[1 + \beta_{ij}(\varepsilon)P_2(\cos\phi)] + \frac{1}{2}\frac{\sigma_{ij}(\varepsilon)}{4\pi}[1 + \beta_{ij}P_2(\cos\phi')] \\ &= \frac{\sigma_{ij}(\varepsilon)}{4\pi}\left[1 + \frac{\beta_{ij}(\varepsilon)}{2}\left(\frac{3\cos^2\phi + 3\cos^2\phi' - 2}{2}\right)\right] \\ &= \frac{\sigma_{ij}(\varepsilon)}{4\pi}\left[1 - \frac{\beta_{ij}(\varepsilon)}{2}\left(\frac{3\cos^2\theta - 1}{2}\right)\right] \\ &= \frac{\sigma_{ij}(\varepsilon)}{4\pi}\left[1 - \frac{\beta_{ij}(\varepsilon)}{2}P_2(\cos\theta)\right] \end{aligned} \tag{67}$$

where we have used the fact that $\cos^2\phi + \cos^2\phi' + \cos^2\theta = 1$. Circularly polarized light gives the same form [Eq. (67)] as unpolarized light (Peshkin, 1970; Jacobs, 1972) since it can be represented as a coherent sum of two linearly polarized beams. Thus, the derivation of Eq. (67) goes through as shown because the coherence or incoherence of the linearly polarized beams does not enter.

The expressions for partially polarized light (Samson, 1969b, 1970) and elliptically polarized light (Schmidt, 1973) are the same (Samson and Starace, 1975) and depend upon one additional parameter, p, where

$$p = \frac{I_y - I_x}{I_y + I_x}, \tag{68}$$

with I_x and I_y being the intensity of light polarized in the photon–photoelectron plane and perpendicular to it, respectively. The photoelectron angular distribution can then be obtained by straightforward geometrical arguments (as above) to be (Samson and Starace, 1975)

$$\frac{d\sigma_{ij}(\varepsilon)}{d\Omega} = \frac{\sigma_{ij}(\varepsilon)}{4\pi} \left[1 - \frac{\beta_{ij}(\varepsilon)}{2} P_2(\cos\theta) - \frac{3}{2} p(\cos^2\phi - \cos^2\phi') \right]. \quad (69)$$

Equations (66), (67), and (69) can be obtained from quite general considerations, depending only on the absorption of radiation going via an electric dipole process (Yang, 1948). Deviation from the form of these equations implies the presence of a nonnegligible amount of absorption via processes other than electric dipole (Yang, 1948). The asymmetry parameter $\beta_{ij}(\varepsilon)$ for the process does, however, depend upon the details of the calculation. It must, however, lie in the range $-1 \leq \beta_{ij}(\varepsilon) \leq 2$ to insure nonnegativity of the differential cross section.

Before proceeding to a discussion of the methods of calculation of the asymmetry parameter, it is worthwhile to discuss the range of validity of the dipole approximation for photoelectron angular distributions. We have previously (Section II,A,1) discussed the application of the dipole approximation to cross sections and indicated that angular distributions were another matter. For angular distributions (Bethe and Salpeter, 1957), the lowest order correction to the dipole approximation goes as $4v/c$, where v is the photoelectron velocity and c is the velocity of light, while for total cross sections this term integrates out to zero and the lowest order correction goes as $(v/c)^2$. Thus, for 1 keV photoelectrons, the correction to the total cross section due to breakdown of the dipole approximation is about 0.2%, while the correction to the angular distribution is approximately 15%. These effects on the angular distribution have been observed experimentally (Krause, 1969) and they have been interpreted theoretically (Cooper and Manson, 1969).

2. Central Field Calculation of the Asymmetry Parameter

In order to calculate the asymmetry parameter, $\beta_{ij}(\varepsilon)$, one needs to generate a continuum wave function which represents a photoelectron of energy ε moving in a specific direction $\psi_{\varepsilon,\mathbf{k}}$, which is a solution to the central field Schrödinger equation before separation of variables. This is in contrast to the cross-section calculation, where the equation was solved in the angular momentum representation. Here, however, angular momentum cannot be a good quantum number since the photoelectron direction is specified and the two are mutually exclusive. In fact, $\psi_{\varepsilon,\mathbf{k}}$ can be expanded in terms of the angular momentum wave functions as (Geltman, 1969; Cooper and Zare, 1969)

$$\psi_{\varepsilon, \mathbf{k}} = \sum_{l, m} i^l \exp{(-i\xi_l)} \frac{P_{\varepsilon l}(r)}{r} Y_l^m(\mathbf{r}) Y_l^m(\mathbf{k})^*, \qquad (70)$$

where $\mathbf{k}$ is the unit vector in the direction of the photoelectron velocity, $P_{\varepsilon l}(r)$ is the single-particle solution of the central-field radial Schrödinger equation (22), and

$$\xi_l(\varepsilon) = \sigma_l(\varepsilon) + \delta_l(\varepsilon) \qquad (71)$$

the total phase shift sum of the Coulomb (σ_l) and non-Coulomb (δ_l) parts from Eq. (25). The $\psi_{\varepsilon, \mathbf{k}}$ has the correct asymptotic form of an outgoing plane wave plus *incoming* spherical waves, often referred to as $\psi_\mathbf{k}^{(-)}$.

Using this continuum wave function, the dipole matrix element for photoionization of an nl electron can be calculated, which, after some angular momentum algebra (Cooper and Zare, 1969) and comparison with Eq. (66), yields (Bethe, 1933; Cooper and Zare, 1968, 1969)

$$\beta_{nl}(\varepsilon) = \frac{l(l-1)R_{l-1}^2(\varepsilon) + (l+1)(l+2)R_{l+1}^2(\varepsilon) - 6l(l+1)R_{l-1}(\varepsilon)R_{l+1}(\varepsilon)\cos{[\xi_{l+1}(\varepsilon) - \xi_{l-1}(\varepsilon)]}}{(2l+1)[lR_{l-1}^2(\varepsilon) + (l+1)R_{l+1}^2(\varepsilon)]} \qquad (72)$$

where the radial dipole matrix elements, $R_{l\pm 1}(\varepsilon)$ are given in Eq. (24). The details of the derivation are given by Cooper and Zare (1969). They also show that this result is generally true for a central field calculation even if all the angular momentum couplings are included with the proviso that the coupling is LS.

Basically the expression for β_{nl} consists of two types of terms, apart from the normalizing denominator. The first two terms in the numerator of Eq. (72) contain squares of the dipole matrix elements, much like the expression for the cross section, Eq. (23). The last term contains the dipole matrix elements themselves as well as a dependence on the relative phases $l + 1$ and $l - 1$ continuum functions. This last term represents interference between the $l \to l + 1$ and $l \to l - 1$ photoionization channels. Since the phases of different continuum waves have different energy dependences, the asymmetry parameter is energy-dependent. For photoionization of an s-subshell electron the only possible transition, within the central potential calculation, is $s \to p$; i.e., only a single continuum wave. Thus, by the above discussion, β_{ns} should be constant; plugging $l = 0$ into Eq. (72) yields $\beta_{ns} = 2$, confirming this point.

Thus the asymmetry parameter, obtained from angular distribution measurements, provides information on the sign of the dipole matrix elements as well as on the phase shifts of the various continuum waves. The phase shifts are the central ingredient in scattering processes, and, since a close connection exists between spectroscopy and collision theory (Fano, 1970; Lu, 1971; Lee and Lu, 1973; Lee, 1974), they are closely related to structure as well. It is only in an interference experiment, such as photoelectron angular distributions, that one gets direct information on phase shifts.

As pointed out in Section II,A,3, central-field calculations contain some serious inaccuracies. These can be important for cross sections, as discussed, but for angular distributions the inexact treatment of exchange can be crucial. The emission of a photoelectron can be considered a two-step process. First, the photon is absorbed by a bound state electron with the accompanying transfer of angular momentum to the photoelectron. Second, the photoelectron escapes. During this escape, however, angular momentum can be transferred between the photoelectron and the ion core. This second part of the process is completely neglected in the above treatment. Fortunately, angular momentum conservation abnegates the second step for closed-shell, $^1S_0$, systems (if fine structure is neglected) so Eq. (72) for β_{nl} is relevant to the noble cases where most of the experimental work has been done. For open shell atoms, however, it will be shown in the next section that anisotropic interactions can make the second step of the photoionization process quite important indeed (Dill et al., 1974; Dill et al., 1975).

3. General Calculation of the Asymmetry Parameter

The general theory of the calculation of the asymmetry parameter has been formulated in various ways (Lipsky, 1967; Jacobs, 1972; Fano and Dill, 1972; Dill and Fano, 1972; Dill, 1973; Dill et al., 1975) which are equivalent to each other. In this section we shall use the angular momentum transfer expansion framework of Dill et al. since the essential points emerge most clearly in this formulation. In addition, the use of this formalism allows one to pinpoint the fundamental differences with the central field results of the previous section.

The ejection of a photoelectron from an *unpolarized* atom X by electric dipole interaction with a photon $h\nu$ may be represented schematically as

$$X(J_0 \pi_0) + h\nu(j_\gamma = 1, \pi_\gamma = -1) \to X^+(J_c \pi_c) + e^-[lsj, \pi_e = (-1)^l] \quad (73)$$

with π the parity of the various components. The differential cross section for this process can be separated into contributions characterized by alternative magnitudes of the angular momentum transfer

$$\mathbf{j}_t = \mathbf{j}_\gamma - \mathbf{l} = \mathbf{J}_c + \mathbf{s} - \mathbf{J}_0 \quad (74)$$

provided no measurement is made of either the spin of the photoelectron or the orientation of the residual ion. The vector j_t is the angular momentum transferred between the *unobserved* initial and final angular momenta in the photoionization process, Eq. (73), i.e., between the total angular momentum J_0 of the atom X and the combined angular momenta of the ion X^+ and the photoelectron spin s which we denote by $J_{cs} = J_c + s$. The possible values of j_t in the photoionization process are determined by the conservation of angular momentum **J** and parity π:

$$J = J_0 + j_\gamma = J_c + s + l = J_{cs} + l, \tag{75}$$

$$\pi = \pi_\gamma \pi_0 = -\pi_0 = \pi_c(-1)^l. \tag{76}$$

The resolution of the cross section into contributions corresponding to alternative values of j_t requires that first one find the allowed values of j_t from Eqs. (74)–(76). Then each value of j_t is characterized as being parity-favored or parity-unfavored (Dill and Fano, 1972), corresponding to whether the parity change of the atom X, $\pi_0 \pi_c$, is equal to $+(-1)^{j_t}$ or $-(-1)^{j_t}$, respectively. Thus, from Eq. (76), which shows that $\pi_0 \pi_c = -(-1)^l$, we find that parity-favored transitions occur for $l = j_t \pm 1$ and parity-unfavored for $l = j_t$; angular momentum conservation, Eq. (75), shows that no other values are possible. The importance of this separation is that the parity-favored contribution to the differential cross section has a complicated asymmetry parameter (discussed below), while the asymmetry parameter of the parity-unfavored part is determined only by angular momentum and parity consideration

$$\beta(j_t)_{unf} = -1 \tag{77}$$

independently of any dynamical considerations.

The cross section for the photoionization process may then be expressed as a sum over the contributions of the various values of j_t (Dill, 1973),

$$\sigma = \sum_{J_{cs}, j_t} \sigma(j_t), \tag{78}$$

and the asymmetry parameter is given by a weighted average of the various $\beta(j_t)$

$$\beta = \left[\sum_{J_{cs}} \left(\sum_{j_t}^{fav} \sigma(j_t)_{fav} \beta(j_t)_{fav} - \sum_{j_t}^{unf} \sigma(j_t)_{unf} \right) \right] / \sigma \tag{79}$$

with the minus sign on the parity-unfavored contribution coming from using the explicit value of $\beta(j_t)_{unf}$, Eq. (77). Note that Eqs. (77) and (78) have sums over J_{cs} but the dependence of $\beta(j_t)$ and $\sigma(j_t)$ on J_{cs} is not indicated. This dependence is hidden in the scattering matrix amplitudes $\bar{S}_l(j_t)$, in terms of which $\sigma(j_t)$ and $\beta(j_t)$ are given by (Dill, 1973)

$$\sigma(j_t)_{\text{fav}} = \pi\lambda^2 \frac{2j_t + 1}{2J_0 + 1}[|\bar{S}_+(j_t)|^2 + |\bar{S}_-(j_t)|^2] \tag{80}$$

$$\sigma(j_t)_{\text{unf}} = \pi\lambda^2 \frac{2j_t + 1}{2J_0 + 1} |\bar{S}_0(j_t)| \tag{81}$$

$$\beta(j_t)_{\text{fav}} = \frac{(j_t + 2)|\bar{S}_+(j_t)|^2 + (j_t - 1)|\bar{S}_-(j_t)|^2 - 3[j_t(j_t + 1)]^{1/2}[\bar{S}_+(j_t)\bar{S}_-(j_t)^+ + \text{cc}]}{(2j_t + 1)[|S_+(j_t)|^2 + |S_-(j_t)|^2]} \tag{82}$$

In these equations λ is $(2\pi)^{-1}$ times the photon wavelength and cc stands for complex conjugate. The subscripts "$\pm$" in the parity-favored cross sections and amplitudes denote $l = j_t \pm 1$. For the parity-unfavored cross sections, the subscript "0" denotes $l = j_t$. The scattering matrix amplitudes are related to the dipole matrix elements and continuum wave phase shifts. Before proceeding with a discussion of them, however, it is of interest to consider the physical significance of the angular momentum transfer.

At the end of the previous section, photoionization as a two-step process was introduced, the steps being absorption of the photon and the subsequent escape of the photoelectron. The angular momentum transfer is always equal to the difference between the angular momentum input to the atom ($j_\gamma = 1$ for the electric dipole interaction) and the angular momentum output from the atom (the photoelectron's orbital angular momentum, l). Thus the angular momentum transfer, $\mathbf{j}_t = \mathbf{j}_\gamma - \mathbf{l}$ is the net angular momentum deposited in the target atom by the electric dipole photoionization process. (Note that since we consider the situation where the photoelectron spin s is unobserved, s is included as part of the angular momentum of the residual target.) The allowed values of j_t are, however, different for the two stages of the photoionization process.

In the initial stage the photon imparts angular momentum $j_\gamma = 1$ to the photoelectron, which has initial orbital angular momentum $\mathbf{l}_0$ (in an independent-particle model), yielding a final orbital angular momentum $\mathbf{l}' = \mathbf{l}_0 + \mathbf{j}_\gamma$. Thus, in the first stage, the angular momentum transfer is

$$\mathbf{j}'_t = \mathbf{j}_t - \mathbf{l}' = -\mathbf{l}_0, \tag{83}$$

so that j'_t has the single magnitude l_0. Owing to parity conservation, $l' = l_0 \pm 1$; hence $j'_t = l_0$ is parity-favored. In the central-field calculation, then, where this first stage is the only part of the process considered, not only are parity-unfavored angular momentum transfers excluded, but any j_t other than $j_t = l_0$ are excluded as well.

During the subsequent escape of the photoelectron, additional angular momentum transfers can arise, within the allowed range determined by Eq. (74), from anisotropic (noncentral) interactions between the photoelectron and the ion core; these interactions are obviously excluded from a central-field calculation. If spin-dependent forces are neglected, the interaction is between $\mathbf{l}'$ and the initial orbital angular momentum of the core, $\mathbf{L}'_c$. Owing to the resulting angular momentum exchange, $\mathbf{k}$, between photoelectron and core, only the total orbital angular momentum $\mathbf{L}$ is conserved. In particular, the photoelectron orbital angular momentum can change from $\mathbf{l}'$ to $\mathbf{l}$ during the departure of the photoelectron from the atom, in which case the angular momentum transfer is no longer $\mathbf{j}_t = -\mathbf{l}_0$ but

$$\mathbf{j}_t = \mathbf{j}_\gamma - \mathbf{l} = \mathbf{j}'_t - \mathbf{k}. \tag{84}$$

Note that even if the magnitudes of $\mathbf{l}'$ and $\mathbf{L}'_c$ remain unchanged, a precession about $\mathbf{L}$ is sufficient to produce a change in magnitude of j_t.

The details of the S-matrix elements as well as the implementation of the j_t analysis are best discussed within the context of a specific example. To this end we consider the photoionization of the ground $3p^4\ ^3P$ state of atomic sulfur (Dill *et al.*, 1974; Dill *et al.*, 1975)

$$h\nu + S(3p^4\ ^3P) \rightarrow S^+(3p^3\ ^4S,\ ^2P,\ ^2D) + e^- \tag{85}$$

in LS coupling. For a given photoionization channel, i.e., a specific $S_c L_c$ of the ionic core, a number of transitions are possible corresponding to the possible values of the photoelectron angular momentum, l, and the various ways a given εl can couple to $S_c L_c$. Considering the ^2D channel, if $J_0 = 2$ and $J_c = \frac{5}{2}$, angular momentum and parity conservation, Eq. (74)–(76), give the result that the possible values of $l = 0, 2, 4, 6$. Note that the central field calculation allows only $l_0 \pm 1$ or 0 and 2 for $l_0 = 1$. The extra possibilities, $l = 4, 6$, are the result of the second stage of the photoionization process where l' goes to l via angular momentum exchanges with the ion core.

If a HF calculation is carried out (Section II,A,4) the matrix elements for the $l = 4$ and 6 values vanish, not because of any selection rules, but rather because of the approximations inherent in the single-particle HF method. Even within HF, however, this analysis shows deviations from Eq. (72). Consider, for example, the ^2D term of the ion where Eqs. (74)–(76) show that for $l = 0$, a total final state (ion plus photoelectron) ^3D corresponding to $j_t = 1$ (parity favored) is allowed while for $l = 2$ total final states ^3S, ^3P, and ^3D are allowed corresponding to admixtures of $j_t = 1$ (parity favored), 2 (parity unfavored), and 3 (parity favored). The values of $j_t = 2, 3$ are a result of the various possible couplings of l' and $\mathbf{J}'_{cs}$ as discussed above and, thus, are absent from Eq. (72). The S-matrix elements turn out to be (Dill et al., 1975)

$$\bar{S}_s(1) = (C/3) \exp\{i[\sigma_s + \delta_s(^3D)]\} R_{\varepsilon s}(^3D) \tag{86a}$$

$$\bar{S}_d(1) = C(2^{1/2}/5) e^{i\sigma_d} [\tfrac{1}{3} e^{i\delta_d(^3S)} R_{\varepsilon d}(^3S)$$
$$+ \tfrac{3}{4} e^{i\delta_d(^3P)} R_{\varepsilon d}(^3P) + \tfrac{7}{12} e^{i\delta_d(^3D)} R_{\varepsilon d}(^3D)] \tag{86b}$$

$$\bar{S}_d(2) = C(2^{1/2}/5) e^{i\sigma_d} [-\tfrac{1}{3} e^{i\delta_d(^3S)} R_{\varepsilon d}(^3S)$$
$$- \tfrac{1}{4} e^{i\delta_d(^3P)} R_{\varepsilon d}(^3P) + \tfrac{7}{12} e^{i\delta_d(^3D)} R_{\varepsilon d}(^3D)] \tag{86c}$$

$$\bar{S}_d(3) = C(2^{1/2}/5) e^{i\sigma_d} [\tfrac{1}{3} e^{i\delta_d(^3S)} R_{\varepsilon d}(^3S)$$
$$- \tfrac{1}{2} e^{i\delta_d(^3P)} R_{\varepsilon d}(^3P) + \tfrac{1}{6} e^{i\delta_d(^3D)} R_{\varepsilon d}(^3D)]. \tag{86d}$$

In Eqs. (86) C denotes constant factors which are common to all channels and will drop out of the expression for β, Eq. (82), s and d refer to $l = 0$ and 2, respectively, and $R_{\varepsilon d}(^3S)$, for example, denotes the dipole matrix element to the $(^2D\varepsilon d)^3S$ final state. The σ_l are the Coulomb phase shifts of Eq. (71).

The asymmetry parameter for photoionization to the ^2D term of the ion is then obtained by combining Eqs. (86) with Eqs. (77) through (82)

$$\beta = \frac{3|\bar{S}_d(1)|^2 - 3 \cdot 2^{1/2} [\bar{S}_d(1)\bar{S}_s(1)^\dagger + cc] - 5|\bar{S}_d(2)|^2 + 2|\bar{S}_d(3)|^2}{3|\bar{S}_s(1)|^2 + 3|\bar{S}_d(1)|^2 + 5|\bar{S}_d(2)|^2 + 7|\bar{S}_d(3)|^2}, \tag{87}$$

where the $j_t = 2$ and 3 contributions are clearly indicated. Note that the non-Coulomb phase shifts δ_l in Eqs. (86) depend upon the total final state coupling, as do the dipole matrix elements. This dependence is the result of the anisotropic interactions. In a central field calculation all of the $R_{\varepsilon d}$ and all of the δ_d are equal, so that $\bar{S}_d(2)$ and $\bar{S}_d(3)$ vanish, all the terms in $\bar{S}_d(1)$ can be combined, and Eq. (87) reduces to Eq. (72), the central field result.

Note that Eq. (87) gives terms depending on the cosines of *all* of the differences between the four phase shifts $\delta_s(^3D)$, $\delta_d(^3D)$, $\delta_d(^3P)$, and $\delta_d(^3S)$. When the differences between the δ_d's are small the $j_t = 2, 3$ contributions are small and Eq. (72) is then a good approximation. Physically this means that when we have four distinct continuum waves they can all interfere with one another. When anisotropic interactions are absent, as in a central-field calculation or a closed shell atom, or when they are quite small, only two distinct continuum waves are in evidence. Thus by comparing δ_d for the various total final states, one has an *a priori* criterion for whether or not the terms with $j_t \neq l_0$ are important. In addition, comparing the δ_d for the different channels yields *a priori* information on how different the β's will be, in this case for the ^4S, ^2P, and ^2D channels.

C. Relationship of Ionization by Fast Charged Particles to Photoionization

In this section we briefly discuss the ionization of atoms by fast charged particles; by *fast* is meant that the first Born approximation applies to the process (Bethe, 1930). The cross section for the ionization of an atom in state $|i\rangle$ to a final state $|f\rangle(=|j, \varepsilon\rangle)$ by a fast charged particle of mass M, charge ze, and velocity v is given in the first Born approximation by (Bethe, 1930; Livingston and Bethe, 1937; Inokuti, 1971)

$$\sigma_{ij}(\varepsilon) = \frac{4\pi a_0^2 z^2}{T/R} \int_{K_{\max}}^{K_{\min}} \frac{|F_{if}(K)|^2}{(Ka_0)^2} \, d\ln(Ka_0)^2, \tag{88}$$

where a_0 is the Bohr radius, $T = mv^2/2$ with m the electron mass, R is the Rydberg energy, and $\hbar K$ is the momentum transfer. The inelastic scattering form factor is defined as

$$|F_{if}(K)|^2 = \frac{1}{g_i} \sum_{i,f} \left| \langle i | \sum_j e^{i\mathbf{K} \cdot \mathbf{r}_j} | f \rangle \right|^2, \tag{89}$$

with $\mathbf{r}_j$ the position vector of the jth target electron and the sum going over all degenerate initial and final states. The limits of integration of Eq. (88) can be obtained from the relation between the momentum transfer and the scattering angle of the projectile, θ:

$$(Ka_0)^2 = \left(\frac{M}{m}\right)^2 \left\{ 2\frac{T}{R} - \frac{\Delta E}{R}\frac{m}{M} - 2\left[\frac{T}{R}\left(\frac{T}{R} - \frac{\Delta E}{R}\frac{M}{m}\right)\right]^{1/2} \cos\theta \right\}, \tag{90}$$

where $\Delta E \, (= \varepsilon + I_{ij})$ is the energy lost by the projectile. $K_{\min}$ occurs for forward scattering $(\theta = 0)$ and $K_{\max}$ for backscattering $(\theta = \pi)$. If one expands Eq. (88) in inverse powers of T, employing Eqs. (89) and (90), the result is (Miller and Platzman, 1957; Inokuti, 1971)

$$\sigma_{ij}(\varepsilon) = \frac{4\pi a_0^2 z^2}{T/R} \left[|M_{if}|^2 \ln\left(\frac{4T}{R}\right) + b_{if} + O\left(\frac{\Delta E}{T}\right) \right], \tag{91}$$

where M_{if} is the dipole matrix element, Eq. (8), and b_{if} is T-independent. Thus for large incident energies, $T/R \gg 1$ and $\Delta E/T \ll 1$, only the first term of Eq. (91) is significant. In that case, the cross section for charged particle ionization is *proportional* to the square of the dipole matrix element and, thus, the photoionization cross section.

The above analysis can also be carried out for the ejected electron angular distribution. If we now denote the angle the ejected electron direction makes with the incident projectile direction by θ, the expansion in inverse powers of T yields (Kim, 1972)

$$\frac{d\sigma_{ij}(\varepsilon)}{d\Omega} = \frac{4\pi a_0^2 z^2}{T/R}\left[\frac{1}{4\pi}|M_{if}|^2\left\{1 - \frac{\beta_{if}}{2}P_2(\cos\theta)\right\}\ln\left(\frac{4T}{R}\right) + \gamma(\theta) + O\left(\frac{\Delta E}{T}\right)\right]. \quad (92)$$

Thus we see that under the conditions which make the first term of Eq. (92) the only significant one, we have an angular distribution of ejected electrons just proportional to the angular distribution of photoelectrons [Eq. (67)]. Note that $\gamma(\theta)$ is T-independent.

Another connection between ionization by protons and fast charged particles is through the differential generalized oscillator strength (GOS), which is given by (Miller and Platzman, 1957; Inokuti, 1971)

$$\frac{df_{if}}{d(\varepsilon/R)}(K) = \frac{\Delta E}{R}\frac{|F_{if}(K)|^2}{(Ka_0)^2}. \quad (93)$$

The GOS has the property that it approaches the differential optical oscillator strength $df_{if}/d(\varepsilon/R)$ in the limit of $K \to 0$. The optical oscillator strength is related to the photoionization cross section σ_{if} by (Fano and Cooper, 1968)

$$\sigma_{if}(Mb) = 8.07\frac{df_{if}}{d(\varepsilon/R)}. \quad (94)$$

Thus if the projectiles are detected in the forward direction in coincidence with the ejected electrons, the result is a measurement of the GOS at $K_{\min}$. If, further, the projectiles are sufficiently swift so that $K_{\min}$ is quite small, one is essentially obtaining optical information, namely, the angular distribution of "photoelectrons." This technique has been used in several experiments to infer photoelectron angular distributions and cross sections (van der Wiel and Wiebes, 1971; El-Sherbini and van der Wiel, 1972; van der Wiel and Brion, 1972a, 1972b, 1973; Wight and Brion, 1974; Backx and van der Wiel, 1974).

III. Experimental Techniques

The experimental arrangement for photoelectron spectroscopy essentially consists of a photon source, an ionization chamber, and a photoelectron analyzer and detector. A great many recent monographs, reviews, and conference proceedings on the experimental situation in photoelectron spectroscopy have appeared recently. Thus, rather than duplicate them, presented herein is a brief guide to this extensive literature; essentially an annotated bibliography.

The books by Siegbahn et al. (1967, 1969) give a great deal of detail about the collision chamber and photoelectron spectrometers, both magnetic and electrostatic, for X-ray photoelectron spectroscopy. In an earlier work Siegbahn (1965) discussed magnetic spectrometers in great detail along with some of their difficulties and, more recently, Fadley et al. (1972) has proposed ways of overcoming the difficulties.

An extensive and inclusive discussion of both line and continuum photon sources in the UV is given by Samson (1967). In addition such subsidiary matters as vacuum techniques and filters and windows are treated in detail. Zaidel and Shreider (1970) also discuss photon sources in the UV extensively. In both of the above works detailed reference lists are provided. One particular photon source, synchrotron radiation, has been reviewed in great detail by Godwin (1969) due to its importance as a continuous source from the UV to the X-ray region. In this review the problems and possibilities of synchrotron radiation are focused on particularly.

The subject of UV emission from hot plasmas and an overview of the instrumentation used in photoelectron spectroscopy are treated in the volume edited by Damany et al. (1974). The sections on photoelectron energy analysis and energy resolution are particularly well done. Eland (1974) gives a fairly concise review of experimental techniques which is particularly suitable as an elementary introduction to experimental photoelectron spectroscopy.

Krause (1975) discusses some soft X-ray lines in the region between 100 eV and 1 keV which are about the only photon sources in this range other than synchrotron radiation. In addition, he gives a good description of the instrumentation for electron detection. For photoelectrons of very low energy the time-of-flight method (Toburen and Wilson, 1975) seems attractive. Although it has not yet been used for photoelectrons, it has been successfully employed in analyzing electrons ejected from atoms by charged particle impact ionization (Toburen and Wilson, 1975).

Recent conference proceedings have also contained a wealth of information on the experimental methods of photoelectron spectroscopy. The proceedings of the International Conference on Electron Spectroscopy held in 1971 and edited by Shirley (1972) and the one held in 1974 and edited by Caudano and Verbist (1974) contain many papers on all phases of the experimental situation, particularly on photoelectron analysis and detection. These works are indispensible to anyone who wishes to be informed on the latest developments. In addition the proceedings of the International Conference on Vacuum Ultraviolet Radiation Physics (Nakai, 1971; Koch et al., 1974) also provide much detailed information, especially on new photon sources in the UV.

Acknowledgments

This work was supported by the National Science Foundation and the U.S. Army Research Office. The author gratefully acknowledges Dr. Mitio Inokuti for his critical reading of the manuscript.

References

Åberg, T. (1969). *Ann. Acad. Sci. Fenn., Ser. A6:* Physica, No. 308.
Åberg, T. (1973). *In* "Inner Shell Ionization Phenomena" (R. W. Fink, S. T. Manson, J. M. Palms, and P. V. Rao, eds.), Conf-720404, p. 1509. USAEC, Oak Ridge, Tenn.
Altick, P. L., and Glassgold, A. E. (1964). *Phys. Rev.* **133**, A632.
Altick, P. L., and Moore, E. N. (1966). *Phys. Rev.* **147**, 59.
Amusia, M. Ya. (1971). *In* "Atomic Physics" (G. K. Woodgate and P. G. H. Sandars, eds.), Vol. 2, p. 249. Plenum, New York.
Amusia, M. Ya., Cherepkov, N. A., Chernysheva, L. V., and Sheftel, S. I. (1969). *Sov. Phys.-JETP* **29**, 1018.
Amusia, M. Ya., Cherepkov, N. A., and Chernysheva, L. V. (1971). *Sov. Phys-JETP* **33**, 90.
Backx, C., and van der Wiel, M. J. (1974). *In* "Vacuum Ultraviolet Radiation Physics" (E. E. Koch, R. Haensel, and C. Kunz, eds.), p. 137. Vieweg, Braunschweig.
Bagus, P. S. (1965). *Phys. Rev.* **139**, A619.
Bahr, J. L. (1973). *Contemp. Phys.* **14**, 329.
Bates, D. R. (1946). *Mon. Not. Roy. Astr. Soc.* **106**, 432.
Bethe, H. A. (1930). *Ann. Phys. (Leipzig)* **5**, 325.
Bethe, H. A. (1933). *In* "Handbuch der Physik" (H. Geiger and K. Sheel, eds.), Vol. 24/1, p. 482. Springer-Verlag, Berlin.
Bethe, H. A., and Salpeter, E. E. (1957). "Quantum Mechanics of One- and Two-Electron Atoms," pp. 247–323. Springer-Verlag, Berlin.
Brueckner, K. A. (1955). *Phys. Rev.* **97**, 1353.
Brueckner, K. A. (1959). "The Many-Body Problem." Wiley, New York.
Burke, P. G. (1968). *Advan. At. Mol. Phys.* **4**, 173.
Burke, P. G., and Moores, D. L. (1968). *J. Phys. B* **1**, 575.
Burke, P. G., and Robb, W. D. (1972). *J. Phys. B* **5**, 44.
Burke, P. G., and Robb, W. D. (1975). *Advan. At. Mol. Phys.* **11**, 143.
Burke, P. G., and Seaton, M. J. (1971). *Methods Comput. Phys.* **10**, 2.
Burke, P. G., and Webb, T. G. (1970). *J. Phys. B* **3**, L131.
Burke, P. G., McVicar, D. D., and Smith, K. (1963). *Phys. Rev. Lett.* **11**, 559.
Burke, P. G., Berrington, K. A., Le Dourneuf, M., and Vo Ky Lan (1974). *J. Phys. B* **7**, L531.
Callaway, J., Oberoi, R. S., and Seiler, G. J. (1970). *Phys. Lett.* **31A**, 547.
Carlson, T. A. (1973). *In* "The Physics of Electronic and Atomic Collisions" (B. C. Cobic and M. V. Kurepa, eds.), Invited Lectures of the Eighth ICPEAC, p. 205. Inst. of Physics, Belgrade.
Caudano, R. and Verbist, J., eds. (1974). "Electron Spectroscopy." North-Holland, Amsterdam.
Chang, T. N., and Fano, U. (1975). *Phys. Rev. A* **12** (in press).
Clementi, E. (1965). *IBM J. Res. Develop.* **9**, 2 and supplement.
Clementi, E., and Roetti, C. (1974). *At. Data Nucl. Data Tables* **14**, 177.
Combet Farnoux, F., and Heno, Y. (1967). *C. R. Acad. Sci., Ser. B* **264**, 138.
Cooper, J., and Zare, R. N. (1968). *J. Chem. Phys.* **48**, 942.

Cooper, J., and Zare, R. N. (1969). *In* "Lectures in Theoretical Physics" (S. Geltman, K. Mahanthappa, and W. Brittin, eds.), Vol. IIC, p. 317. Gordon and Breach, New York.
Cooper, J. W. (1962). *Phys. Rev.* **128**, 681.
Cooper, J. W., and Manson, S. T. (1969). *Phys. Rev.* **177**, 159.
Dalgarno, A., Stewart, A. L., and Henry, R. J. W. (1964). *Planet. Space Sci.* **12**, 235.
Damany, N., Romand, J., and Vodar, B. (1974). "Vacuum Ultraviolet Radiation Physics." Pergamon, Oxford.
Damburg, R., and Karule, E. (1967). *Proc. Phys. Soc., London* **90**, 637.
Dill, D. (1973). *Phys. Rev. A* **7**, 1976.
Dill, D., and Fano, U. (1972). *Phys. Rev. Lett.* **29**, 1203.
Dill, D., Manson, S. T., and Starace, A. F. (1974). *Phys. Lett.* **32**, 971.
Dill, D., Starace, A. F., and Manson, S. T. (1975). *Phys. Rev. A* **11**, 1596.
Dubau, J., and Wells, J. (1973). *J. Phys. B* **6**, 1452.
Einstein, A. (1905). *Ann. Phys. (Leipzig) Ser. IV* **17**, 132.
Eland, J. H. D. (1974). "Photoelectron Spectroscopy." Wiley, New York.
El-Sherbini, Th. M., and van der Wiel, M. J. (1972). *Physica* **59**, 433.
Fadley, C. S., Healey, R. N., Hollander, J. M., and Miner, C. E. (1972). *In* "Electron Spectroscopy" (D. A. Shirley, ed.), p. 121. North-Holland, Amsterdam.
Fano, U. (1961). *Phys. Rev.* **124**, 1866.
Fano, U. (1970). *Phys. Rev. A* **2**, 353.
Fano, U., and Dill, D. (1972). *Phys. Rev. A* **6**, 185.
Fano, U., and Lee, C. M. (1973). *Phys. Rev. Lett.* **31**, 1573.
Fano, U., and Cooper, J. W. (1965). *Phys. Rev.* **137**, A1364.
Fano, U., and Cooper, J. W. (1968). *Rev. Mod. Phys.* **40**, 441.
Fano, U., and Prats, F. (1964). *J. Nat. Acad. Sci. India* **33**, 553.
Fermi, E. (1928). *Z. Phys.* **48**, 73.
Fischer, C. F. (1968). "Some Hartree-Fock Functions for the Atoms Helium to Radon," Dept. of Math., University of British Columbia Report (unpublished).
Geltman, S. (1969). "Topics in Atomic Collision Theory," p. 44. Academic Press, New York.
Godwin, R. P. (1969). *Springer Tracts Mod. Phys.* **51**, 1.
Goldstone, J. (1957). *Proc. Roy. Soc., Ser. A* **239**, 267.
Grant, I. P. (1974). *J. Phys. B* **7**, 1458.
Grant, I. P., and Starace, A. F. (1975). *J. Phys. B* **8**, 1999.
Hall, H. (1936). *Rev. Mod. Phys.* **8**, 358.
Harris, F. E., and Michels, H. H. (1971). *Methods Comput. Phys.* **10**, 144.
Hartree, D. R. (1928). *Proc. Cambridge Phil. Soc.* **24**, 111.
Hartree, D. R. (1946). *Rep. Prog. Phys.* **11**, 113.
Hartree, D. R. (1957). "The Calculation of Atomic Structures." Wiley, New York.
Heitler, W. (1954). "Quantum Theory of Radiation." Oxford Univ. Press, London.
Hercules, D. M., and Carver, J. C. (1974). *Anal. Chem. Ann. Rev.* **46**, 133R.
Herman, F., and Skillman, S. (1963). "Atomic Structure Calculations." Prentice-Hall, Englewood Cliffs, New Jersey.
Hotop, H., and Mahr, D. (1975). *J. Phys. B.* **8**, L301.
Huzinaga, S., and Arnau, C. (1970). *Phys. Rev. A* **1**, 1285.
Inokuti, M. (1971). *Rev. Mod. Phys.* **43**, 297.
Ishihara, T., and Poe, R. T. (1972a). *Phys. Rev. A* **6**, 111.
Ishihara, T., and Poe, R. T. (1972b). *Phys. Rev. A* **6**, 116.
Jacobs, V. L. (1972). *J. Phys. B* **5**, 2257.

Jacobs, V. L., and Burke, P. C. (1972). *J. Phys. B* **5**, L67.
Kelly, H. P. (1964). *Phys. Rev.* **136**, B896.
Kelly, H. P. (1971). In "Atomic Physics" (G. K. Woodgate and P. C. H. Sandars, eds.), Vol. 2, p. 227. Plenum, New York.
Kelly, H. P. (1975). In "Atomic Inner-Shell Processes" (B. Crasemann, ed.), Vol. I, p. 331. Academic Press, New York.
Kelly, H. P., and Ron, A. (1972). *Phys. Rev. A* **6**, 1048.
Kennedy, D. J., and Manson, S. T. (1972). *Phys. Rev. A* **5**, 227.
Kim, Y. K. (1972). *Phys. Rev. A* **6**, 666.
Koch, E. E., Haensel, R., and Kunz, C., eds. (1974). "Vacuum Ultraviolet Radiation Physics." Vieweg, Braunschweig.
Kohn, W. (1948). *Phys. Rev.* **74**, 1763.
Krause, M. O. (1969). *Phys. Rev.* **177**, 151.
Krause, M. O. (1975). In "Atomic Inner Shell Processes" (B. Crasemann, ed.), Vol. II, p. 33. Academic Press, New York.
Krause, M. O., and Wuilleumier, F. (1972). *J. Phys. B* **5**, L143.
Latter, R. (1955). *Phys. Rev.* **99**, 510.
Lee, C. M. (1974). *Phys. Rev. A* **10**, 584.
Lee, C. M., and Lu, K. T. (1973). *Phys. Rev. A* **8**, 1241.
Lefebvre, R., and Moser, C., eds. (1969). *Advan. Chem. Phys.* **14**, 1.
Lin, C. D. (1974a). *Phys. Rev. A* **9**, 181.
Lin, C. D. (1974b). *Phys. Rev. A* **10**, 1986.
Lipsky, L. (1967). In "Fifth International Conference on the Physics of Electron and Atomic Collisions: Abstracts of Papers," p. 617. Nauka, Leningrad.
Livingston, M. S., and Bethe, H. A. (1937). *Rev. Mod. Phys.* **9**, 282.
Lu, K. T. (1971). *Phys. Rev. A* **4**, 579.
McGuire, E. J. (1968). *Phys. Rev.* **175**, 20.
Mann, J. B. (1968). "Atomic Structure Calculations," Los Alamos Sci. Lab. Rept. Nos. LA-3690, 1 (unpublished).
Manson, S. T. (1976). *J. Electron Spectrosc. Relat. Phenom.* **7** (in press).
Manson, S. T., and Cooper, J. W. (1968). *Phys. Rev.* **165**, 126.
Manson, S. T., Kennedy, D. J., Starace, A. F., and Dill, D. (1974). *Planet. Space Sci.* **22**, 1535.
Marr, G. V. (1967). "Photoionization Processes in Gases." Academic, New York.
Mies, F. (1968). *Phys. Rev.* **175**, 164.
Miller, W. F., and Platzman, R. L. (1957). *Proc. Phys. Soc., London* **85**, 51.
Moore, C. E. (1949). "Atomic Energy Levels," NBS Publ. No. 467. U.S. Govt. Print. Off., Washington, D.C.
Nakai, Y., ed. (1971). "Vacuum Ultraviolet Radiation Physics." Phys. Soc. Japan, Tokyo.
Ormonde, S., Whitaker, W., and Lipsky, L. (1967). *Phys. Rev. Lett.* **19**, 1161.
Ormonde, S., Smith, K., Torres, B. W., and Davies, A. R. (1973). *Phys. Rev. A* **8**, 262.
Percival, I. C., and Seaton, M. J. (1957). *Proc. Cambridge Phil. Soc.* **53**, 654.
Peshkin, M. (1970). *Advan. Chem. Phys.* **18**, 1.
Price, W. C. (1974). *Advan. At. Mol. Phys.* **10**, 131.
Rountree, S. P., Smith, E. R., and Henry, R. J. W. (1974). *J. Phys. B* **7**, L167
Salpeter, E. E., and Zaidi, M. H. (1962). *Phys. Rev.* **125**, 248.
Samson, J. A. R. (1967). "Techniques of Vacuum Ultraviolet Spectroscopy." Wiley, New York.
Samson, J. A. R. (1969a). *Phys. Rev. Lett.* **22**, 693.
Samson, J. A. R. (1969b). *J. Opt. Soc. Amer.* **59**, 356.
Samson, J. A. R. (1970). *Phil. Trans. Roy. Soc. London, Ser. A* **268**, 141.

Samson, J. A. R., and Starace, A. F. (1975). *J. Phys.* **B 8**, 1806.
Schmidt, V. (1973). *Phys. Lett.* **A 45**, 63.
Seaton, M. J. (1951). *Proc. Roy. Soc., Ser.* **A 208**, 418.
Sevier, K. D. (1972). "Low Energy Electron Spectrometry." Wiley-Interscience, New York.
Shirley, D. A. (1972)., ed. (1972). "Electron Spectroscopy." North-Holland, Amsterdam.
Shirley, D. A. (1973). *Advan. Chem. Phys.* **23**, 85.
Siegbahn, K. (1965). *In* "Alpha-, Beta-, and Gamma-Ray Spectroscopy" (K. Siegbahn, ed.), Vol. I, p. 79. North-Holland, Amsterdam.
Siegbahn, K., Nordling, C., Fahlman, A., Nordberg, R., Hamrin, K., Hedman, J., Johansson, G., Bergmark, T., Karlson, S. E., Lindgeren, I., and Lindberg, B. (1967). *Nova Acta Regiae Soc. Sci. Upsal. Ser., IV* **20**, 1.
Siegbahn, K., Nordling, C., Johansson, G., Hedman, J., Heden, P. F., Hamrin, K., Gelius, U., Bergmark, T., Werme, L. O., Manne, R., and Baer, Y. (1969). "ESCA—Applied to Free Molecules." North-Holland, Amsterdam.
Silverstone, H. J., and Yin, M. L. (1968). *J. Chem. Phys.* **49**, 2076.
Slater, J. C. (1930), *Phys. Rev.* **36**, 57.
Slater, J. C. (1951). *Phys. Rev.* **81**, 385.
Slater, J. C. (1960). "Quantum Theory of Atomic Structure." McGraw-Hill, New York.
Smith, K. J. (1971). "The Calculation of Atomic Collision Processes." Wiley-Interscience, New York.
Smith, K., Henry, R. J. W., and Burke, P. G. (1967). *Phys. Rev.* **157**, 51.
Starace, A. F. (1970). *Phys. Rev.* **A 2**, 118.
Starace, A. F. (1971). *Phys. Rev.* **A 3**, 1242.
Starace, A. F. (1973). *Phys. Rev.* **A 8**, 1141.
Starace, A. F., Manson, S. T., and Kennedy, D. J. (1974). *Phys. Rev.* **A 9**, 2453.
Thomas, L. H. (1927). *Proc. Cambridge Phil. Soc.* **23**, 542.
Toburen, L. H., and Wilson, W. E. (1975). *Rev. Sci. Instrum.* **46**, 851.
Truhlar, D. G., Abdallah, J., Jr., and Smith, R. L. (1974). *Advan. Chem. Phys.* **25**, 211.
Turner, D. W., Baker, A. D., Baker, C., and Brundle, C. R. (1970). "Molecular Photoelectron Spectroscopy." Wiley-Interscience, New York.
van der Wiel, M. J., and Brion, C. E. (1972a). *J. Electron Spectrosc. Relat. Phenom.* **1**, 309.
van der Wiel, M. J., and Brion, C. E. (1972b). *J. Electron Spectrosc. Relat. Phenom.* **1**, 443.
van der Wiel, M. J., and Brion, C. E. (1973). *J. Electron Spectrosc. Relat. Phenom.* **1**, 439.
van der Wiel, M. J., and Wiebes, G. (1971). *Physica* **53**, 225.
Vilesov, F. I., Kurbatov, B. C., and Terenin, A. N. (1961). *Dokl. Akad. Nauk. SSSR* **138**, 1320.
Vilesov, F. I., Kurbatov, B. C., and Terenin, A. N. (1962). *Sov. Phys.-Dokl.* **8**, 883.
Weiss, A. W. (1973). *Advan. At. Mol. Phys.* **9**, 1.
Wendin, G. (1973). *J. Phys.* **B 6**, 42.
Wick, G. C. (1950). *Phys. Rev.* **80**, 268.
Wight, G. R., and Brion, C. E. (1974). *J. Electron Spectrosc. Relat. Phenom.* **3**, 191.
Wigner, E., and Eisenbud, L. (1947). *Phys. Rev.* **72**, 29.
Yang, C. N. (1948). *Phys. Rev.* **74**, 764.
Zaidel, A. N., and Schreider, E. Ya. (1970). "Vacuum Ultraviolet Spectroscopy." Ann Arbor-Humphrey, Ann Arbor, Michigan.

Emission of Polarized Electrons from Solids

M. CAMPAGNA,* D. T. PIERCE,† F. MEIER,‡
K. SATTLER,‡ AND H. C. SIEGMANN‡

I. Introduction 113
II. Techniques and Apparatus 115
   A. Photoemission of Polarized Electrons 115
   B. Apparatus for Spin-Polarized Photoemission 118
   C. Field Emission of Polarized Electrons 123
   D. Apparatus for Spin-Polarized Field Emission 124
   E. Other Techniques 128
III. Results of Spin-Polarized Photoemission 130
   A. 4f Semiconductors and 4f Insulators 130
   B. 3d Semiconductors and 3d Semimetals 136
   C. 3d Metals 144
   D. Optically Magnetized Solids 150
IV. Results of Other Techniques 157
   A. Field Emission 157
   B. Superconducting Tunneling 159
   C. Electron Capture by Deuterons 160
V. Conclusion 161
   References 162

I. Introduction

The electron is the earliest known, most useful, and most common elementary particle. In 1925 Goudsmit and Uhlenbeck postulated that the electron has a spin. For a long time, the bulk of the evidence confirming this hypothesis was somewhat indirect, since the evidence was concerned only with electrons bound in solids or atoms. The first conclusive proof of the existence of free polarized electrons was provided in a double-scattering experiment performed by Shull *et al.* in 1943 (*1*), although, as Kuyatt recently pointed out (*2*), Davisson and Germer observed polarization effects in scattering as early as 1929 (*3*) without recognizing them as such. It has become clear only very recently that polarized electron beams can be as intense as unpolarized beams if the electrons are extracted from a solid under suitable circumstances. The present paper gives an account of experi-

---

* Bell Telephone Laboratories, Murray Hill, New Jersey.
† National Bureau of Standards, Washington, D.C.
‡ Laboratorium für Festkörperphysik der ETH, Zurich, Switzerland.

ments performed in several laboratories in the past seven years revealing this fact. The purpose of these research efforts was mainly to understand the physical properties of solids and the nature of the electron emission process. The discovery of a new generation of sources of polarized electrons was an important by-product. Hence the problems discussed in this paper will mainly pertain to solid state and surface physics. But the authors have attempted to emphasize the basic ideas involved in the experiments with the hope of reaching a larger audience than would be possible with a more narrowly focused presentation.

An ensemble of electrons is said to be spin-polarized if the electron spin has a preferred direction in space. The polarization vector is defined as $\mathbf{P} = \{\langle\sigma_x\rangle, \langle\sigma_y\rangle, \langle\sigma_z\rangle\}$, where σ_x, σ_y, σ_z are the Pauli matrices. In practice, coordinates can be chosen such that $\langle\sigma_x\rangle = 0$ and $\langle\sigma_y\rangle = 0$. $P = \langle\sigma_z\rangle$ is called the degree of polarization along the z axis, or the electron spin polarization (ESP). In this work we are dealing primarily with magnetic materials and have chosen to define P in terms of the electron magnetic moment (note that the spin moment and magnetic moment are antiparallel in the case of electrons). We have $-1 \leq P \leq +1$, the limits corresponding to ensembles of electrons with magnetic moments completely antiparallel and parallel to the z direction, respectively. $\mathbf{P}$ is an axial vector. Therefore, if one wants to extract polarized electrons from a solid, one has to define a z axis in the solid by an axial vector, such as a magnetic field or circularly polarized light. An electric field or linearly polarized light is not suitable.

The principle of the experiment is simple. One defines the z direction in a solid by applying a magnetic field or by determining the direction of incidence of a circularly polarized light beam. Then one extracts electrons from the solid by field emission or photoemission. The emitted electrons are collimated into a beam, the beam is accelerated to a suitable energy, and the polarization is measured by observation of the asymmetry in elastic scattering from heavy nuclei. This technique of detecting the ESP is known as Mott-scattering because it is based on the theoretical work of Mott, published in 1929. Kessler has written an extensive review on Mott-scattering (4).

Figure 1 shows the principle of the experiment, without the complex apparatus actually needed to perform it. The solid is a cylinder in which magnetization $\mathbf{M}$ is generated by a magnetic field. The electrons are emitted from the lower face and deflected electrostatically at an angle of 90°. In an electric field the polarization vector does not change as long as the electrons are nonrelativistic [see Ref. (5) for a review]. The scattering occurs in a plane perpendicular to $\mathbf{M}$ and therefore measures the ESP along the direction of $\mathbf{M}$ (see Section II). The fact that the direction of the z axis must be defined by an axial vector can be seen in Fig. 1. As soon as z is specified and the emitted electrons have an ESP different from zero, we can distinguish be-

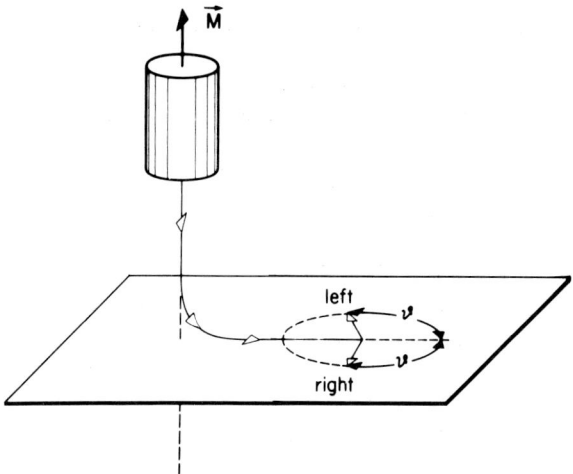

FIG. 1. Principle of the experiment. The z axis (dashed line) is defined by the magnetization **M**. The scattering plane is perpendicular to **M**.

tween right and left. If the preferred direction of the spin is antiparallel to **M**, more electrons are scattered to the left than to the right. This statement is valid for scattering angles $\zeta = \pm 120°$ at electron energies of 100 keV.

One difficulty in this experiment is the preparation and preservation of atomically clean surfaces. Another problem arises if one wants to work with magnetic materials. The magnetic field needed to generate a magnetization has unfavorable effects on the electron optics. If its direction is in the plane of the emitting surface, the Lorentz force inhibits emission. The direction of the magnetic field H must be mainly perpendicular to the surface. Even then, the magnitude of H has a substantial effect on the shape and position of the extracted electron beam. This is perilous when the scattering asymmetry is being measured, because large asymmetries that have nothing to do with ESP may be introduced by the electron optics. These are the main reasons why, despite many previous attempts, the definite experimental proofs for the existence of photo-ESP and field-ESP were adduced only in 1969 (6) and 1972 (7), respectively. A review of the development of polarized electron beams in atomic physics has been given by Farago (8).

II. Techniques and Apparatus

A. Photoemission of Polarized Electrons

In this experiment the electrons are emitted from the sample by irradiation with light. The apparatus used is described below in detail, but a short explanation of the physical principles of the ESP measurement and of photoemission is appropriate at this point.

1. Detection of Spin-Polarization by Mott-Scattering

We consider elastic scattering of electrons from atoms with zero magnetic moment. There are two types of interactions. The first is the electrostatic Coulomb interaction, which alone would give rise to the familiar Rutherford scattering. In the nonrelativistic limit ($v/c = \beta \ll 1$), the differential scattering cross section is (9)

$$d\sigma/d\Omega = (Z\alpha/2m\beta^2)^2(1/\sin^4(\theta/2)), \qquad (1)$$

where $\alpha = 1/137$, Z is the charge of the nucleus, m the mass of the electron, and θ the angle between the wave vectors $\mathbf{k}_1$ of the incoming electron and $\mathbf{k}_2$ of the scattered electron. The second interaction, which allows one to detect the polarization, is due to the spin–orbit interaction. From symmetry one can predict which components of the spin can lead to additional scattering. Since σ is an axial vector and $\mathbf{k}_1$ and $\mathbf{k}_2$ are polar vectors, $\sigma \cdot \mathbf{k}_1$ and $\sigma \cdot \mathbf{k}_2$ are pseudoscalars that change sign under the parity transformation $(x, y, z) \rightarrow (-x, -y, -z)$. This change is not possible in electromagnetic interactions. Consequently, only the spin components that are perpendicular to both $\mathbf{k}_1$ and $\mathbf{k}_2$ (that is, perpendicular to the scattering plane) can be detected in Mott-scattering. In the coordinate system in which the electron is at rest we have a magnetic field due to the electric field F' of the nucleus:

$$\mathbf{B}' = (1/c)(\mathbf{v}' \times \mathbf{F}'); \qquad (2)$$

$\mathbf{v}'$ is the relative velocity.

In the frame in which the atom is at rest and the electron moves, one obtains for the energy of the spin in that magnetic field

$$E = -\mu_B \cdot \mathbf{B} = \text{const } (1/r) \cdot dV(r)/dr \cdot \mathbf{l} \cdot \mathbf{S}. \qquad (3)$$

μ_B is the magnetic moment of the electron. The electric field was replaced by the derivative of the potential $V(r)$, and $\mathbf{v}$ by the angular momentum $\mathbf{l} = m\mathbf{r} \times \mathbf{v}$. The sign of E depends on the sign of $\mathbf{l}$, or, in other words, on whether the electron is scattered to the right or to the left. In one case the spin–orbit force dE/dr adds to the Coulomb force, and in the other case it subtracts from it.

The actual calculation of cross sections is rather lengthy. For applications it is sufficient to know that the component of $\mathbf{P}$ which is perpendicular to the scattering plane, $P_z = (n\uparrow - n\downarrow)/(n\uparrow + n\downarrow)$, is related to the observed scattering asymmetry $A = (N_R - N_L)/(N_R + N_L)$ by $A = P \cdot S$; $n\uparrow$ and $n\downarrow$ are the number of electrons with magnetic moments parallel and antiparallel to z, respectively. N_R is the number of electrons scattered into an angle θ to the right and N_L the number scattered into an angle $-\theta$ to the left. S, which is called the Sherman function, depends on the energy of the electrons, the scattering angle θ, and the strength of spin–orbit coupling, which is different for each atom. Fairly complete tables of S have been calculated, for instance, by Holzwarth and Meister (10) for Au and Hg atoms. At electron energies of 100 keV, S has a flat maximum of $S = 0.39$ centered at $\theta = 120°$. This is commonly used for detection of ESP.

2. Some Models of Photoemission

If photons with energy $\hbar\omega$ fall on the surface of a solid, electrons with various kinetic energies are emitted. The energy distribution curves (EDCs) of photoelectrons are specific for the solid and for the state of its surface as well. The object of photoemission is to establish which properties the solid and its surface must possess in order to produce the observed EDCs. Until now, this problem has been only partially solved because, as Caroli et al. (11) point out, it is impossible to treat from first principles all the phenomena that might occur. However, there are a few limiting cases in which one knows which approximations are good. One considers the energy balance in photoemission:

$$\hbar\omega + E(N) = E_v(N - 1) + e\Phi(\infty) + E_{kin}. \quad (4)$$

$E(N)$ is the energy of the solid with its N electrons, $E_v(N - 1)$ is the energy of the solid left behind in the state v after emission of one photoelectron, $e\Phi(\infty)$ is the potential energy of the photoelectron at infinite distance from the solid, and E_{kin} is its kinetic energy. Now the photothreshold Φ is introduced. It is the minimum photon energy $\hbar\omega_0$ required to remove an electron from the solid and bring it to infinite distance with $E_{kin} = 0$. Minimum excitation energy requires that the solid be left behind in the ground state $v = 0$. This yields

$$\hbar\omega_0 + E(N) = E_0(N - 1) + e\Phi(\infty). \quad (5)$$

Inserting $e\Phi(\infty)$ from Eq. (5) into Eq. (4) yields with $\hbar\omega_0 = \Phi$:

$$\hbar\omega = E_v(N - 1) - E_0(N - 1) + \Phi + E_{kin}. \quad (6)$$

One knows $\hbar\omega$, and one measures the intensity of photoelectrons $N(E)$ with a certain kinetic energy. Φ can also be determined by the experiment; it is the threshold photon energy $\hbar\omega_0$ at which photoemission sets in. The main drawback is that one does not know in which state v the solid was left behind.

The approximation $E_v(N - 1) - E_0(N - 1) \simeq$ const is valid for photoemission at low photon energies from extended band states in semiconductors like Si, Ge, or GaAs. $N(E)$ can be predicted from the band structure of the bulk crystal if anisotropic matrix elements for the optical transitions are taken into account. This has been demonstrated, for instance, by Grobman and Eastman with Ge (12), and the photo-ESP measurements on GaAs reported in Section III,D of this chapter are another example of how well the results can be understood in terms of the band structure.

Another approximation can be made for photoemission from electron states that are strongly localized on one atom. One particular electron in such a localized atomic state will interact most strongly with those electrons that are localized in the same shell, whereas its interaction with the electrons in neighboring atoms and in other atomic shells will be comparatively weak. N in Eq. (6) is then the number of electrons in that shell instead of the total number of electrons in the solid. $E_v(N-1) - E_0(N-1)$ is the excitation spectrum of the ion core left behind after photoemission. This spectrum of a single ion in a crystal field (SICF) is well known from atomic theory. The SICF model has been shown to apply to photoemission from 4f states regardless of whether the solid is metallic or insulating [see, for example, Ref. (13)]. This is not surprising, since the 4f states behave like atomic states even when conduction electrons are present. The interesting question is what happens in the case of 3d states. It seems that in simple oxides with poor electrical conductivity, like MnO or NiO, the SICF model is still applicable. This was shown recently by Wertheim et al. (14) and by Eastman and Freeouf (15). What goes on in the case of 3d metals is not known. It is possible that the SICF model is still important, but the multiplets are smeared and distorted by the screening action of the conduction electrons on the localized hole in a 3d shell (16,17). The older view of photoemission from 3d metals is that the band-structure effects are the most important.

The measurement of photo-ESP from magnetic materials adds new facets to these intriguing problems, as will be shown in Sections III,A, B, and C. The final goal of the investigations is to understand the behavior of d electrons. This knowledge in turn will provide a better physical basis for the description of magnetic and catalytic phenomena, to name only two fields in which there will be a major impact.

B. Apparatus for Spin-Polarized Photoemission

In all types of photoemission experiments some common problems have to be overcome. The distinguishing characteristic in spin-polarized photoemission is that the emitted electrons are collimated to form a beam that is directed onto a polarization detector, in our case based on Mott-scattering. The following devices are required.

1. An electron-optical system to form a beam. The expected ESP must be perpendicular to the beam because Mott-scattering detects transverse polarization only (see Section II,A,1).

2. An electron accelerator to bring the beam to an energy at which a suitable spin asymmetry occurs in Coulomb scattering (100 keV).

3. A counting system that measures the electrons scattered from heavy nuclei (Au).

4. For magnetic materials, a magnetic field that generates the magnetization in the sample.

5. A cooling system to keep the sample below the magnetic ordering temperature or to avoid spin–lattice relaxation in nonmagnetic materials.

Figure 2 shows the scheme of the apparatus. The magnetic field H, up to 50 kOe, is produced by the superconducting coil (No. 4). The sample is

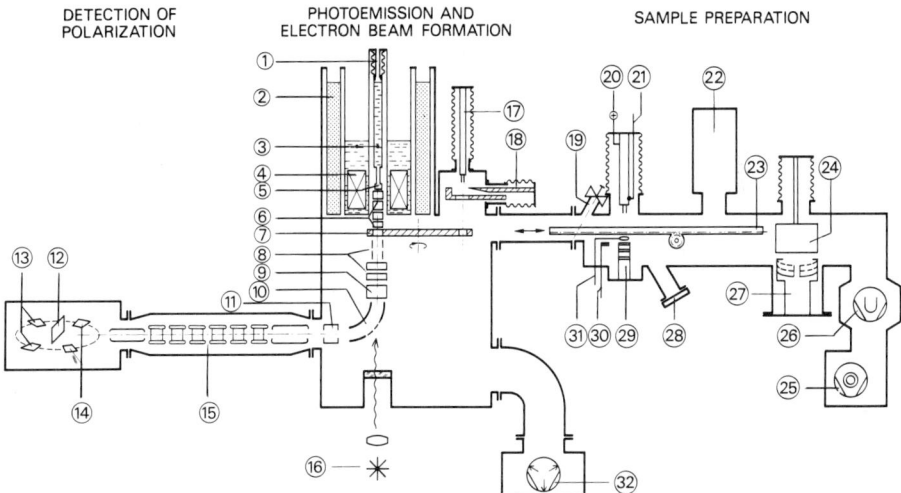

FIG. 2. Apparatus for spin-polarized photoemission. (1) Gripper with inner cryostat, (2) liquid-nitrogen Dewar, (3) outer cryostat, (4) superconducting coil, (5) sample in measuring position, (6) accelerating electrodes, (7) wheel and crystal magazine, (8) parallel beam shifters, (9) plane condenser, (10) cylindrical condenser, (11) diaphragm, (12) gold-foil scattering target, (13) forward detectors, (14) backward detectors, (15) accelerator, (16) light source with monochromator and optics, (17) gripper, (18) cleaving mechanism, (19) gate valve, (20) feedthrough to sample, (21) thermocouple, (22) mass spectrometer, (23) cog rail, (24) mirror for observing LEED pattern, (25) turbomolecular pump, (26) Ti sublimation pump, (27) LEED-Auger system, (28) window, (29) sputtering gun, (30) cesium oven, (31) electron gun, (32) ion getter pump.

located in the center of the coil. H is uniform in the absence of a sample. The field acting on the spins in the sample is H minus the correction due to demagnetization. Since the geometry of the sample is not perfectly defined and since surfaces are rarely smooth on the scale of the escape depth of photoelectrons, which is only a few lattice constants, the strength of the field acting on the spins is generally not known precisely.

The electrons emitted from the crystal are extracted electrostatically in a direction parallel to the magnetic field. The voltage on the rings (No. 6) is adjusted in such a way that the intensity of the beam is a maximum. The stray magnetic field outside the coil is minimized by an iron shield. The expected polarization of the beam is parallel or antiparallel to the beam direction (longitudinal ESP). At an energy of 4.5 keV, the electron beam passes through a series of deflector plates (Nos. 8, 9) that allow correction of the beam angle and beam position. The cylindrical deflector (No. 10) deflects the beam by 90°. It has three functions: (1) to transform the longitudinal ESP into a transverse one; (2) to deflect the electron beam out of the light beam; (3) to discriminate photoelectrons produced at the acceleration stages by stray light. For the last function, the diaphragm (No. 11) is also essential, since it stops all the spurious electrons from entering the six-stage 100-keV electrostatic accelerator (No. 15). Other diaphragms at the end of the accelerator define the angle and point of incidence of the electron beam on the scattering target, which is a self-supported Au foil. The vertical component of the earth's magnetic field is compensated by a coil. All these precautions are necessary because the scattering into a counter depends very strongly on where the beam hits the target and at what scattering angle θ the intensity is measured [see Eq. (1)]. The scattering chamber is maintained at 100 keV against ground potential.

Two of the four detectors are installed at $\theta = \pm 45°$ (No. 13), where the spin asymmetry $A_s = S(\theta) \cdot P = 0$ because $S(\theta) = 0$. The purpose of these counters is to monitor the beam position, that is, to determine the instrumental scattering asymmetry A_i that has nothing to do with ESP. The two remaining counters measure the electrons scattered into cones with 5° openings centered at $\theta = 120°$. Here $S(\theta)$ has a maximum, and one measures A_s. The effect of A_i on the size of A_s can thus be taken into account. Single electrons are counted with Si surface-barrier detectors. The electronics is the standard type. A rate meter indicates the intensity of the scattered beam, which helps in adjusting the electron optics. The detectors have a resolution of ~ 10 keV FWHM, which ensures that low-energy secondary electrons produced when the beam strikes a diaphragm are not counted.

1. *Calibration of the Polarization Detector*

Instrumental scattering asymmetries are introduced by different sensitivities of the detectors, by oblique incidence or inhomogeneity of the electron beam, and by wrinkles or holes in the thin gold-foil scattering target. These instrumental asymmetries are eliminated by reversing the beam polarization. This is no problem with nonmagnetic materials, because one simply switches from left- to right-circularly polarized light, which does not affect the electron beam optics. With magnetic materials one has to reverse the direction of the magnetic field. It is not possible to have the magnetic field so perfectly aligned with the axis of the extraction system that the position and angle of the electron beam will be unaffected. After the direction of H is reversed, one has to readjust the beam until the forward counters indicate identical scattering conditions. The remaining uncertainty in the measurement of the ESP is still of the order of 1%. It arises because the scattering greatly depends on θ [see Eq. (1)].

To calculate the absolute value of the ESP, one needs to correct the theoretical value S_0 of the Sherman function for the effects of multiple scattering in a real target of finite thickness d. Multiple scattering is proportional to the square of the number of scattering centers and therefore will decrease faster with d than single scattering decreases. Greenberg et al. (18) have established the following empirical relation:

$$A_0/A(d) = 1 + \alpha d, \tag{7}$$

where A_0 and $A(d)$ are the scattering asymmetries for targets with thickness $d = 0$ and d, respectively. If one measures $A(d)$ with a beam of constant polarization P_0, one obtains from Eq. (7) the relative Sherman function $S_0/S(d) = 1 + \alpha d$.

For this calibration, the scattering chamber contains a wheel with six Au foils with thicknesses varying from 150 μg/cm^2 to 460 μg/cm^2 as determined by a vibrating quartz monitor during fabrication of the foils in a commercial evaporation system. More accurate values for the effective thicknesses were obtained *in situ* by letting an electron beam of constant intensity fall on the foils and then measuring the intensity of electrons back-scattered into the counters at $\theta = \pm 120°$. For thin foils, there is a linear relationship between foil thickness and counting rate. Figure 3 shows the result of the measurement. One sees that Eq. (7) is valid only for foils with

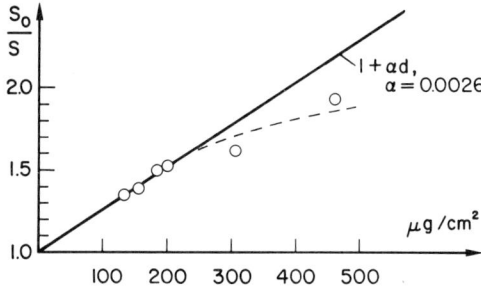

FIG. 3. Relative Sherman function as a function of foil thickness.

$d < 200$ μg/cm². One obtains $\alpha = 0.0026$. The effective Sherman function $S(d)$ is calculated under the assumption that the theoretical S_0 is correct.

It is difficult to estimate the uncertainties in the measurement of the ESP. In most papers only the statistical uncertainties are given. However, there is an uncertainty in calibration, and there are also remaining effects of the instrumental asymmetry that are especially large with magnetic substances. One could estimate that the uncertainty in calibration introduces an uncertainty of $\pm 5\%$ of the absolute values of the ESP, but additional systematic errors can be much larger, depending on how carefully instrumental asymmetries are eliminated.

2. Further Details of the Photoemission Apparatus

This subsection treats features of the apparatus shown in Fig. 2 that are not specifically related to spin-polarized photoemission.

Because of the sensitivity of photoemission to the state of a surface, the experiment must be done in ultrahigh vacuum. In the section of the apparatus where photoemission and electron beam formation occur, the pressure of 2×10^{-10} torr is maintained by a 200-liter/sec ion getter pump (No. 32 in Fig. 2) and by the pumping action of the walls of the liquid He cryostat (No. 3). This section also contains a crystal magazine (No. 7), as well as a knife with anvil (No. 18) to obtain atomically clean surfaces by cleaving single crystals mounted with individually shaped silver spacers on stainless-steel holders. The holders can be caught by grippers (Nos. 1, 17). The gripping works on the banana plug principle. For a measurement, the crystal is selected from the magazine by rotation of the wheel and is caught by a gripper (No. 17). After it has been cleaved it is put back in the wheel, and the wheel is rotated until the other gripper (No. 1) can catch it and pull it up into measuring position.

The light source is either a Hg–Xe high-pressure arc sealed in a quartz envelope or a hydrogen-discharge lamp. The Hg–Xe arc covers the photon energy range up to 6.5 eV. The H discharge is used for $6 \leq \hbar\omega \leq 11.2$ eV, and is operated with a heated cathode for increased intensity (19). Monochromatic light is obtained by a 1-m McPherson spectrograph. The pressure of 10^{-5}–10^{-6} torr in the optical system is maintained by an oil diffusion pump. The optical system is separated by a LiF window from the measuring system.

The sample preparation chamber is used for obtaining defined surfaces with crystals that cannot be cleaved. It is connected by an interlock to the measuring chamber. The vacuum in the sample preparation chamber is maintained by a Ti sublimation pump and by a bakeable turbomolecular pump. The chamber contains (i) an argon ion gun for sputtering away dirty surface layers, (ii) a cesium oven for lowering the work function, (iii) an adjustable leak valve for admission of various gases (not shown), (iv) a mass spectrometer, (v) an electron gun for heating, and (vi) a LEED–Auger unit for monitoring the state of the surface. The crystal under investigation is positioned on a cog rail and can thus be transported to all the different stages. If the surface has been well prepared and examined, the crystal is moved into the measuring system and put on the crystal magazine wheel, from which it can be caught by a gripper (No. 1) for measurement. In the sample preparation chamber it is possible to perform even complex surface-preparation techniques, such as obtaining a negative electron affinity on GaAs, and it is possible to use Cs and Ar without contaminating the measuring system. For putting in new samples, it is necessary to open and bake out only the preparation chamber, not the whole system.

In conventional photoemission it is usually not important to control the temperature of the sample during the measurement. In contrast, the photo-ESP reflects the state of magnetization, and temperature becomes an important parameter. The sample can be cooled to about $10°K$, and brought up to $300°K$ by a heater in the gripper (No. 1). The heater is connected by sliding contacts as soon as the gripper is in measuring position. For $T > 60°K$, the temperature is measured by a Cu–constantan thermocouple and kept constant by an automatically regulated current through the heater.

C. Field Emission of Polarized Electrons

The technique of field emission was developed to a much more sophisticated level than photoemission by the early 1960s (20,21). The process of making and cleaning a field-emission tip cannot be applied to a material that does not withstand the necessary tensile stress. [For a review of the status of this technology see Ref. (22).] Of the magnetic materials of interest for a spin-polarized field-emission experiment, only less refractory metals such as Ni, Co, or Gd and their alloys can be obtained in a suitable tip form without too many difficulties. Semiconducting compounds can only be evaporated onto the tip of a refractory metal such as W or Mo. The first apparently successful experiment involving emission of polarized electrons from solids was accomplished with a Gd field-emission cathode in 1967 (23). Ni was later investigated and polarization effects of about 10% were reported (24). These results could not be easily understood for two simple reasons: (1) in the same experiment comparable polarizations could be detected in field emission from W, and (2) the sign of the polarization varied with the strength of the magnetic field.

On the other hand, Müller and co-workers (7) reported the observation of no spin-polarization effects from W but a very large effect (up to 90%) for

EuS evaporated on W. The status of spin-polarized field emission, then, is less satisfactory than that of photoemission. This is surprising since the physics involved in the field-emission process, tunneling, is thought to be less complicated than that of the photoemission process.

D. *Apparatus for Spin-Polarized Field Emission*

In view of the difficulties with the interpretation of the observations in Refs. 23 and 24, we describe here a recently constructed apparatus (25) that differs in a few important features from those used so far in spin-polarized field emission (FE) studies. In Fig. 4 a schematic diagram of the source part

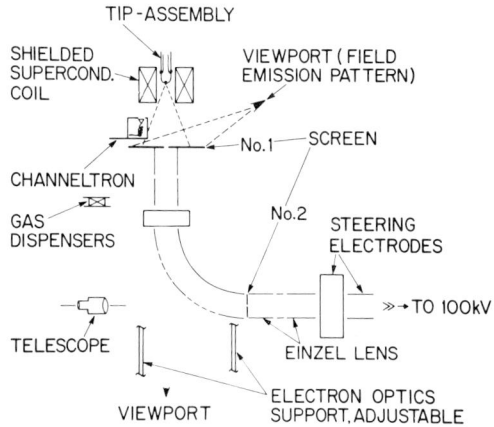

FIG. 4. Schematic diagram of the field emission source. The electron beam is bent by 90° to obtain transverse ESP as in photoemission.

is shown. The detection part is the same as the one used in photoemission and is based on Mott-scattering. Fundamental for an ESP measurement in FE as a function of the external magnetic field and for various (*hkl*) crystallographic directions are the following capabilities: (1) a controlled way of cleaning the tip surface, (2) static operation of the magnetic field at the tip position, and (3) an adjustable screen probe-hole arrangement for selecting the desired (*hkl*) direction.

The first capability is the most crucial, since it is known that tips of 3d metals like Ni are usually covered with an oxide layer that cannot be desorbed by simple thermal treatment. Gomer showed in his pioneering work (26) how one can obtain clean Ni tips by heat treatment in O_2 and H_2 atmosphere. After the introduction of field ion microscopy, field evaporation with the use of an imaging gas was the technique most used. The present approach is different because it involves a controlled field evaporation without the use of an imaging gas. This technique has the advantage of keeping the system absolutely clean. Field evaporation is achieved by moving the channeltron in Fig. 4 toward the axis of the apparatus and moving the tip vertically until the acceptance angle of the channeltron is 45-60°. A positive voltage is applied to the tip while the channeltron is at about -2.60 kV. The channeltron is connected to an amplifier and also to an oscilloscope and counter-timer. The rapid onset of field evaporation greatly depends on the tip temperature and the tip radius (27), but the process can be readily monitored in this way. The tip is supported by a W loop with Pt leads for measurement of the temperature. The field evaporation is followed by annealing of the supporting W loop at 950-1000°C, in a vacuum better than 1×10^{-10} torr. Best results in field evaporation have been obtained when the Ni tip is at a temperature near 150-200°K. Evaporation rates of 2000 Ni ions/sec can be safely achieved. Within 100 to 500 sec a few layers of material can be evaporated and an atomically clean field-evaporated end form is then available for the ESP experiment.

The result of a field evaporation can be directly seen on the screen (No. 1 in Fig. 4), which can be viewed at an angle of about 30-45°. The channeltron is retracted in the meantime from the axis of the Dewar electron optics system, while the necessary negative voltage for the FE mode is supplied to the tip. The result of a cleaning procedure outlined above can be seen in Fig. 5, A-C. Figure 5A shows the field emission pattern of a (001)-oriented Ni tip, which has been previously field-evaporated and kept for a few days in the 10^{-10} torr vacuum. No symmetry is apparent from the picture. After field evaporation the fourfold symmetry around the (001) direction is clearly visible, despite the slightly inconvenient viewing angle (Fig. 5B). To allow easier identification of the main (*hkl*) directions, we have indexed the FE pattern crystallographically.

For the ESP measurement, the tip is then retracted into the center of the superconducting coil. The importance of having a static magnetic field and a screen probe-hole arrangement available for a meaningful ESP measurement in FE from various (*hkl*) directions can be best appreciated by considering Fig. 5, C-G. Figure 5C shows a FE pattern of the field-evaporated tip when it is annealed at about 100°C. The shape of the tip and consequently the field distribution near the apex have drastically changed. When positioned into the center of the coil, part of the FE image is cut out by the presence of the inner walls of the Dewar so that, although the total FE current has not changed, the well-developed (001) dark plane now covers most of the

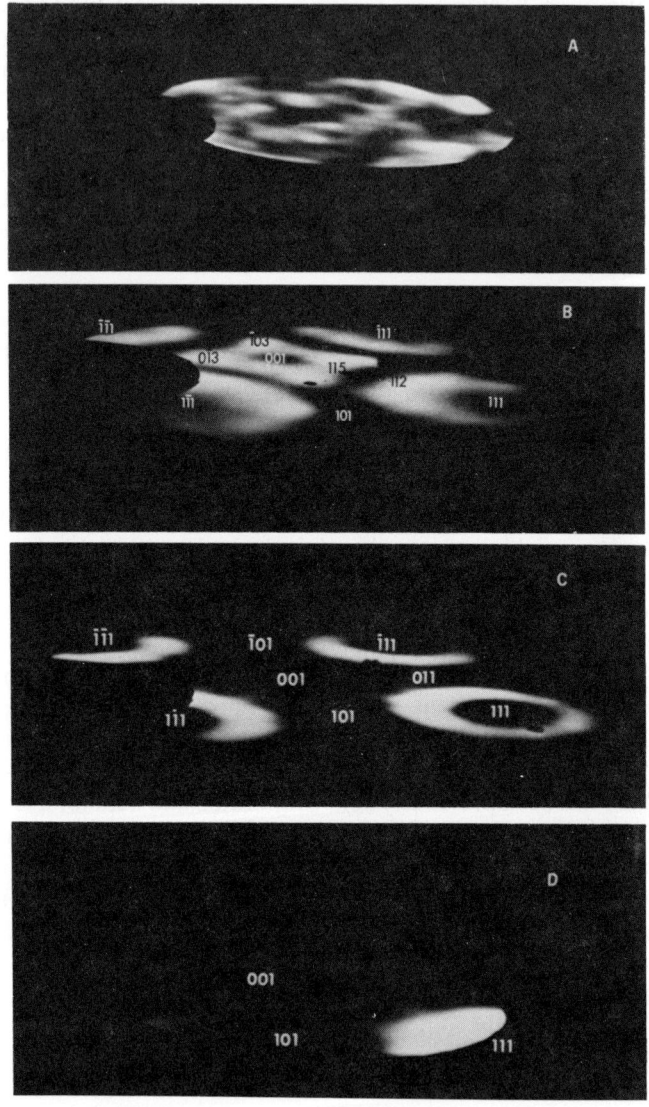

FIG. 5, A–D. Caption on facing page.

screen. Sweeping the external magnetic field is known to generate, besides a rotation, a drastic distortion (compression) of the field-emission image.

At lower fields the distortions are more important for electrons leaving the tip at large angles ϕ ($\phi = 0$ for emission parallel to the tip axis). They depend on the strength of the external magnetic field and on the potential

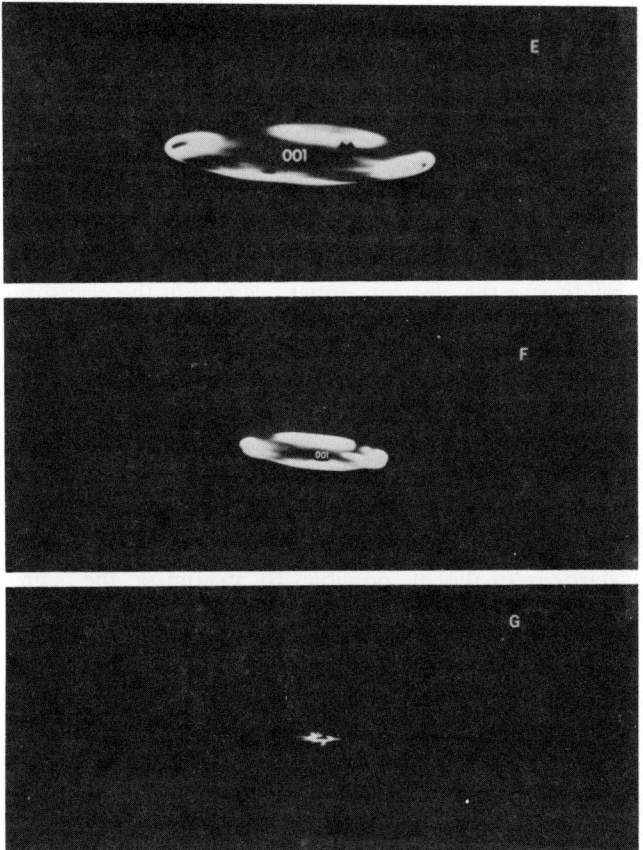

FIG. 5. The effects of cleaning and of a strong external magnetic field on FE patterns. (A) "Dirty" Ni tip, (001)-oriented; (B) clean, field evaporated end form of (001) Ni; (C) image obtained after annealing at 100°C; (D) tip positioned in the center of the superconducting coil; (E, F, G) fields of 750, 1000, and 1400 Oe, respectively, are applied.

used to extract the electrons from the magnetic field. In our arrangement, since no immersion electrode is located near the tip, the extraction voltage is the tip voltage itself. This is of the order of a few hundred volts, up to 2–3 kV depending on tip radius, the upper bound being dictated by voltage requirements. (The electron optics cannot be operated at voltages higher than +15 kV.) For example, with a tip voltage of −1714 V and total FE current of 1.6×10^{-7} A, one can progressively focus the FE in the center of the screen by increasing the H field to 1.4 kOe (see Fig. 5, E–G). The image expands and will be recompressed again for increasing field values.

All this illustrates clearly how careful one has to be to measure the polarization vs. magnetic field H for a given crystallographic direction. The ideal situation for a precise $P(H)$ measurement is to have the geometrical axis of the electron optics, the axis of the magnetic field, and the tip axis completely aligned. This is technically difficult to achieve, especially because each tip has its own geometry, which may change under thermal treatment. But only in this case will the center of the FE image when positioned onto the probe hole not move by sweeping H. Instead of making the tip and the Dewar with superconducting coil completely adjustable relative to each other and relative to the electron optics, we mount the electron optics assembly on a system allowing x, y, and z movement *and* tilting relative to the tip and Dewar, while the tip-to-screen and coil-to-screen distances can be independently adjusted. With tips mounted as centrally as possible on the axis of the coil, we have found that small adjustments of the electron optics (tilting angles $< 5°$) are usually sufficient to bring the desired (hkl) direction onto the probe hole. After passing through the probe hole, the beam of electrons is focused onto a screen (No. 2) at the entrance of the 100-keV accelerator, which can be viewed through a telescope. An einzel lens and steering electrodes make it possible to maximize the signal in the Mott-scattering.

E. Other Techniques

There are ways of emitting electrons from solids other than photoemission and field emission. There are also alternative methods of measuring the ESP. In the tunneling experiment (Section II,E,1), the electrons are not emitted into a vacuum, but into a superconductor or from a superconductor into a ferromagnetic metal. The ESP can be measured because the density of states in a superconductor depends on the spin state if a magnetic field is applied parallel to the junction. In the second experiment, discussed in Section II,E,2, the electrons at the surface of the solid are captured by deuterons passing by the surface. Their ESP is detected in a nuclear reaction.

1. *Superconducting Tunneling*

In a series of impressive experiments Tedrow and Meservey (*28*) observed that when a large magnetic field is applied to an $Al-Al_2O_3-Al$ junction the quasiparticle energy states of Al are split by the interaction of the quasiparticle magnetic moment μ with the field H, where the relative displacement for up- and down-spins is $2\mu H$. A careful analysis of the tunneling conductance made it possible even to observe spin-state mixing in superconductors, i.e., the process whereby, because of spin–orbit interaction, a small fraction of the spin-up states have the same energy as the spin-down states. These experiments also showed that spin-flip tunneling can be neglected. When, on one side of the junction, Al is replaced by a ferromagnetic metal, the available tunneling density of states is polarized, provided an external field is applied so as to align the magnetic domains. The asymmetry resulting in the tunneling conductance depends on the polarization of the carriers; the information one obtains concerns the polarization of electron states very near to the Fermi energy (± 1 meV).

2. Electron Capture at Surfaces of Ferromagnets

Owing to their efforts to obtain an efficient way of producing polarized deuteron beams, Kaminsky (29) and Rau and Sizmann (30) developed a new method of emitting electrons from magnetized surfaces. The principle of the technique is shown schematically in Fig. 6. A beam of charged deuterons

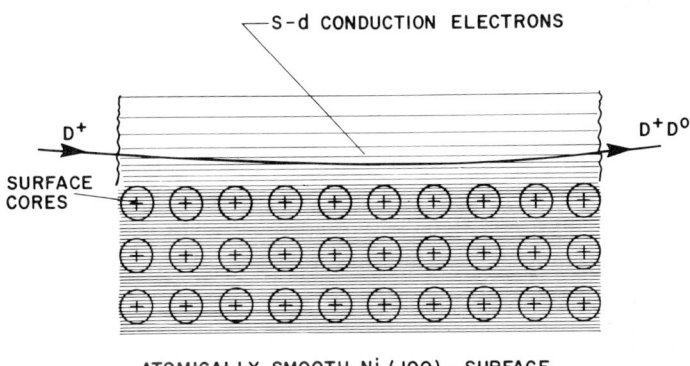

FIG. 6. Principle of the experiment involving capture of polarized electrons by deuterons from the surface of magnetic metals (Ni).

(150 keV) is scattered at grazing incidence from a Ni single crystal with a smooth surface. The deuterons are partially neutralized by capture of polarized electrons from the surface. The electron polarization is then transformed via hyperfine interaction to the deuteron nuclei, and this polarization is used to extract values of the electron polarization. The physics involved in the process of electron capture by fast charged particles is not yet understood in detail.

III. Results of Spin-Polarized Photoemission

A. 4f Semiconductors and 4f Insulators

This section treats diatomic compounds that contain a rare earth element as one component. [For a review, see Ref. (*31*).] For photoemission from the 4f states of the rare earth, the model of a single ion in a crystal field (SICF) is applicable, and the structure in optical absorption and photoemission spectra is dominated by the excitation spectrum of the ion core left behind after Eq. (6). With seven electrons, the 4f shell is half-occupied and the magnetic moment of the ion has the maximum possible spin-only value of $7\mu_B$. This situation is realized either with divalent Eu or trivalent Gd. Compounds with these elements are simplest from the viewpoint of spin-polarized photoemission because the excitation spectrum of the $4f^7$ ion core is the ^7S multiplet. Further, energy distribution curves (EDCs) of photoelectrons have been published for many of these compounds, and are understood in their basic features.

The materials investigated are: (1) the ferromagnetic insulators EuO, EuS, EuSe, Eu_3P_2, Eu_3As_2, and GdN with Curie temperatures of 69, 16.8, 4.2, 24, 18, and 70°K; and (2) the antiferromagnetic semiconductors or semimetals EuTe, GdP, GdAs, and GdSb with Néel temperatures of 9.8, 15, 19, and 28°K. Additionally, the effect of extra electrons introduced by *n*-type doping of EuO and EuS with several trivalent elements (La, Gd, Ho, Sc) and the effects of structural disorder introduced by depositing thin films on substrates cooled to 4.2°K were studied.

Three types of electronic levels are distinguished in the experiments: (1) the occupied valence bands formed by covalent mixing of p-wave functions of the anions and 5d states of the cations, (2) the localized magnetic states of the rare earth cations ($4f^7$), and (3) the conduction states, which can be intrinsic or extrinsic. These three types of states behave differently in photoemission: for emission from $4f^7$, the SICF model is applicable, whereas for emission from extended valence or conduction states the band model may be a reasonable approximation. The ionization energy of 4f states may depend on the presence of conduction electrons, because the hole in the $4f^6$ final state can be screened by the conduction electrons in the sense of the impurity model of Friedel (*32,33*). The mechanism exists independently from the magnetic order. Another interesting phenomenon that depends on magnetic order has been observed in EuO when conduction electrons are introduced on *n*-type doping. One detects a large reduction of the photoelectric work function ($\Delta\Phi \cong 0.5$ eV) upon cooling below the Curie temperature (*34*).

This phenomenon can be understood if one takes into account that the conduction electrons are spin-polarized via the exchange interaction with the localized spins in the 4f shell (34).

The measurements described in this section include:

1. Yield measurements. The yield Y is the number of photoelectrons emitted per incident photon $\hbar\omega$.

2. Photoelectric magnetization curves (PMC). Here the photo-ESP is measured as a function of the strength of the magnetic field at constant photon energy and constant temperature.

3. Spectra of spin polarization (SSP). The magnetic field strength and the temperature are kept constant. The photo-ESP is measured as a function of the photon energy $\hbar\omega$. A review of many of the results is provided in Ref. (35).

1. *Photoelectric Magnetization Curves (PMCs)*

The main point of measuring PMCs is to obtain information on the magnetic properties of the surface. Since in most cases the escape depth of photoelectrons is only a few lattice constants, the photo-ESP reflects the magnetic behavior of a very thin sheet at the surface. If we neglect for a moment spin flips in the photoemission process, the photo-ESP should be proportional to the magnetization: $P = f(\hbar\omega) \cdot M(T, H)/M_0$. $M(T, H)$ is the magnetization at the temperature T and the magnetic field H; M_0 the saturation magnetization if all the atoms have their magnetic moments aligned parallel. $f(\hbar\omega)$ is a factor that depends on the initial states from which the photoelectrons were emitted. $f = 1$ for emission from $4f^7$ shells; $f \cong 0$ for emission from doubly occupied valence states; and $0 \leq f \leq 1$ for emission from conduction states. The latter holds because the exchange coupling of conduction states to the 4f electrons is much larger than kT_0, as has been determined by optical studies, especially the red shift of the absorption edge in the ferromagnetic compounds (31). T_0 is the magnetic ordering temperature.

The magnetization curves of the bulk are well known. If the PMCs are proportional to the bulk magnetization curves independently of whether the electrons are emitted from 4f states or conduction states, we can conclude that the surface sheet has magnetic properties identical to the bulk. This was found to apply, within the accuracy of the experiment, to thin films of the antiferromagnets EuTe, GdP, GdAs, and GdSb (36). The PMCs are straight lines whose slopes decrease with increasing T_0 just as the bulk magnetization curves do. As an example, Fig. 7 shows the measurements on EuTe. The slope of the curve also depends on the spectrum of photon energies incident

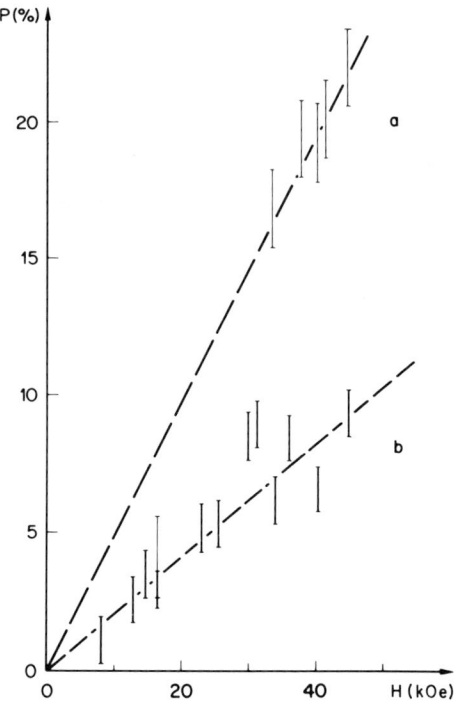

FIG. 7. Dependence of photo-ESP on magnetic field strength for a thin-film sample of EuTe at $T \cong 4.2°$K; points were measured (a) with photon energies near photoelectric threshold and (b) with the full spectrum of the Hg-Xe arc.

on the surface, because with increasing $\hbar\omega$ more unpolarized valence electrons are admixed and the ESP decreases.

In contrast, the ferromagnetic compounds EuO, EuS, and Eu_3P_2 exhibit PMCs that are different from the bulk magnetization curves (37). Even at temperatures far below T_0 the photo-ESP does not saturate at high-magnetic field strengths. Figure 8 shows measurements on pure EuO and on EuO doped with Gd. Saturation of the PMCs should occur at $H \cong 10$ kOe in these single-crystal samples, which clearly is not the case. Also, the photo-ESP should reach the value $P = 100\%$, since with the photon energy chosen emission occurs predominantly from $4f^7$ states. There is only a small kink in the PMCs when H reaches the bulk saturation value.

These observations can be understood if one assumes that the magnetic coupling in the surface sheet is very weak, actually close to zero, and if the effect of this single sheet on the total photo-ESP is enhanced by surface-spin exchange scattering of the escaping photoelectrons (38). Both these assumptions are conceivable. The weakening of the exchange coupling of surface-

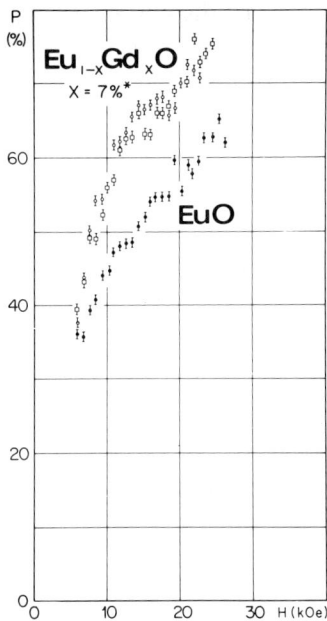

FIG. 8. Dependence of photo-ESP on magnetic field strength for single-crystal samples of pure and doped EuO at $T \cong 10°$K. Photon energy was chosen so that emission occurs predominantly from $4f^7$ states.

spin moments might be due to the fact that the number of nearest neighbors is reduced, and the fact that at the surface of ionic crystals relaxation takes place (displacement of anions and cations in opposite directions perpendicular to the surface). Superstructure or reconstruction in the top surface sheet may also be present (33). As soon as the surface spins are free to change their z component, there must be substantial spin-exchange scattering with electrons emerging from layers deeper in the bulk. Numerical estimates show that one single nonsaturated layer at the surface can reduce the initial polarization of 100% characteristic of electrons from deeper $4f^7$ states, to 50% (38). The spin-exchange scattering greatly enhances the effect of a single surface sheet. Even if the escape depth of photoelectrons is very great, say 10 atomic layers, we can in this way still understand why the photo-ESP is only 50%.

The conditions for the existence of a nonsaturated surface sheet in Heisenberg ferromagnets have been further investigated. Introducing conduction electrons by doping enhances the coupling between surface and bulk spins. This is seen from the results on $Eu_{1-x}Gd_xO$ displayed in Fig. 8. The photo-ESP is higher at a given H compared to that of pure EuO. The increase of the photo-ESP also occurs on doping with La, Ho, and Sc (33). Hence it does not depend on whether the dopant has a 4f spin or not. It has long been known that the presence of conduction electrons enhances the ferromagnetic part of the exchange (31), and that it does not matter how the conduction electrons are introduced. An enhanced surface-to-bulk coupling can apparently be generated by introducing "conduction" electrons in the surface only. This has been shown by depositing a few atomic layers of metallic Cs at the surface of EuO (39). The nonsaturated surface sheet seems to be a property of only the insulating 4f ferromagnets. In magnetite, the atomic moments are due to the unfilled 3d shell of Fe, and the PMCs saturate at magnetic fields of ~ 6 kOe as expected (40). The same is true for the PMCs of the ferromagnetic 3d metals Fe, Co, and Ni (41).

2. *Spectra of Spin Polarization (SSPs)*

The SSP enables one to identify the character of the initial state from which the photoelectron was emitted. Photoelectrons from 4f states are spin-polarized, whereas the average ESP of those from valence states is zero. Figure 9 shows as an example the SSP of EuO (42). The photo-ESP is high

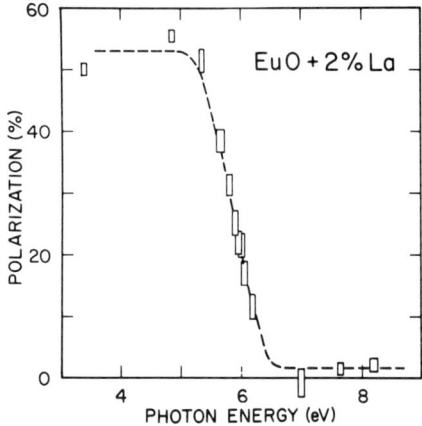

FIG. 9. Dependence of photo-ESP on photon energy for a single crystal of EuO at $T \cong 10°K$ and at $H = 12$ kOe.

near threshold and decreases to zero as the emission of unpolarized oxygen 2p electrons sets in. This is readily understood in terms of the energy-level scheme of EuO established by optical studies (43). The $4f^7$ states are the highest occupied electron states, and the threshold for 2p emission is ~ 5.5 eV. The emission strength from 2p states is substantial compared to that of 4f states because the matrix elements for optical excitation of 4f electrons are very weak near threshold owing to the centrifugal barrier. This explains why the ESP decreases to zero although 4f electrons are still present even when the threshold of 2p electrons is surpassed. This effect is important when the ESP measurement is applied to other oxides, for which the distinction between 2p electrons and magnetic electrons is often crucial in the interpretation of spectroscopic data.

An interesting example of an application is provided in a study of the extra electrons introduced in EuO single crystals by doping (33). These extra electrons have lower ionization energy than the 4f electrons. Measurement of the photoelectric yield and of the SSP for $\hbar\omega$ near threshold shows that two types of electron states are created on doping. Figure 10 shows the measurements on Gd-doped EuO. The yield Y shows a kink at $\hbar\omega = 4$ eV. At $\hbar\omega \leq 3$ eV, the emission occurs mainly from impurity states that have been introduced on doping with Gd. The two different curves shown in Fig. 10 have been observed in many cases. They occur with the same crystal,

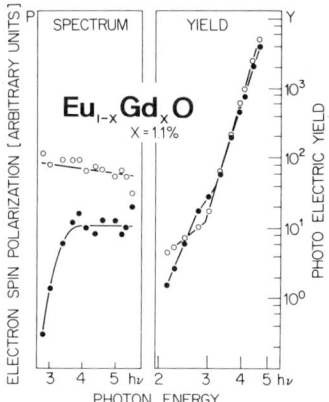

FIG. 10. Dependence of photo-ESP and photoelectric yield on $\hbar\omega$ at $T \cong 10°$K for EuO single crystals nominally doped with 1.1% Gd. The SSP was measured with H = const. The two curves were measured on two different cleave surfaces of the same crystal.

but with two different surfaces obtained in successive cleavings. Yield and SSP are correlated. When Y increases slowly with $\hbar\omega$, the ESP of the impurities is as high as that of the $4f^7$ electrons. When Y rises steeply with $\hbar\omega$, the impurities are weakly polarized. The latter kind of impurity is related to lattice distortions and the extra electrons are then localized. The first kind of impurity, with high ESP, is present when a conduction band is formed. It is a well-known fact that the concentration of conduction electrons in EuO cannot be simply calculated from the amount of Gd that has been substituted for Eu, and the ESP measurement shows directly that "doping" of

EuO is very delicate and not homogeneous. This result has been corroborated by the optical work of Schoenes and Wachter (44).

Another interesting application of SSPs is concerned with the effects of structural disorder in the NaCl lattices of 4f magnetic semiconductors. Structural disorder was introduced in ferromagnetic EuS and EuSe, and in antiferromagnetic GdP, GdAs, and GdSb as well, by depositing thin films on substrates kept at liquid-He temperature (45). The disorder was removed by annealing. Extra electron states are introduced with disordering, and their ESP can be measured directly. The extra states generally have very large effects on the magnetic interactions. With GdP, the effect is dramatic: at the same temperature it is paramagnetic in the disordered state but antiferromagnetic in the ordered state. By a series of measurements on the same sample in the disordered and ordered states and by a comparison of the systematic variation of properties in a series of related materials, it has been possible to understand many of the very interesting changes in electronic and magnetic properties that occur on disordering.

B. 3d Semiconductors and 3d Semimetals

Materials containing atoms with incomplete 3d shells are in many respects the antipodes of those with incomplete 4f shells. The extension of the 3d shell is much larger, and direct interatomic overlap of 3d states is possible. This leads to high magnetic ordering temperatures and in many cases conduction band formation cannot be excluded. But these bands are generally quite narrow. The question is: How should one describe the d electrons? Neither a collective band nor a purely atomic model can account for the variety of phenomena observed in 3d materials. It has been pointed out in Section II,A,2 that photoemission can help one decide whether optical excitations should be interpreted in the framework of an atomic model, that is, the model of a single ion in a crystal field (SICF), or whether a description in terms of initial band states is more appropriate. We show that spin-polarization measurements can answer this question even in quite complex cases where other methods are ambiguous or do not work at all.

1. Magnetite and Some Simple Ferrites

Magnetite, Fe_3O_4, is the oldest magnetic material known and the prototype of the ferrites, which form the basis of a large industry. The general formula is $MeFe_2O_4$, where Me is mostly a divalent metal ion. The ferrites crystallize in the spinel crystal structure, in which the comparatively large O ions form an fcc lattice. The metal ions Me^{2+} and Fe^{3+} are located either in the tetrahedral or the octahedral interstitials called A and B sites, respectively. The number of occupied B sites is twice the number of A sites. Which site Me^{2+} occupies depends on the kind of atom and very often on the thermal treatment of the crystal. The full formula is: $Me_\delta^{2+}Fe_{1-\delta}^{3+}[Me_{1-\delta}^{2+}Fe_{1+\delta}^{3+}]O_4$. The ions in brackets occupy B sites. For magnetite (MAG), $Me^{2+} = Fe^{2+}$ and $\delta = 0$; that is, at A sites we have Fe^{3+} only, whereas at B sites Fe^{2+} and Fe^{3+} coexist. For magnesioferrite (MGF), $Me^{2+} = Mg^{2+}$, and the δ parameter can be varied within $0 \leq \delta \leq 1/3$ by heat treatment; $\delta = 1/3$ corresponds to a statistical distribution of the Mg ions onto the two available lattice sites. Lithium ferrite (LIF) has the formula $Fe^{3+}[Li_{0.5}^{+}Fe_{1.5}^{3+}]O$. The net magnetization of these simple ferrites is given by $M = M_B - M_A$, where M_B is the B-sublattice and M_A the A-sublattice magnetization. All the A-ion spins are parallel among themselves and are antiparallel to the B-ion spins. Fe^{3+} has a magnetic moment of $5\mu_B$ (Bohr magnetons), and Fe^{2+} has one of $4\mu_B$; Mg^{2+} and Li^+ have, of course, no magnetic moment. Hence the magnetization M per formula unit is $4\mu_B$, $\delta \cdot 10\mu_B$, and $2.5\mu_B$ for MAG, MGF, and LIF, respectively.

For the photoemission experiment, it is important to note that with ferrites the photoelectric threshold can be as high as 6.5 eV, and that the 3d states are the highest occupied intrinsic states. Therefore, at threshold, one has pure d-state emission except for the impurities, and one must go at least to photon energies in the vacuum ultraviolet. It turns out that MGF is the ideal material to test the applicability of the SICF model. First, the excitation levels of the ion cores are very simple. Mg^{2+} does not emit any photoelectrons at low $\hbar\omega$, because the closed inner shells have high binding energy. Further, there exists Fe^{3+} only, which has a spherically symmetric $^6A_1$ configuration, and all the spins in the 3d shell of this ion are parallel. If we neglect spin flips, the photoelectrons are either up-spin or down-spin, depending on whether they have been emitted from a B or an A site. The Fe^{4+} left behind can be in only two excited states, depending on whether a t_{2g} or an e_g electron has been emitted. These two states are ^5E and $^5T_2$ in the cubic crystal field of the ferrites. The nomenclature is the one used by Sugano et al. (46). The energy separation Δ of ^5E and $^5T_2$ is given by the strength of the crystal field acting on the ion core. It turns out that this crystal field is comparatively strong, so that the line separation is large and the two lines can be resolved.

One can calculate the ratio of the crystal-field splitting on B sites compared to that on A sites, Δ_B/Δ_A, from crystal-field theory (47) without knowing the size of the charge on the oxygen ions. The main point, which is most important for the interpretation of photo-ESP, is that the sign of the crystal field on B sites is opposite to that on A sites. This means that the sequence of the ion core multiplets is inverted; at the A site, $^5T_2$ is the ground state of Fe^{4+}, but at the B site it is ^5E.

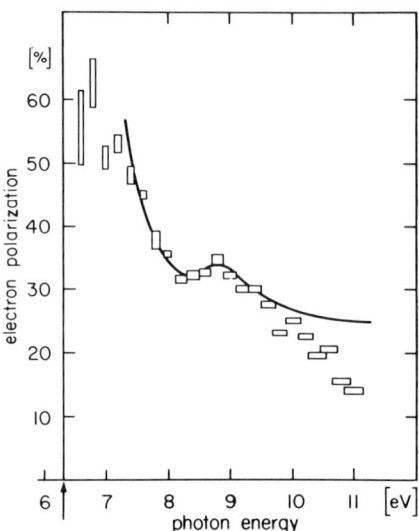

FIG. 11. Dependence of photo-ESP on photon energy for a single crystal of $MgFe_2O_4$ at $T \cong 60°K$, $\delta = 0.24$, and $H = 16\,kOe$. The solid line was calculated with the SICF model. Arrow indicates photoelectric threshold Φ.

Figure 11 shows the ESP of photoelectrons from a surface of MGF as a function of photon energy (SSP). The ESP is high at photoelectric threshold and decreases with increasing photon energy after passing through a relative maximum at $\hbar\omega \sim 9$ eV. The energy position and height of this maximum do not depend on the thermal treatment of the crystals, that is, on the size of the δ parameter. The maximum plays a crucial role in the interpretation. It would not be there if an interpretation based on the initial band states were correct. The initial state (Fe^{3+}) is spherically symmetric and cannot be split by either cubic crystal field or by the exchange, since Fe^{3+} contains only up-spins. In the SICF model, on the other hand, multiplet structure is expected. It corresponds to the emission of an electron from Fe^{3+} at a B site leaving behind Fe^{4+} in the $^5T_2$ excited state. The full curve in Fig. 11 is an SSP calculated on the SICF model. There is good agreement with the observed values of the ESP over a large range of $\hbar\omega$. Note that the absolute value of $P(\hbar\omega)$ must agree; that is, no background can be subtracted. With one and the same set of five parameters it has been possible to fit all the $P(\hbar\omega)$ curves measured on ten MGF surfaces obtained by cleaving single crystals with different thermal treatment and different values of the δ parameter (48). For $\delta = 0$, we have an A–B antiferromagnet, i.e., the net magnetization $M = 0$. Yet polarized photoelectrons are obtained. This shows that B sites emit electrons at lower $\hbar\omega$ than A sites. There are two significant deviations of the experimental SSPs from the calculated curves; near threshold Φ and for $\hbar\omega > 10$ eV, the observed ESP is too low. The deviations near Φ can be attributed to photoemission from unpolarized impurity states lying in the forbidden energy band. The deviation at $\hbar\omega > 10$ eV is most likely due to the onset of emission from unpolarized oxygen bands.

Figure 12 illustrates the simple model that was used to calculate the SSPs. The lower part shows the assumed excitation spectrum of Fe^{4+} at B

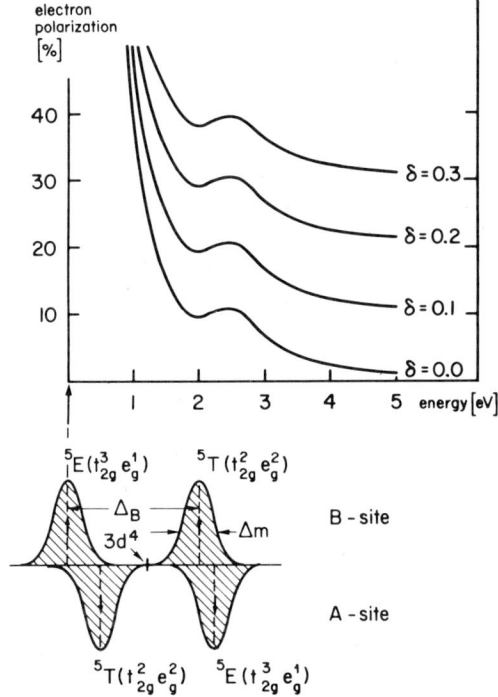

FIG. 12. Calculated dependence of photo-ESP on $\hbar\omega$-Φ for different distributions δ of Mg^{2+} over the lattice sites. The energy levels of Fe^{4+} ion cores left behind on A and B sites are also shown. $3d^4$ means center of gravity of the $3d^4$ shell. Vertical arrow indicates threshold Φ.

and A sites. The energy levels have a finite width $\Delta m = 0.5$ eV to account mainly for the finite lifetime of Fe^{4+}. The relative strength η of the transitions $^6A_1 \to {}^5E$ and $^6A_1 \to {}^5T$ should be 2 : 3 at high $\hbar\omega$. It has been found that η must be an adjustable parameter at low $\hbar\omega$. The value of η = 1.00 is in agreement with the experiment. The center of gravity (CG) of $3d^4$ divides the splitting Δ_B in the ratio 3 : 2 and Δ_A in the ratio 2 : 3 because there are three t_{2g} orbitals and two e_g orbitals in which the hole created by emission of a photoelectron can be located. Since the relative occupancies of A and B sites can be influenced by thermal energies of 100 meV, one can assume that the center of gravity of $3d^4$ has the same energy at both lattice sites in MGF: $\Delta CG = 0$. In general, ΔCG is another parameter. Finally one has to take into account that the electrons must escape from the material over the surface barrier potentials. When the threshold Φ_v for leaving behind the ion core in a state v is reached, the intensity of photoelectrons will increase with $\hbar\omega$ according to $i_v = \exp(-(\hbar\omega - \Phi_v)/k)$. The escape constant k is the fourth parameter. It has been found that $k = 1.5$ eV, which is reasonable (49). The upper part of Fig. 12 shows the SSPs calculated for different values of the distribution δ of Mg^{2+} over the two lattice sites. One salient feature of the observations is reproduced: the position and height of the relative maximum do not depend on δ. The other striking result of the experiment, namely, that the ESP is high at threshold and that its sign corresponds to emission from B sites, also finds a natural explanation. It is due to the fact that the sign of the crystal field on A sites is opposite to the sign of that on B sites. Note that the average energy of the 3d electrons on both sublattices was assumed to be the same by setting $\Delta CG = 0$.

The most convincing argument for the validity of this model is that it also explains the SSPs of LIF and MAG. Figure 13 shows the measurements on LIF (40) together with the excitation states of Fe^{4+} assumed to calculate the SSP (full line). Except for deviations at low and high $\hbar\omega$ for the same reasons as in MGF, there is a perfect fit. The parameters η, Δm, and k have identical values in LIF and MGF, but the crystal field splitting Δ_B and ΔCG have to be different. The value $\Delta CG = 0.25$ eV in LIF may reflect the fact that the Li ions never occupy A sites.

In MAG, Fe^{2+} is present in addition to Fe^{3+}. The final state multiplet of singly ionized Fe^{2+} contains six lines, which makes the calculation a bit more difficult. Figure 14 shows the measurements on MAG (40) together with the SSP calculated on the SICF model. The main features of this spectrum are: (1) from $\hbar\omega = 6.5$ eV on, the shape is very similar to the SSPs observed with ferrites containing only Fe^{3+} (LIF and MGF); (2) at threshold the ESP becomes negative. The SICF model correctly predicts these features, especially that the presence of Fe^{2+} must generate negative ESP at threshold. This can be seen directly: Fe^{2+} has one electron more than Fe^{3+}. This extra electron must be a down-spin because all up-spin orbitals are already occupied. The emission of the extra electron requires lowest energy because the ion core left behind is in the ground state. Hence the emission of the extra electron occurs at photoelectric threshold. Additional parameters can be obtained from the ESP measurement on MAG, especially the intraconfiguration energy U_{eff}, which is the energy difference between the CG of $3d^4$ and $3d^5$.

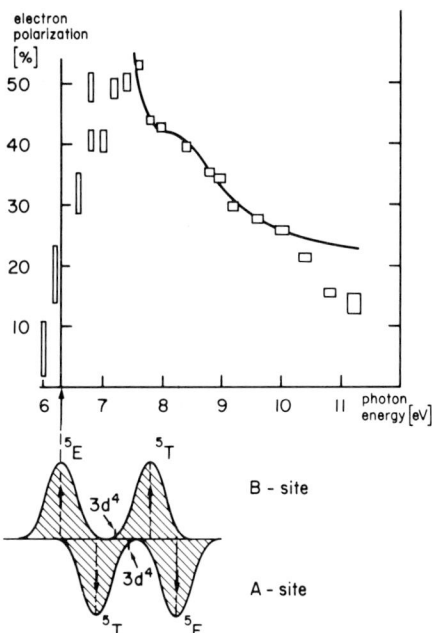

FIG. 13. Dependence of photo-ESP on $\hbar\omega$ for a single crystal of $Li_{0.5}Fe_{2.5}O_4$ at $T = 60°K$ and $H = 8.5$ kOe. The energy levels of Fe^{4+} left behind as assumed for the calculation (full curve) are also shown.

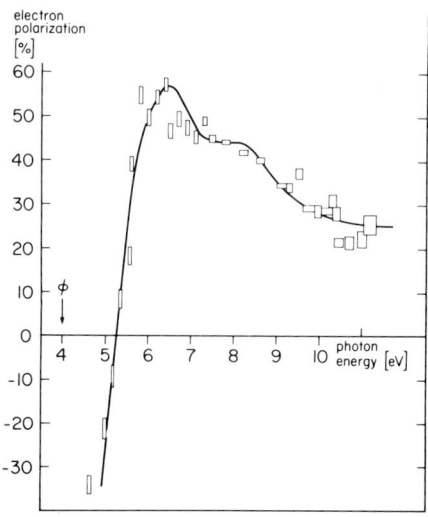

FIG. 14. Dependence of photo-ESP on $\hbar\omega$ for a single crystal of Fe_3O_4 at $T = 10°K$ and $H = 8.5$ kOe.

One can see, then, that the SICF model explains the electronic excitations in ferrites. With few parameters and a plausible choice of their magnitude, one can fit all the data. This result was obtained because simple ferrites that contain only Fe^{3+} have been investigated, and because the ESP measurement allows one to distinguish whether an excitation has occurred on the A site or on the B site. Earlier measurements on magnetite have been erroneously interpreted on the basis of initial one-electron energy levels generated by crystal field effects and exchange interactions (40,50,51).

2. *The Catalyst* $La_{1-x}Pb_xMnO_3$

The metallic-like oxides of the type $La_{1-x}Pb_xMnO_3$ crystallize in the perovskite structure. They are ferromagnetic for $x \cong 0.3$ and may replace Pt-based catalysts for oxidation of pollutants in automobile exhaust (52). The active centers for catalysis are probably the high-spin-state Mn ions located in octahedral oxygen coordination. Mn^{3+} and Mn^{4+} coexist in the ratio $(1 - x)/x$. The Mn ions are also responsible for the peculiar behavior of the resistivity near the ferromagnetic transition temperature which is $T_0 = 340°K$ for $x = 0.3$ (53). An understanding of these phenomena implies an understanding of the behavior of the d electrons. Both a generalized LCAO scheme (54) (that is, a collective behavior) and a localized picture (55) have been put forward, but experimental evidence supporting either of these approaches has been very scarce. This is not surprising, since Pb 6s electrons, O^{2-} 2p electrons, and the Mn 3d electrons coexist at about the same energy, and with the conventional techniques it is practically impossible to disentangle these different contributions. With the ESP measurement, one can distinguish between spin-polarized 3d emission and nonpolarized 6s and 2p emission. A structure has been observed in the 3d emission on a large background, and by combining the ESP results with X-ray and UV-photoemission data it has been possible to show that the collective picture is appropriate for $LaPbMnO_3$ (56).

Figure 15 shows the spectrum of spin polarization (SSP) and the energy distribution curves (EDCs) of photoelectrons at various photon energies in the ultraviolet, both measured on surfaces of single crystals obtained by cleaving. Considering first the EDCs, one sees that the density of states is very low at the Fermi energy E_F: in fact, the discontinuity at E_F appears only after amplification. $LaPbMnO_3$ contradicts the rule that a catalyst should have a high density of states at E_F. The general shape of the EDCs does not depend on photon energy up to photon energies of $\hbar\omega \cong 1500$ eV (X-rays), except that the intensity of emission from states near E_F is somewhat stronger in comparison to the big hump at ~ 3 eV below E_F. This suggests that the Mn 3d states are located near E_F, since the matrix elements for d emis-

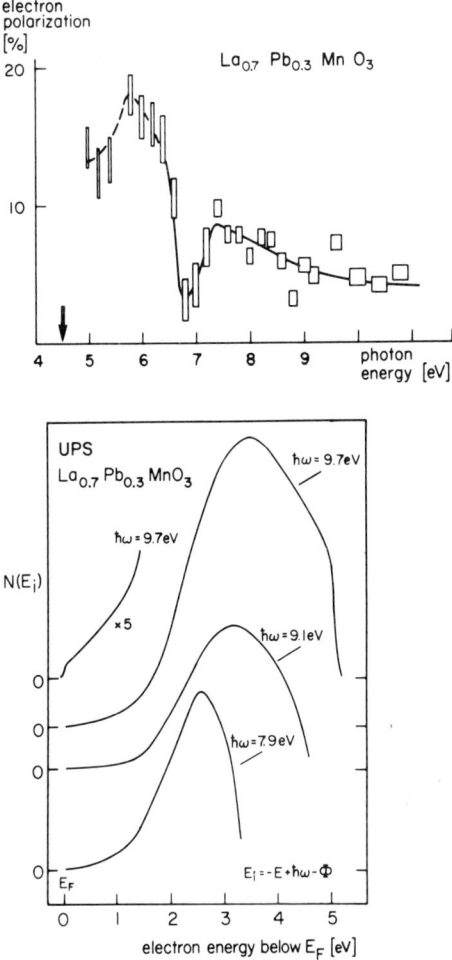

FIG. 15. (Top) dependence of photo-ESP on $\hbar\omega$ for a LaPbMnO$_3$ single crystal at $T \cong 10°$K and $H = 16.4$ kOe. Arrow indicates photoelectric threshold. (Bottom) energy distribution of photoelectrons $N(E_i)$ in arbitrary units at various photon energies. $T = 300°$K.

sion increase with photon energy because of the centrifugal barrier. The SSPs show that the d states are indeed at E_F, since the ESP is highest near threshold. The first decrease of the ESP at $\hbar\omega - \Phi \cong 1.6$ eV coincides with the start of the pronounced shoulder in the EDC. This implies that the onset of the shoulder is due to emission of unpolarized electrons, i.e., either 6s or 2p electrons. There is a second rise in the ESP at $\hbar\omega - \Phi \cong 2.2$ eV before it decreases to a rather low value. For pure 3d emission the ESP should be

100%. The fact that the ESP is quite low at all $\hbar\omega$ shows that d emission is superimposed on a large unpolarized background. This explains why the structure of the d emission is not seen in the EDCs. At $\hbar\omega = 9$ eV, for instance, the contribution of all the d electrons to the total photocurrent is only a few percent.

For the interpretation of the structure in the SSP, one has to discard the SICF model because it is in conflict with the low intensity of emission from states near E_F observed in the EDCs. The SICF model predicts the following: on the fast time scale of photoemission ($< 10^{-16}$ sec) both Mn^{4+} and Mn^{3+} are found to coexist, in analogy to the two ionic configurations of the Sm^{2+} and Sm^{3+} ions in the metallic phase of SmS (57). Then the two peaks in the SSP, whose thresholds are separated by ~ 2.2 eV, are identified as the final states of the transitions $Mn^{3+} \to Mn^{4+}$ + photoelectron and $Mn^{4+} \to Mn^{5+}$ + photoelectron. The difference in the thresholds is the effective intra-ionic correlation energy, $U_{eff} \cong 2.2$ eV. But the total intensity in the two transitions should correspond to the density of Mn^{4+} and Mn^{3+}. The X-ray photoemission studies show that this is not the case, and provide support for the collective picture of Mattheiss (54). Here the structure is due to an e_g band at E_F filled with 0.7 electrons per Mn atom and a full and narrow t_{2g}^3-derived band ~ 2 eV below E_F.

All this shows the promising aspects of the ESP measurement for investigation of the behavior of electrons in the critical but interesting intermediate range, where neither the uncorrelated band model nor the strong correlated limit of the Hubbard model is the correct approach.

C. 3d Metals

The 3d ferromagnetic transition metals, Fe, Co, and Ni, are the center of controversy in the history of spin-polarized photoemission measurements. Although in earlier experiments (58,59,60) no polarization of the photoemitted electrons was observed, measurements in the last five years at the ETH in Zurich (61,41,62,63) definitely show that the electrons photoemitted from Fe, Co, and Ni are polarized. The failure to observe a spin polarization in the earlier measurements was most likely due to poor vacuum conditions and unfavorable electron optics, especially the application of the magnetic field parallel to the emitting surface.

The measurement of sizable polarizations from the 3d ferromagnets only heightened the controversy, because these measurements, unlike the results from the magnetic semiconductors, cannot be readily explained within existing theories of magnetism and photoemission. In particular, Fe, Co, and Ni are expected to be itinerant ferromagnets, and the widely used band theory of ferromagnetism developed by Slater (64), Stoner (65), and Wolfarth (66) should be applicable. In this theory, the exchange interaction between the itinerant electrons produces an average Hartree–Fock molecular field that is different for the two electron spin states. The band is formally divided into two pure spin subbands split by the exchange energy ΔE_{ex}, thought to be about 0.4 eV for Ni and 1.1 eV for Co at $T = 0°K$. For Ni and Co, one subband is filled (majority spins) and the Fermi level cuts through the other partially filled band (minority spins). Electrons emitted from states near E_F in Ni and Co are expected to give rise to a negative polarization. In Fe, both bands are only partially filled. The majority electrons dominate at E_F; so, in contrast to Ni and Co, the measured polarization is expected to be positive.

In fact, a positive polarization was measured for all three metals. This discrepancy with the band theory of ferromagnetism led to measurements beyond the initial measurements on polycrystalline films to films coated with Cs in order to probe states that lie energetically deeper below E_F, and to very thin films in order to test the relative importance of bulk and surface contributions. These results are discussed below along with their implications for theories of magnetism in Fe, Co, and Ni. Since our understanding of 3d ferromagnets is by no means complete, spin-polarized photoemission measurements are currently being extended to single-crystal surfaces and photon energies in the vacuum ultraviolet.

1. Clean Polycrystalline Films

Fe, Co, and Ni films (61,41) were evaporated by electron gun onto stainless steel substrates held at $T = 400°K$ (type I) or $T = 4.2°K$ (type II). The pressure rose to 10^{-8} torr during evaporation but fell immediately afterwards to the 10^{-10} torr range. The annealing of the film and the measurements were performed at a pressure of 2×10^{-10} torr. No difference was observed between a type-II film annealed to 400°K and a type-I film.

The first measurement was that of the ESP of electrons photoemitted from states near E_F. The work functions of Fe, Co, and Ni, which range from 4.6 to 5.2 eV, the photoelectron escape function, and the decreasing intensity of the Hg–Xe lamp between 5 and 6 eV all combine to limit the photoelectrons to those originating from states within a few tenths of an eV from E_F. The saturation ESP was found to be $+54\%$, $+21\%$, and $+15\%$ for polycrystalline films (type-I and annealed type-II) of Fe, Co, and Ni at 4.2°K.

The dependence of the ESP on applied magnetic field, the photoelectric magnetization curve (PMC), is shown for Fe in Fig. 16. As expected, an applied field about equal to the saturation magnetization of 21.5 kG is necessary to reach saturation because of the large demagnetizing field in this

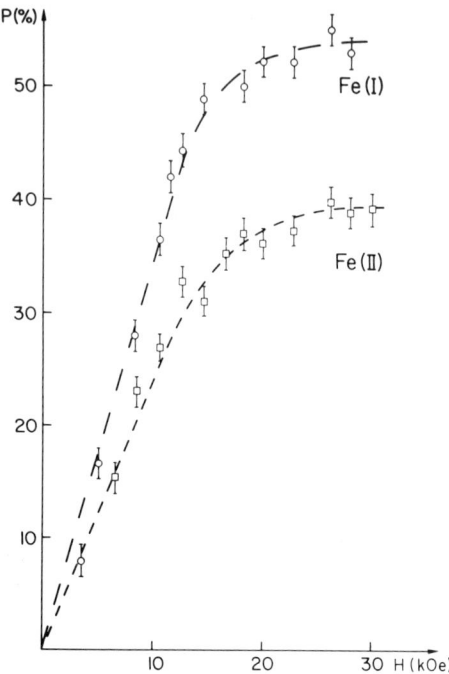

FIG. 16. Dependence of photo-ESP on magnetic field strength at 4.2°K for polycrystalline (I) and noncrystalline (II) films of Fe with the full spectrum of a Hg-Xe arc.

geometry. Fe, Co, and Ni all exhibited PMCs that saturated like that for a bulk ferromagnet, in contrast to the PMCs observed from the 4f ferromagnetic semiconductors (Section III,A,1). The saturation ESP of a noncrystalline type-II film is lower than that for a type-I film. Magnetization measurements (67) have also been made for noncrystalline Fe and Ni. The provocative observation is that the decrease in ESP near E_F scales with the decrease in M; for Fe, $M_I/M_{II} = P_I/P_{II} = 0.7$ and for Ni, $M_I/M_{II} = P_I/P_{II} = 0.6$. This correlation between the ESP near E_F and the magnetization and the positive ESP observed for Ni and Co are both important experimental facts to be explained by a theory of magnetism.

2. Spectra of ESP from Cesiated Films

In order to obtain sharper experimental criteria to be met by a new theory of ferromagnetism, the ESP measurements were extended to films where the work function was lowered by deposition of Cs. The saturation ESP measured for electrons from near E_F in Fe, Co, and Ni is higher than the average ESP of all the band electrons: $\bar{P} = n_B/n = +26\%, +17\%$, and

+5% for Fe, Co, and Ni, respectively. Here n_B is the number of Bohr magnetons and n is the number of s and d electrons. It is clearly of interest to determine how the ESP varies as electrons are emitted from states lying deeper below E_F. The work functions ϕ of Fe, Co, and Ni polycrystalline films prepared as in Section III,C,1 above were varied from 3.8 to 2.2 eV by depositing Cs from a zeolite source. In this way it was possible to measure $P(\hbar\omega)$, the spectrum of spin polarization (SSP), with the Hg–Xe light source.

The photon energy dependence of the ESP for cesiated Fe, Co, and Ni films (62,63) with a work function of 3.25 ± 0.1 eV is shown in Fig. 17. In all

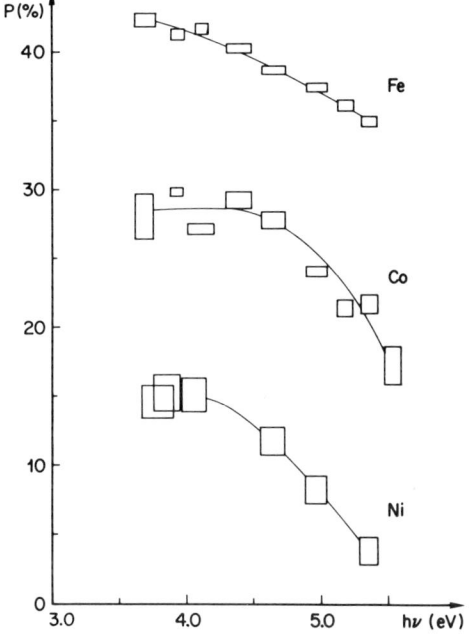

FIG. 17. Dependence of photo-ESP on photon energy for cesiated Fe, Co, and Ni films with work function $\Phi = 3.25 \pm 0.1$ eV.

three curves, the ESP is highest at threshold, that is, for electrons originating near E_F. The maximum value of the ESP decreases with increasing number of d electrons (and decreasing Bohr magneton number). The spectral width of the region of high ESP also decreases from Fe to Ni and is correlated with, but less than, the d bandwidths of 3.8, 3.6, and 3.3 eV of Fe, Co, and Ni, respectively, determined from ultraviolet photoemission measurements (68) at $\hbar\omega = 40.8$ eV. The polarization P approaches the average polarization $\bar{P} = 17\%$ and 5% for Co and Ni, and is tending to $\bar{P} = 26\%$ for Fe, although measurements at higher $\hbar\omega$ are needed to verify this trend.

The exact shape of the spectra predicted by the band model of ferromagnetism depends also on the model assumed for the photoemission process. The nondirect transition model, in which the photocurrent is proportional to the product of the initial and final density of states, predicts a negative P at threshold for Ni and Co [up to $\hbar\omega - \phi = 2$ eV for Co (62)], which increases to $\bar{P}$ at larger $\hbar\omega$. Smith and Traum (69) have used the direct transition model, in which conservation of $\mathbf{k}$ is important in the optical excitation process, for calculating the variation of P with changing $\hbar\omega$ or ϕ. They have shown that under certain circumstances a positive P is expected because no final state is available for minority electrons. When the work function is altered, the final state for threshold photoelectrons is changed and will not continue to discriminate against minority electrons over a range of, say, 1 eV. At some point the emission of minority electrons will be reflected in a negative ESP.

Representative spectra for cesiated Co with a work function ranging from 2.5 to 3.3 eV are shown in Fig. 18. The ESP is positive and at higher photon energies approaches the expected $\bar{P} = 17\%$. The band theory of ferromagnetism with either a direct or nondirect transition model of photoemission is inadequate to explain the data.

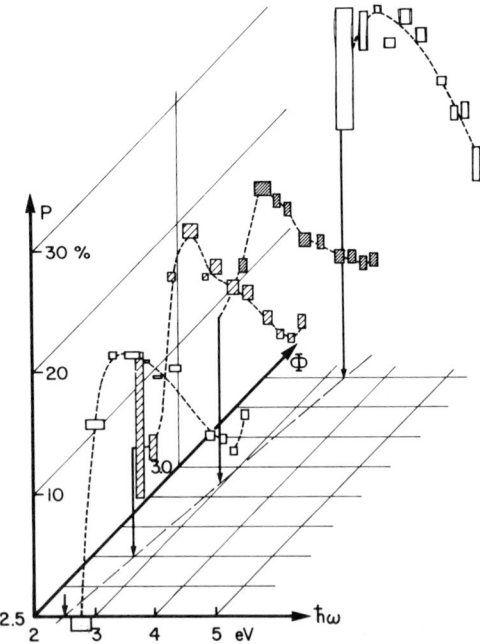

FIG. 18. Dependence of photo-ESP on photon energy and on work function Φ for cesiated Co films at $T = 4.2°$K and $H = 17$ kOe.

For higher Cs coverages, $\phi \lesssim 3$ eV, a depolarization is observed near threshold. For low Cs coverages, the Cs 6s electron goes to the metal and a dipole layer is formed by the Cs ions. For higher coverage, there is a higher charge density in the neighborhood of the Cs ion that results in more atomic-like Cs on the surface in addition to ions. The photo-ESP may change owing to (1) direct emission of the Cs 6s electrons, (2) spin exchange collisions of the photoexcited electrons with the Cs 6s electrons, or (3) an adatom-induced change in the position of the d level that results in a decrease of the surface magnetization (70). The spin-exchange-scattering cross section increases rapidly for electrons with kinetic energies less than 0.5 eV (71). This is an attractive explanation of the depolarization, but it is curious that no such effect was observed in Fe. Analysis of the spin-polarized photoemission from GaAs indicates that spin exchange might be a very important depolarization mechanism.

3. ESP Dependence on Film Thickness

The conflict with the band theory could be resolved by assuming that the surface layer has different magnetic properties from the bulk and that emission from the surface layer dominates the photocurrent. It has been suggested that a magnetically dead layer exists at the surface of Ni (72,73) or that there is an antiferromagnetic coupling of the surface spins to the bulk (74). The photoemission probing depth in Ni and its dependence on photoelectron energy have been determined by Eastman (49), using the overlayer technique and measuring kinetic energies of electrons. For electrons with energy 5 eV above E_F, the mean free path for inelastic scattering λ_e is about 10 Å; λ_e is longer for the lower kinetic energies encountered with cesiated films.

The question arises, however, whether the mean free path λ_σ for the conservation of spin is shorter than the mean free path λ_e for the conservation of energy. Recall that in EuO, for example, $\lambda_\sigma < \lambda_e$ and the effect of the surface was enhanced in the photo-ESP data. Measurements were made of the rate of buildup of the photo-ESP for successively thicker films of Ni deposited on Cu to get a measure of λ_σ (75). In effect, the ESP "marked" the electrons from the Cu substrate or from the Ni overlayers. The polarization increased exponentially from $P_{Cu}(\infty) = 0$ to $P_{Ni}(\infty) = 15\%$ according to $P(x) = P_{Ni}(\infty)\{1 - \exp[-(x - x_0)/\lambda]\}$, where x is the Ni film thickness and x_0 allows for a possible nonzero intercept. A least squares fit to the experimental points gave $x_0 = 1.2 \pm 1$ Å; so it is not possible to rule out one dead layer (~ 2 Å). It appears that the second layer is already ferromagnetic and that above two layers ferromagnetism is certainly present at $T = 80°K$. λ was determined to be 11 ± 2 Å for electrons 5.4 ± 0.3 eV above E_F. This value is in good agreement with that obtained by Eastman, who measured electron kinetic energies indicating that inelastic electron-electron scattering is dominant in determining the photoemission probing depth and $\lambda_\sigma \geq \lambda_e$. No evidence for a thickness-dependent depolarization mechanism is observed. This is in agreement with estimates of electron-magnon scattering cross sections (76).

A spin-dependent mean free path, caused, for instance, by a much stronger scattering of the minority spins, could explain the observed positive ESP, but the positive ESP observed even in very thin films argues strongly against this. The measurement of the thickness dependence of the ESP also shows that one magnetically dead layer or antiferromagnetically coupled layer cannot change the sign of the ESP. At the photon energies of the ESP experiments, the contribution of the surface layer to the total photocurrent is at most 20%.

We are left with the fact that the experimental data are not explained by the band theory of ferromagnetism in combination with either the direct or nondirect transition model of photoemission. We have been assuming that the 3d metals are sufficiently itinerant that photoemission from the valence band probes the band properties of the electrons. If the hole created in photoemission is highly localized, as in the case of the ferrites, which is highly unlikely in Fe, Co, or Ni, one observes the energy spectrum of the hole state left behind and it would not make sense to compare the results with the band theory of ferromagnetism (77).

Several theories have invoked many-body effects to explain positive polarization (78). While there is no doubt about the existence of many-body effects, it is difficult at present to assess their quantitative importance, though the discrepancies suggest they may be substantial. The gathering body of experimental data we have discussed imposes stringent requirements on any new theory that is proposed.

D. Optically Magnetized Solids

Thus far we have discussed photo-ESP from magnetic materials. The electrons are polarized by the exchange interaction or the external magnetic field. Now we turn to photo-ESP from solids that have no spin moments. It is possible that a preferred orientation of the photoemitted electrons is established by the photoexcitation process. This process occurs when electrons are excited by circularly polarized light from or to spin-orbit split bands.

Measurements of the photo-ESP of nonmagnetic solids have so far been made only for the semiconductor GaAs and, under very different conditions, for the alkali metals. The measurements of GaAs were made on cleaved single crystals in ultrahigh vacuum (79,80,81). The preliminary measurements of the alkali metals (82) were on polycrystalline samples continuously evaporated in a vacuum of 10^{-5} torr. The results from these two very different systems—semiconductor and alkali—are discussed below.

1. GaAs

The production of spin-oriented electrons in the conduction band of GaAs at the Γ point is illustrated in Fig. 19. The spin-orbit interaction splits the sixfold degenerate p states of the valence band into a fourfold degenerate

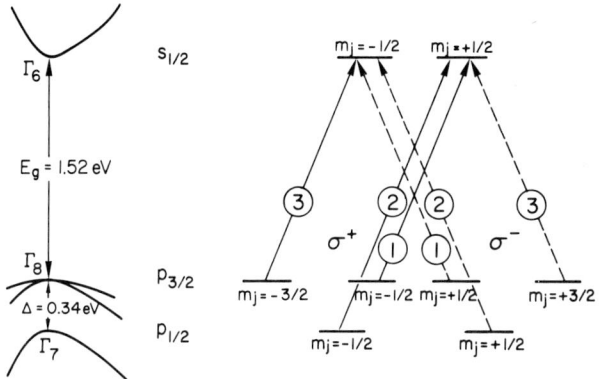

FIG. 19. Energy bands of GaAs at $k = 0$ (left) and transitions from the $P_{1/2}$ and $P_{3/2}$ levels to the $S_{1/2}$ states. Solid and broken lines are for σ^+ and σ^- light, respectively. The circled numbers give the relative transition intensities.

$P_{3/2}$ level and a twofold degenerate $P_{1/2}$ level separated by $\Delta_{LS} = 0.34$ eV. On the right side of the figure, transitions are shown from the states labeled by their m_j quantum numbers. For σ^+ circularly polarized light, angular momentum in light direction **k**, transitions with $\Delta m_j = m_f - m_i = 1$ are allowed. If the photon energy is chosen to excite electrons just across the gap E_g, three times as many electrons are excited to the $S = -\frac{1}{2}$ state (positive magnetic moment) as to the $S = +\frac{1}{2}$ state for a polarization $P = (3 - 1)/(3 + 1) = 50\%$. Note that the spin is not changed in the optical transitions. The relative transition probabilities are obtained, given the initial and final state wavefunctions ψ_i and ψ_f, by calculating the matrix element $\langle \psi_f | H_{int} | \Psi_i \rangle$, where $H_{int} \propto X \pm iY$ for $\sigma \pm$ polarizations.

The spin-orbit splitting of the valence band is essential to obtain an ESP in the conduction band. As $\hbar\omega$ is increased above $E_g + \Delta_{LS}$, excitations from the $P_{1/2}$ level come in with a relative strength of 2 and the ESP decreases to zero. Optical orientation of electron spins in the conduction band was first observed by Lampel (83). The photoexcited electrons are emitted into vacuum if their energy is higher than the vacuum level E_∞.

Fortuitously, in GaAs the electron affinity E_a (the energy difference between E_∞ and the conduction band minimum in the bulk) can be decreased to zero, as was first pointed out by Scheer and van Laar (84), and can even be made negative by treatment with Cs and O (85). While a negative electron affinity is not required for the observation of spin-dependent band structure at points other than the conduction band minimum, it has important consequences for GaAs as a high-intensity source of spin-polarized electrons. The key feature of the negative electron affinity (NEA) is that photoemission is now limited by the diffusion length of the order of 1 μ for thermalized electrons rather than by the hot electron scattering length of $\sim$ 100 Å.

The p-type GaAs (1.3×10^{19} cm^{-3} Zn) single crystals were cleaved in ultrahigh vacuum along the (110) plane. The cleaved crystal was transferred to the sample preparation chamber for cesiation and oxidation (see Fig. 2). The quantity of Cs evaporated from the cesium chromate channels was chosen so as to maximize the photocurrent. An exposure of 0.6×10^{-6} torr-sec constituted a typical oxygen treatment. Successive treatments will be denoted as GaAs + Cs(OCs)^n. The base pressure in the preparation chamber was 6×10^{-9} torr and that in the cleaving and measurement chamber was 3×10^{-10} torr. A polaroid linear polarizer and variable $\lambda/4$ platelet of the Soleil-Babinet type produced the circular polarization of the light.

The spectrum $P(\hbar\omega)$ measured on a sample at temperature $T \cong 10°$K is shown in Fig. 20. The rectangular fields represent the uncertainty in the

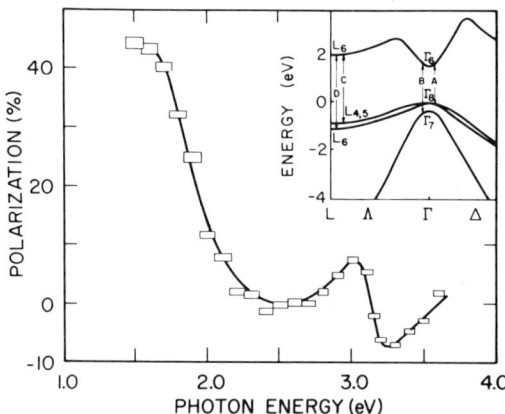

FIG. 20. Dependence of photo-ESP on the energy of circularly polarized light for GaAs with negative electron affinity at $T \cong 10°$K. The insert shows the relevant portion of the band structure. After Zucca et al. (88).

measured points, the width being the monochromator resolution and the height the statistical uncertainty from electron counting and the estimated uncertainty in the light polarization. The polarization has a maximum of 45% at threshold, which agrees with the luminescence measurements of Ekimov and Safarov (86,87). The spectrum is for a sample to which Cs was first applied, followed by oxygen and again Cs, i.e., GaAs + Cs(OCs)[1]. Measurements of samples with more Cs and O showed lower ESP, possibly because of spin-exchange scattering in the Cs–O layer. The maximum P in the case of GaAs + Cs(OCs)[5] was only 33%.

For $\hbar\omega > E_g + \Delta$, the polarization decreases to zero as emission from the split-off valence band with $P = -100\%$ is mixed in. The decrease is not abrupt, because the onset of the transitions from the split-off band is $\propto (\hbar\omega - E_g - \Delta)^{1/2}$, whereas the probability of the transitions at the top of the valence band is $\propto (\hbar\omega - E_g)^{1/2}$.

The spectrum exhibits a positive and negative peak at 3.0 and 3.2 eV, which can be understood in terms of transitions near the L point. The relevant part of the GaAs energy-band diagram of Zucca et al. (88) is shown in the insert of Fig. 20. The rise of ESP at 3.0 eV is due to transitions from L_4 and L_5 to L_6 (transition C in Fig. 20). The transitions from L_6 to L_6 occur with equal probability but go to the other spin state and cause the decrease at 3.2 eV (transition D in Fig. 20).

The structure arising from L-point transitions has a different shape from the structure due to transitions at Γ. This is because even though the bands bend away from each other as we move away from the L point in a direction perpendicular to [111] there is a large region where the energy surfaces are approximately parallel. Thus, as the photon energy is increased, the transitions C give rise to the positive peak, and then the transitions D dominate and cause the negative excursion. In contrast, at Γ, transitions A from the upper valence band continue to be competitive as transitions B from the split-off band begin.

A detailed analysis of the spectrum involves calculations with the appropriate wave functions (81). Two main points are: (1) transitions are not isolated along a given symmetry axis, and (2) to calculate transition probabilities, one must express the wave functions at some point in the Brillouin zone in a coordinate system defined by the light angular momentum. For example, it might be expected that if somehow it were possible to pick out transitions from the heavy hole band near Γ with σ^+ light, a polarization $P = 100\%$ would result. In fact, this is true only along the quantization axis. But when one integrates over solid angle as required for NEA GaAs, a maximum $P = 50\%$ is obtained. A similar situation arises at the L point, where, if a calculation is made for transitions at critical points on the light direction taken as [111], $P = \pm 100\%$. However, there are six other directions no longer equivalent to [111] and [1̄1̄1̄], and when all are taken into account, $P = 50\%$ as at Γ. One must then determine which of the excited electrons are emitted.

For the spectrum of Fig. 20, light is incident normal to the (110) crystal face. Transitions are from an initial state **k** to a final state $\mathbf{K} = \mathbf{k} + \mathbf{G}$, where **G** is a reciprocal lattice vector. The vectors $\mathbf{G}_{111}$ and $\mathbf{G}_{11\bar{1}}$ are the only ones that conserve energy and momentum and still give rise to emission; the polarization resulting from the respective **K-G** initial states is $P = 35\%$. The experimental value $P = 10\%$ at 3.0 and 3.2 eV is lower owing to transitions at other places in the Brillouin zone that occur at the same photon energy. Transitions elsewhere in the Brillouin zone may go to final states with kinetic energy different from the L transition.

The spectra from two of the seven surfaces studied behaved anomalously near threshold. Two new features were observed, as shown in Fig. 21. First,

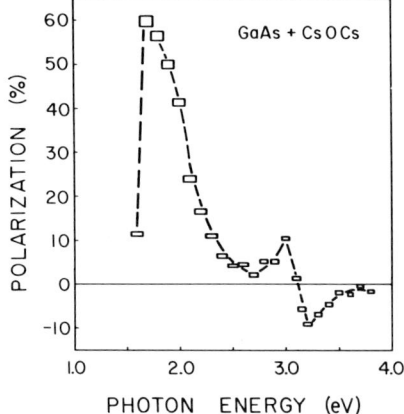

FIG. 21. Spectrum of spin polarization from GaAs at $T \cong 10°K$. The electron affinity is probably slightly positive.

there is the low value of the ESP right at threshold. Second, the maximum attains a value of 60%, higher than the theoretical maximum, and in addition occurs at a higher photon energy than it does in Fig. 20. This behavior is consistent with these two surfaces having a slightly positive electron affinity (PEA) rather than a negative electron affinity. In the case of NEA, electrons in the conduction band minimum can escape into vacuum and are in fact accelerated as they pass through the band-bending region. In contrast, with PEA, electrons at the conduction band minimum cannot escape.

For a PEA surface, the photoelectrons emitted at threshold travel with zero kinetic energy through the Cs–O layer. It is well known that spin-exchange scattering from alkali atoms is very large and increases rapidly as the electron energy decreases (71). If the Cs–O activation layer is not stoichiometric Cs_2O but has a Cs excess as suggested by Clark (89), these cesium atoms provide the means for spin-exchange scattering of the low-energy electrons leading to the depolarization observed in Fig. 21. This depolarization and the fact that photon energies greater than E_g are required for emission explain the shift in the maximum of the polarization to higher energy in the PEA case. The higher polarization $P = 60\%$ also results from the PEA, which emphasizes transitions from the heavy hole band, limits the emission to an escape cone about the light direction, and thus produces a polarization between the 50% of the NEA case and the 100% expected if the emission is due only to transitions on the light quantization axis.

The measurements of the photo-ESP from GaAs demonstrate the sensitivity of spin-polarized photoemission to the effects of spin-orbit coupling on the electronic structure. A negative electron affinity is not required for these studies; all that is necessary is that the final state lie above the vacuum level. It is expected that the most interesting applications will be realized in materials where the band structure is unknown.

Negative electron affinity GaAs is an excellent source of polarized electrons. The photocurrent is the product of the photoelectric yield Y and the incident flux of photons. For an optimized cathode, $Y = 0.5$, and with any commercial high-pressure arc the photocurrents will be limited by space-charge effects as in thermal cathodes. An important feature of the GaAs source is that the ESP can be reversed rapidly by reversing the polarization of the light, without affecting the width and position of the polarized electron beam. Lock-in techniques can therefore be used to detect even small spin-dependent effects in an interaction.

The NEA GaAs source also has a high brightness and is quite monochromatic. To achieve brightness, the light must be focused to give a small emitting area. The emittance has been estimated to be 2 mrad-cm for 1-eV electrons (81). The initial energy and energy spread of the electrons emitted near threshold in the region of high polarization are only 0.1–0.2 eV. This source should have many applications in physics and technology.

2. Alkali Metals

Heinzmann et al. (82) deposited the alkali metals continuously at a rate of 100 atomic layers/sec to maintain fresh surfaces in a vacuum of 10^{-5} torr. Their results on Cs and Rb are shown in Fig. 22. A maximum polarization of 4.5% was observed for Cs at about 4500 Å or 2.8 eV. The maximum polarization decreases with decreasing atomic number. $P(Cs) = -4.5 \pm 0.7\%$, $P(Rb) = -2.4 \pm 0.7\%$, $P(K) = -1.1 \pm 0.7\%$, and within experimental uncertainty $P = 0$ for Na and Li. The low-energy edge of the

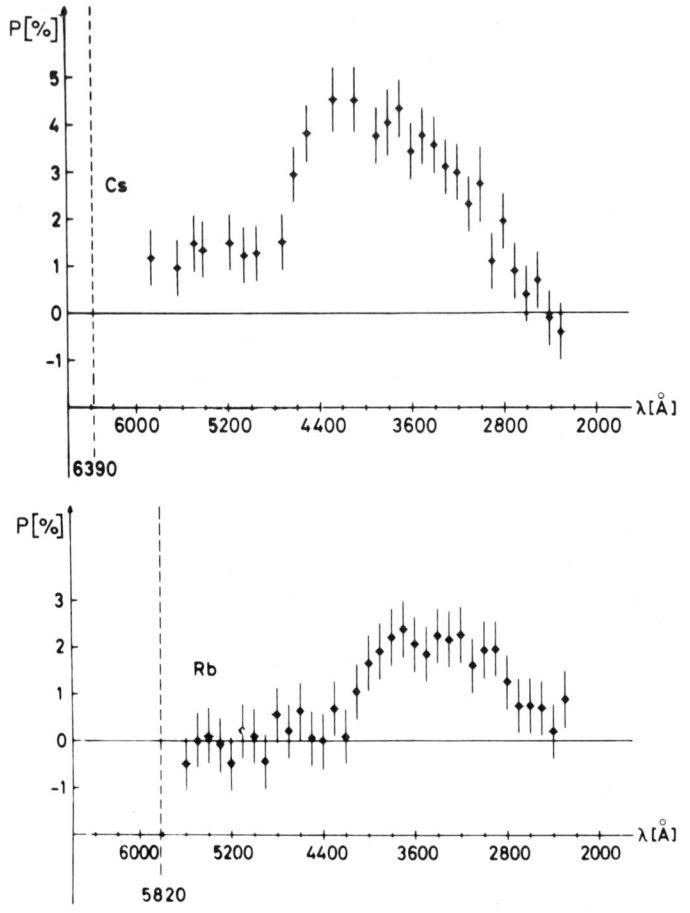

FIG. 22. Spectra of spin polarization from Cs and Rb. After Heinzmann et al. (82).

peak shifts to higher photon energy in the case of lighter elements, while the high-energy edge remains approximately the same. The peak is consistent with transitions from spin-orbit-split bands. The decreasing width of the peak on going to lighter elements can be understood by the reduced strength of the spin-orbit interaction.

As predicted by Fano (90), photoionization of free alkali atoms by circularly polarized light also results in emission of polarized electrons. The spectra are very different from those of the solid. In the case of Cs atoms (91), the peak ESP of -100% at 4.3 eV decreases to zero at 4.7 eV and reaches $+40\%$ at ~ 5 eV. In the solid, the ESP not only is much lower but also occurs at photon energies below the threshold of the atom.

For a detailed interpretation of the spectra, the probabilities of the transitions to the different spin states as well as the factors that might change the polarization as the electrons are emitted must be considered. Koyama and Merz (92) have neglected these difficult aspects, and have considered the angular part of the wave functions along a few symmetry directions in bcc Cs. They point out that in the [100] direction, transitions to Δ_7 yield $P = 100\%$ and those to Δ_6 yield $P = -100\%$. However, even if it were possible to select emission due to transitions at Δ, by kinetic energy analysis, for instance, high ESP can be expected only if (1) the light quantization axis is in the [100] direction, and (2) only the electrons emitted from the critical point in the [100] direction are accepted (i.e., there is no overlap with emission cones because of excitations at critical points along other Δ directions not on the quantization axis).

In the polycrystalline films of the experiment (82), the total ESP is the sum of the contributions from each microcrystallite. This greatly reduces the polarization. A further loss in ESP occurs if the photoelectrons undergo spin-flip scattering in escaping from the metal. This cannot be excluded, considering the way the sample surface had to be prepared. Nevertheless, these initial experiments demonstrate that photo-ESP measurements can be used to study the spin-orbit splittings in the band structure of metals. But one must remember that the short mean free path of hot electrons in metals, unlike that in semiconductors, causes a greater smearing of **k** and E and hence a broadness of the structure in optical spectra and photoemission energy distribution curves rather than the sharp structure observed at critical points in semiconductors. In particular, for Cs, mean free paths of ~ 2 Å have been measured (93) in the photon energy range of the ESP measurements, whereas for GaAs mean free paths of ~ 100 Å have been measured. For the short mean free paths in Cs, it is unrealistic to expect to obtain information that can be compared with confidence to bulk band-structure calculations.

IV. Results of Other Techniques

A. Field Emission

In view of the observations and remarks we made about the behavior of a FE pattern under the influence of an external magnetic field and especially about the difficulties in obtaining clean surfaces of less refractory metals, we will not discuss here results obtained in earlier experiments with metals (23,24). We will instead discuss new FE measurements of Ni using the apparatus of Fig. 4 (25) and analyze the results of the FE measurements of EuS evaporated on W (7).

With the experimental setup schematically shown in Fig. 4, reproducible results for Ni (001) oriented tips have been obtained very recently, and investigations involving tips with other orientations and other materials are in progress. We have found that the ESP in FE from Ni is small, i.e., $\lesssim 5\%$, independently of the crystallographic direction selected by the probe hole. In a careful experiment the dependence $P(H)$ has been investigated from $H \sim 3$ kOe down to about 200 Oe by slightly adjusting the probe hole in such a way that the same crystallographic direction, (103)–(102), was tested. The ESP did not depend on H within experimental accuracy of $\pm 0.42\%$. For these directions, $P = +3.58 \pm 0.42\%$, a value obtained from different runs with two different tips, one made by us and one kindly supplied by L. W. Swanson, of the FE Corporation, Oregon. For emission along (113)–(112), $P = -2.90 \pm 0.62\%$. The results were obtained within 20–40 min from a given field evaporation, at $T = 100°$K and with total field emission currents in the range 0.5–1×10^{-7} A.

A rather small ESP detected in the total current measurement, i.e., without energy selection, shows that the FE current is primarily due to s electrons, as has been found in a field emission energy distribution (FEED) for Ni (110) (94). No structure attributable to surface states or presence of d bands could be detected in the FEED study. This is in agreement with theoretical estimates (95,96) of the ratio r of the tunneling probability for d and s electrons, expected to be in the range 10^{-1}–10^{-2}. It is also in agreement with preliminary results obtained by Müller and co-workers (97).* We can attribute the ESP to d electrons, since the polarization of the s electrons is expected to be smaller than 1%. The observation of positive ESP is of crucial importance for the understanding of the results of another experiment, the superconducting tunneling (see Section IV,B). From the simple band-theoretical description of magnetism in Ni, one does not expect any positive ESP, since the majority spin d bands do not contribute to the Fermi surface (96). This means that the band approach cannot be applied in a straightforward way to explain the results. Field emission is known to give information on the density of states normal to the surface, because to a good approximation only states with $k_\perp \approx 0$ contribute to the FE current (98). The exact position in space where the transition from ψ_{in} to ψ_{out} and the matching of $d\psi_{in}/dz$, $d\psi_{out}/dz$ takes place is crucial for the results; here ψ_{in} and ψ_{out} are the wave functions describing the tunneling electron in and outside the metal. The position of the matching plane determines the possibility of detecting any d-band structure effects in FEED because it critically influences the ratio r. Politzer and Cutler (99) showed that for the (001) Ni surface and for energies $E \approx E_F$, $6 \times 10^{-3} \leq r \leq 0.15$, the smaller limit

---

* There is an error in the abstract of Ref. (97): the direction investigated for Ni was (120), not (122).

being valid when the matching plane is located one lattice constant from the last layer of ion cores (the upper one when this distance is one-half the lattice constant). The ESP experiment seems to indicate that the proper position for the matching plane is located at a distance near one-half the lattice constant. Furthermore, if the band theory holds, it follows that many-body effects of the type proposed by Anderson (78) for photoemission from strongly correlated magnetic metals are also operative in FE. Hertz and Aoi (100) did a calculation along these lines and were able to produce reasonable numbers for comparison with the tunneling experiment, as we will see. This could be the answer to the discrepancy between the simple prediction of the band theoretical approach and the FE or tunneling experiment. Of crucial importance for confirmation of these ideas will be ESP measurements in FE from dark planes of Ni and Co.

The highest polarization observed today in a field emission experiment is the one reported by Müller and co-workers (7). Values up to 90% obtained by FE from EuS evaporated on W at $T \sim 15°K$ and with a field of a few kOe can be best understood by assuming either emission of completely polarized impurity electrons or emission of 4f electrons. Very recently Kisker et al. (101) have reported field-ESP studies on the W/EuS system that confirm the experiments (7). Additionally, the temperature dependence of the ESP was measured. In agreement with magnetization measurement on bulk EuS, the ESP vanishes at $T \cong 16°K$ if the sample is properly prepared. Interesting experiments in this context would be ESP measurements as a function of the crystallographic direction (with a static H field), as well as measurements of FEED. The role of 4f electrons in FE could then be easily identified. The future of spin-polarized FE looks promising, especially in the case of magnetic semiconductors evaporated on a refractory metal. Interface problems and scattering phenomena during the emission can be studied in principle even if they involve single-atom events.

B. Superconducting Tunneling

In FE, well-developed and atomically smooth single-crystal planes can be obtained, but the anisotropy of the work function coupled to the tip end form may make an ESP measurement for a close-packed, dark plane difficult.

In tunneling with superconducting Al electrodes, the possibility of working with oriented single crystals is so far only a theoretical one; here the anisotropy of the work function does not play a role. Using polycrystalline films of Fe, Co, Ni, and Gd, Tedrow and Meservey (28) obtained results similar to the ones obtained in photoemission. The results are compared in Table I, together with the most recent ones obtained in FE. Both superconducting tunneling and field emission are tunneling processes, in the former case through the oxide into the superconductor and in the latter case into the vacuum. The agreement in Table I between the superconducting tunneling and photoemission measurements rather than between the more closely related superconducting tunneling and field-emission measurements is very surprising. Further experimental and theoretical investigations in spin-polarized field emission, superconducting tunneling, and photoemission from 3d metals are clearly indicated.

TABLE I

COMPARISON OF THREE SPIN-POLARIZATION MEASUREMENTS

Method	Energy range below E_F	Type of sample	Polarization (%)			
			Ni	Fe	Co	Gd
Tunneling	10^{-3} eV	Polycrystalline films	+11	+44	+34	+4.3
Photo-emission	0.4 eV 0.8 eV	Polycrystalline films	+15	+54	+21	+5.7
Field emission	0.2 eV	Single crystal (013)–(012) (115)–(113)	+3.6 −3	— —	— —	— —

C. Electron Capture by Deuterons

The very high polarizations observed in this experiment, as shown in Table II, are definitely not related in any simple way to any of the observations of the other experiments. The neutralization mechanism is not a simple FE process because a dynamic perturbance (the 150 keV deuterons) cannot be simulated by a static electric field. Major disturbances (excitations) of the

TABLE II

SOME VALUES OF THE ESP IN THE ELECTRON-CAPTURE EXPERIMENT ON Ni[a]

Scattering plane	Beam direction	ESP (%)
(110)	[110]	−96.1 ± 2.8
(100)	[110]	−19 ± 2
(111)	[110]	−10 ± 2
(120)	[$\bar{2}$10]	+16 ± 1

[a] After Rau and Sizmann (30).

electrons at the metal surface may be involved and may play a critical role. This novel technique can be a powerful complementary tool for studying magnetism at surfaces, provided some better theoretical insight into the technique itself can be gained.

V. Conclusion

The aim of this paper has been to show the present status of the experiments dealing with the emission of polarized electrons from solids. The field has greatly profited from the rapidly expanding knowledge in surface physics and from the general present-day interest in collecting and understanding spectroscopic data from solid materials.

The main results obtained so far are concerned with establishing the energy order of magnetic and nonmagnetic electron states, obtaining information on magnetism of surfaces and magnetic proximity effects, and getting some insight into the electron states in the critical but interesting intermediate range where neither the band model nor the atomic picture is the correct approach. As for surface magnetism, we have seen that in 4f ferromagnetic insulators there exists a surface sheet that is not magnetically coupled to the bulk. Such a sheet is not present in metals and 3d ferromagnetic semiconductors. This reflects the very special way in which the magnetic moments of the extremely localized 4f shell are coupled in a solid. The proximity of a metal, as well as the presence of conduction electrons introduced by doping, enhances the coupling of the surface sheet to the bulk. Only then does one obtain almost fully polarized electrons either in field emission or in photoemission. Structural disorder also has very drastic effects on both electronic and magnetic properties. The experiments on 4f materials show quite clearly that the electron spin is conserved in the emission process unless one has paramagnetic centers that can give rise to extremely efficient spin-exchange scattering.

As to 3d magnetic materials, we have seen that the electronic excitations in magnetite and other ferrites must be explained by the model of a single ion in a crystal field. Photo-ESP shows unambiguously that there is little relation between initial band density of states and photoemission and optical absorption data. The gross features of the spectra are caused by the spectrum of the hole state left behind.

In the ionic but metallic-like La-Pb perovskites, the ESP measurement can identify the 3d electrons even on the large background of a complex electronic structure. This shows the promising role of ESP measurements in studying surfaces of catalysts.

In the case of ferromagnetic 3d metals, finally, the band model of ferromagnetism does not explain the photo-ESP data from polycrystalline samples. Neither can the results of spin-polarized field emission and tunneling be explained in a straightforward way. We believe that future measurements, involving single crystals, alloys, and higher photon energies will clarify these questions.

The future of spin-polarized electron emission is extremely promising. One branch of research will certainly deal with the properties of 3d electrons, and with the interconnection between surface magnetism and catalysis. An additional stimulus for this branch will come once a better way of detecting spin polarization has been developed. Such a new detector of ESP could be made by employing scattering of low-energy electrons from single crystals. The presently used single scattering from atoms involves the unfavorable condition that the scattering is sensitive only to spin state when the intensity is weak (*4*). This drawback is overcome when constructive interference in scattering from a single crystal enhances the contribution from each atom (*102,103*).

The second branch will make use of the new and efficient sources of polarized electron beams that work on the GaAs principle. These sources make scattering of spin-polarized electrons, for instance from solid surfaces, a very attractive experiment since there is no need to employ a polarization detector. To detect the spin dependence of the scattering, one simply has to measure the change in the scattered intensity when the spin of the primary electron beam is reversed. Lock-in techniques, compared with present-day techniques, will enhance the sensitivity to spin-dependent effects by several orders of magnitude and make a new field of phenomena accessible to investigation.

Acknowledgements

Part of this work was supported by the Schweizerischer Nationalfonds. We wish to thank Prof. G. Busch and G. K. Wertheim for their continued interest and support, and S. F. Alvarado, W. Eib, and M. Erbudak for discussions and suggestions.

References

1. C. G. Shull, C. T. Chase, and F. E. Myers, *Phys. Rev.* **63**, 29 (1943).
2. C. E. Kuyatt, *Phys. Rev.* B **12**, 4581 (1975).
3. C. J. Davisson and L. H. Germer, *Phys. Rev.* **33**, 760 (1929).
4. J. Kessler, *Rev. Mod. Phys.* **41**, 3 (1969).
5. P. S. Farago, *Advan. Electron. Electron Phys.* **21**, 1 (1965).
6. G. Busch, M. Campagna, P. Cotti, and H. C. Siegmann, *Phys. Rev. Lett.* **22**, 597 (1969).
7. N. Müller, W. Eckstein, W. Heiland, and W. Zinn, *Phys. Rev. Lett.* **29**, 1651 (1972).
8. P. S. Farago, *Rep. Progr. Phys.* **34**, 1055 (1971) and references cited therein.
9. H. Ueberall, "Electron Scattering from Complex Nuclei," Part A. Academic Press, New York, 1971.

10. G. Holzwarth and H. J. Meister, "Tables of Asymmetry, Cross Sections and Related Functions for Mott Scattering of Electrons by Screened Au and Hg Nuclei." Univ. of Munich, 1964; *Nucl. Phys.* **59**, 56 (1964).
11. C. Caroli, D. Lederer-Rozenblatt, B. Roulet, and D. Saint-James, *Phys. Rev.* **B 8**, 4552 (1973).
12. W. D. Grobman and D. E. Eastman, *Phys. Rev. Lett.* **33**, 1034 (1974).
13. Y. Baer and G. Busch, *Phys. Rev. Lett.* **31**, 35 (1973) and references cited therein.
14. G. K. Wertheim, H. J. Guggenheim, and S. Hüfner, *Phys. Rev. Lett.* **33**, 1050 (1973).
15. D. E. Eastman and J. L. Freeouf, *Phys. Rev. Lett.* **34**, 395 (1975).
16. P. C. Kemeny and N. J. Shevchik, *Solid State Commun.* **17**, 255 (1975).
17. G. K. Wertheim and S. Hüfner, *Phys. Rev. Lett.* **35**, 53 (1975).
18. J. S. Greenberg, D. P. Malone, R. L. Gluckstern, and V. W. Hughes, *Phys. Rev.* **120**, 1393 (1960).
19. D. E. Eastman and J. J. Donelon, *Rev. Sci. Instrum.* **41**, 1648 (1970).
20. E. W. Müller, *Advan. Electron. Electron Phys.* **13**, 83 (1960).
21. R. Gomer, "Field Emission and Field Ionization," p. 195. Harvard Univ. Press, Cambridge, Mass., 1960.
22. E. W. Müller and T. T. Tsong, "Field Ion Microscopy," p. 314. Elsevier, New York, 1969.
23. M. Hofmann, G. Regenfus, O. Schärpf, and P. J. Kenedy, *Phys. Lett. A* **25**, 270 (1967); **27**, 1066 (1968).
24. W. Gleich, G. Regenfus, and R. Sizmann, *Phys. Rev. Lett.* **27**, 1066 (1971).
25. M. Campagna and T. Utsumi, *J. Vac. Sci. Technol.* **13**, 193 (1976).
26. R. Gomer, *J. Chem. Phys.* **21**, 293 (1973).
27. E. W. Müller and T. T. Tsong, "Progress in Surface Science," Vol. 4, Part 1. Pergamon Press, New York, 1973.
28. P. M. Tedrow and R. Meservey, *Phys. Rev. Lett.* **25**, 1270 (1970); **27**, 919 (1971); **26**, 192 (1971); *Solid State Commun.* **11**, 333 (1972).
29. M. Kaminsky, ANL-Report 7971, p. 178 (unpublished).
30. C. Rau and R. Sizmann, *Proc. Int. Conf. At. Collisions, 5th, Gatlinburg, Tenn., 1973*.
31. S. Methfessel and D. C. Mattis, "Handbuch der Physik," Vol. 18, Springer, Berlin, 1968; P. Wachter, *Crit. Rev. Solid State Sci.* **2**, 189 (1972).
32. J. Friedel, *Comments Solid State Phys.* **2**, 21 (1969).
33. K. Sattler and H. C. Siegmann, *Z. Phys.* **B20**, 289 (1975).
34. W. Eib, F. Meier, D. T. Pierce, and P. Ruchti, *AIP Conf. Proc.* **24**, 401 (1975).
35. H. C. Siegmann, *Phys. Rep.* **17C**, 38 (1975).
36. G. Busch, M. Campagna, and H. C. Siegmann, in "Electron Spectroscopy" (D. A. Shirley, ed.), p. 827. North-Holland, Amsterdam, 1972.
37. M. Campagna, K. Sattler, and H. C. Siegmann, *AIP Conf. Proc.* **18**, 1388 (1974).
38. J. S. Helman and H. C. Siegmann, *Solid State Commun.* **13**, 891 (1973).
39. F. Meier, D. T. Pierce, and K. Sattler, *Solid State Commun.* **16**, 401 (1975).
40. S. F. Alvarado, W. Eib, F. Meier, D. T. Pierce, K. Sattler, H. C. Siegmann, and J. P. Remeika, *Phys. Rev. Lett.* **34**, 319 (1975).
41. G. Busch, M. Campagna, and H. C. Siegmann, *Phys. Rev. B* **4**, 746 (1971).
42. F. Meier, W. Eib, and D. T. Pierce, *Solid State Commun.* **16**, 1089 (1975).
43. P. Cotti and P. Munz, *Phys. Condens. Matter* **17**, 307 (1974); D. E. Eastman, *Phys. Rev. B* **8**, 6027 (1973) and references cited therein.
44. J. Schoenes and P. Wachter, *Phys. Rev. B* **9**, 3097 (1974).
45. M. Campagna and H. C. Siegmann, *Phys. Kondens. Mater.* **15**, 247 (1973); G. Busch, M. Campagna, and H. C. Siegmann, *Int. J. Magn.* **4**, 25 (1973); M. Campagna, D. T. Pierce,

and H. C. Siegmann, *In* "Amorphous and Liquid Semiconductors" (J. Stuke and W. Brenig, eds.), Vol. 2, p. 1379. Taylor and Francis, London, 1974.
46. S. Sugano, Y. Tanabe, and M. Kamimura, "Multiplets of Transition Metal Ions in Crystals," p. 326. Academic Press, New York, 1970.
47. M. T. Hutchins, "Solid State Physics" (F. Seitz and D. Turnbull, eds.) Vol. 16, p. 237. Academic Press, New York, 1964.
48. S. F. Alvarado, W. Eib, H. C. Siegmann, and J. P. Remeika, *Phys. Rev. Lett.* **35**, 860 (1975).
49. D. E. Eastman, *In* "Techniques of Metals Research" (E. Passaglia, ed.), Vol. VI. Interscience, New York, 1972.
50. D. L. Camphausen, J. M. D. Coey, and B. K. Chakraverty, *Phys. Rev. Lett.* **29**, 657 (1972).
51. S. G. Bishop and P. C. Kemeny, *Solid State Commun.* **15**, 1877 (1974) and references cited therein.
52. R. J. H. Voorhoeve, J. P. Remeika, and D. W. Johnson, Jr., *Science* **180**, 62 (1973).
53. C. Zener, *Phys. Rev.* **182**, 403 (1951); A Morrish, B. J. Evans, J. E. Eaton, and L. K. Leung, *Can. J. Phys.* **47**, 2691 (1969).
54. L. F. Mattheis, *Phys. Rev. B* **2**, 3918 (1970); **6**, 4718 (1972).
55. J. B. Goodenough, *Progr. Solid State Chem.* **5**, 145 (1971).
56. S. F. Alvarado, W. Eib, P. Munz, H. C. Siegmann, M. Campagna, and J. P. Remeika, *Phys. Rev. B* June 1 (1976).
57. M. Campagna, E. Bucher, G. K. Wertheim, and L. D. Longinotti, *Phys. Rev. Lett.* **33**, 165 (1974).
58. H. A. Fowler and L. Marton, *Bull. Amer. Phys. Soc.* **4**, 235 (1959).
59. R. L. Long, Jr., V. W. Hughes, J. S. Greenberg, I. Ames, and R. L. Christensen, *Phys. Rev.* **138**, A1630 (1965).
60. A. B. Baganov and D. B. Diatroptov, *Sov. Phys.-JETP* **27**, 713 (1968).
61. U. Bänninger, G. Busch, M. Campagna, and H. C. Siegmann, *Phys. Rev. Lett.* **25**, 585 (1970).
62. G. Busch, M. Campagna, D. T. Pierce, and H. C. Siegmann, *Phys. Rev. Lett.* **28**, 611 (1972).
63. H. Alder, M. Campagna, and H. C. Siegmann, *Phys. Rev. B* **8**, 2075 (1973).
64. J. C. Slater, *Phys. Rev.* **49**, 537 (1936); **49**, 931 (1936).
65. E. C. Stoner, *Proc. Roy. Soc., Ser. A* **154**, 656 (1936); **165**, 372 (1938).
66. E. P. Wohlfarth, *Rev. Mod. Phys.* **25**, 211 (1953).
67. K. Tamura and H. Endo, *Phys. Lett. A* **29**, 52 (1969); W. Felsch, *Z. Phys.* **219**, 280 (1969); *Z. Angew. Phys.* **30**, 275 (1970); M. R. Bennett and J. G. Wright, *Eur. Conf. Phys. Matter, 1st, Florence, 1971*.
68. D. E. Eastman, *J. Phys. (Paris)* **32**, Cl-293 (1971).
69. N. V. Smith and M. M. Traum, *Phys. Rev. Lett.* **27**, 1388 (1971).
70. K. Aoi and K. H. Bennemann, *Proc. Int. Conf. Solid Surfaces, 2nd, Kyoto, 1974*, p. 737.
71. D. M. Campbell, H. M. Brash, and P. S. Farago, *Phys. Lett. A* **36**, 449 (1971).
72. L. Liebermann, J. Clinton, D. M. Edwards, and J. Mathon, *Phys. Rev. Lett.* **25**, 232 (1970).
73. K. Levin, A. Liebsch, and K. H. Bennemann, *Phys. Rev. B* **7**, 3066 (1973).
74. P. Fulde, A. Luther, and R. E. Watson, *Phys. Rev. B* **8**, 440 (1973).
75. D. T. Pierce and H. C. Siegmann, *Phys. Rev. B* **9**, 4035 (1974).
76. U. Bänninger, G. Busch, M. Campagna, and H. C. Siegmann, *J. Phys. (Paris)* **32**, Cl-290 (1971).
77. C. S. Fadley and E. P. Wohlfarth, *Comments Solid State Phys.* **4**, 48 (1972).
78. P. W. Anderson, *Phil. Mag.* **24**, 203 (1971); U. Brandt, *Z. Phys.* **244**, 217 (1971); S.

Doniach, *AIP Conf. Proc.* **5**, 549 (1972); D. M. Edwards and J. A. Hertz, *Phys. Rev. Lett.* **28**, 1334 (1972).
79. D. T. Pierce, F. Meier, and P. Zürcher, *Phys. Lett. A* **51**, 465 (1975).
80. D. T. Pierce, F. Meier, and P. Zürcher, *Appl. Phys. Lett.* **26**, 670 (1975).
81. D. T. Pierce and F. Meier, *Phys. Rev. B* June 15 (1976).
82. U. Heinzmann, K. Jost, J. Kessler, and B. Ohnemus, *Z. Phys.* **251**, 354 (1972).
83. G. Lampel, *Phys. Rev. Lett.* **20**, 491 (1968).
84. J. J. Scheer and J. van Laar, *Solid State Commun.* **3**, 189 (1965).
85. R. L. Bell, "Negative Electron Affinity Devices," p. 37. Clarendon, Oxford, 1973.
86. A. I. Ekimov and V. I. Safarov, *JETP Lett.* **12**, 198 (1970).
87. A. I. Ekimov and V. I. Safarov, *JETP Lett.* **13**, 195 (1971).
88. R. R. L. Zucca, J. P. Waltar, Y. R. Shen, and M. L. Cohen, *Solid State Commun.* **8**, 627 (1970).
89. M. G. Clark, *J. Phys. D* **8**, 535 (1975).
90. U. Fano, *Phys. Rev.* **178**, 131 (1969).
91. U. Heinzmann, J. Kessler, and J. Lorenz, *Z. Phys.* **240**, 42 (1970).
92. K. Koyama and H. Merz, *Z. Phys.* **B20**, 131 (1975).
93. N. V. Smith and G. B. Fisher, *Phys. Rev. B* **3**, 3662 (1971).
94. M. Campagna and T. Utsumi, *AIP Conf. Proc.* **24**, 399 (1974).
95. J. W. Gadzuk, *Phys. Rev. B* **1**, 2110 (1970).
96. A. B. Politzer and P. H. Cutler, *Phys. Rev. Lett.* **28**, 1330 (1972).
97. N. Müller, *Bull. Amer. Phys. Soc.* **20**, 859 (1975).
98. D. R. Penn and E. W. Plummer, *Phys. Rev. B* **9**, 1216 (1974); D. R. Penn, *Phys. Rev. B* **11**, 3208 (1975); J. W. Gadzuk and E. W. Plummer, *Rev. Mod. Phys.* **45**, 487 (1973).
99. A. B. Politzer and P. H. Cutler, *Surface Sci.* **22**, 277 (1970).
100. J. A. Herz and K. Aoi, *Phys. Rev. B* **8**, 3252 (1973).
101. E. Kisker, G. Baum, A. H. Mahan, W. Raith, and K. Schröder, *Phys. Rev. Lett.* **36**, 982 (1976).
102. D. Maison, *Phys. Lett.* **19**, 654 (1966).
103. M. R. O'Neill, M. Kalisvaart, F. B. Dunning, and G. K. Walters, *Phys. Rev. Lett.* **34**, 1167 (1975).

Generation of Images by Means of Two-Dimensional, Spatial Electric Filters

HENNING F. HARMUTH

Department of Electrical Engineering
The Catholic University of America
Washington, D.C.

I. Introduction	168
II. Image Generation by Linear Transformations	172
A. Transformation in the Medium and Inverse Transformation by Tapped Delay Circuits	172
B. Inverse Transformation by Means of the Fourier Transform	179
C. Inverse Transformation by Means of Sampled Storage Circuits	190
III. Focusing for Spherical Wavefronts	197
IV. Reduction of the Array Size	200
A. Array Size for the Classical Limit of Resolution	200
B. Super-Resolution by Means of Sampled Arrays	203
C. Improving the Signal-to-Noise Ratio	211
D. Practical Use of Super-Resolution for Image Generation	213
V. Implementation of Filters by Digital Circuits	216
A. Digital Delay Circuits	216
B. Digital Circuits for the Inverse Fourier Transform	217
C. Synchronous Demodulation as the Key to Practical Circuits	219
VI. Beyond the Capabilities of Photography and Holography	223
A. Range Images and Three-Dimensional Images	223
B. Doppler Images	226
C. Telelens Effect	227
D. Pseudo-Color Images	228
E. Focusing for Spherical Wavefronts Revisited	231
F. Focusing without Approximation	235
G. Time Sharing for High Resolution	239
VII. Experimental Equipment and Test Results	242
Appendix	246
References	247

I. Introduction

There are currently three generally known principles for the generation of images by means of electromagnetic or acoustic waves. The classical method uses lenses. The echo principle, used in radar and sonar, is a second method. Holography is the third and youngest method.

A fourth method, using two-dimensional, spatial electric filters, has been developed theoretically and experimentally during the last few years. Let us first see what such filters do before discussing their use for image generation. A filter in electrical engineering is usually a circuit that modifies a single time-variable input voltage. These are time filters for time-variable signals. The typical time-variable signals are audio signals transmitted by telephone or radio. In addition to audio signals we have video signals. Indeed, in everyday life almost all the information received by us comes in the form of video signals. Video signals are space-variable signals. The shades of gray of a black-and-white photograph vary as a function of two spatial Cartesian coordinates x and y. Let the x and y coordinates be divided into 256 intervals each. The photograph is then divided into $256 \times 256 = 65{,}536$ spatial intervals. The photograph will have about the same resolution as a television picture if each spatial interval has one shade of gray. Let these 65,536 shades of gray be transformed simultaneously into 65,536 voltages. This could be done by an array of 256×256 phototransistors. A two-dimensional, spatial electric filter would accept all the voltages simultaneously at 65,536 input terminals, modify them, and produce simultaneously 65,536 output voltages.

One may readily understand from this example why time filters have been used for half a century while spatial filters have been developed only recently. A time filter has typically one input and one output terminal; in exceptional cases there may be several terminals. The simplest spatial filters have hundreds of terminals, but filters with good spatial resolution need thousands and tens of thousands of terminals. Such numbers were beyond our technology until the arrival of integrated semiconductor circuits.

A moving picture as seen by the eye has two spatial coordinates x and y, as well as a time coordinate t. The simplest filters for such space–time signals consist of a two-dimensional spatial filter with a time filter connected to each of its output terminals. This calls for 65,536 time filters in our previous example. Only one such filter, with 256 time filters, exists at present.

Spatial filters come as time-invariant and time-variable filters. The time-variable filters are always sampling filters, which means that the time-variable elements are switches. More sophisticated forms of time variability are generally too expensive. Of course, the time-variable spatial filters are more powerful than the time-invariant ones, since they make use of the additional variable time. A more detailed discussion of spatial filters may be found in a book by Harmuth (1972).

Let us review the fields of application for the various known imaging methods. The lens is predominantly used in optics, which means for wavelengths of the order of 10^{-5}–10^{-4} cm. In acoustics, the minimum practical wavelength is 1 to 10^{-1} cm, and this leads to very large dimensions of the lens. Sutton (1972) and Rolle (1973) have reported on simple acoustic imaging systems using a lens.

Radar and sonar have both proved useful for the generation of images in the form of shadowgraphs. In contrast to the lens, images are not formed instantly and no moving images can thus be produced.

Holography has been used successfully in optics as well as in acoustics to produce nonmoving images. The excellent features of the optical lens have handicapped the development of optical holography. By the same token, the problem of large size of the acoustic lens has favored the development of acoustic holography. This field has been strongly advanced toward practical use by Booth, Farrah, Fritzler, Keating, Koppelman, Marom, Mueller, Saltzer, Sutton, Zilinskas, and many others. A convenient source for this field is the series of books published under the titles *Acoustical Holography* and *Acoustical Holography and Imaging*. There is also an excellent textbook by Hildebrand and Brenden (1972). We will return to acoustical holography in the section on synchronous demodulation, where its relation to two-dimensional, spatial electric filters will become evident.

Two-dimensional, spatial electric filters can be used in theory for image generation by means of electromagnetic or acoustic waves. In reality, it is very hard to build electronic circuits for response times below 1 nsec or frequencies above 10^9 Hz. This eliminates the field of optics. Acoustics, on the other hand, is ideally suited, since the frequencies used do not exceed a few hundred kilohertz. Within the range of application determined by the response time of electric circuits and their cost, the electric filters are by far the most general and powerful means for image generation. This is due to the variety of operations that can be performed. A lens can produce delays and summations. An electric filter can in addition produce multiplications, integrations, time-variable linear operations, and nonlinear operations, to name only a few. As an example, let us observe that a lens can produce a two-dimensional, moving color image, but an electric filter can produce a genuine three-dimensional, moving color image, and do so in two different ways. The relation between electric filters and holography is more complicated. The two methods cannot be compared in optics since electric filters do not work at optical frequencies. In acoustics, holography, in its currently used practical form, is the process of synchronous demodulation that may or may not be used with electric filters.

The principle of image generation by spatial electric filters is illustrated in Fig. 1. A sound projector insonificates the object plane. Each point in the object plane returns a scattered wave. A reflected wave will also be returned

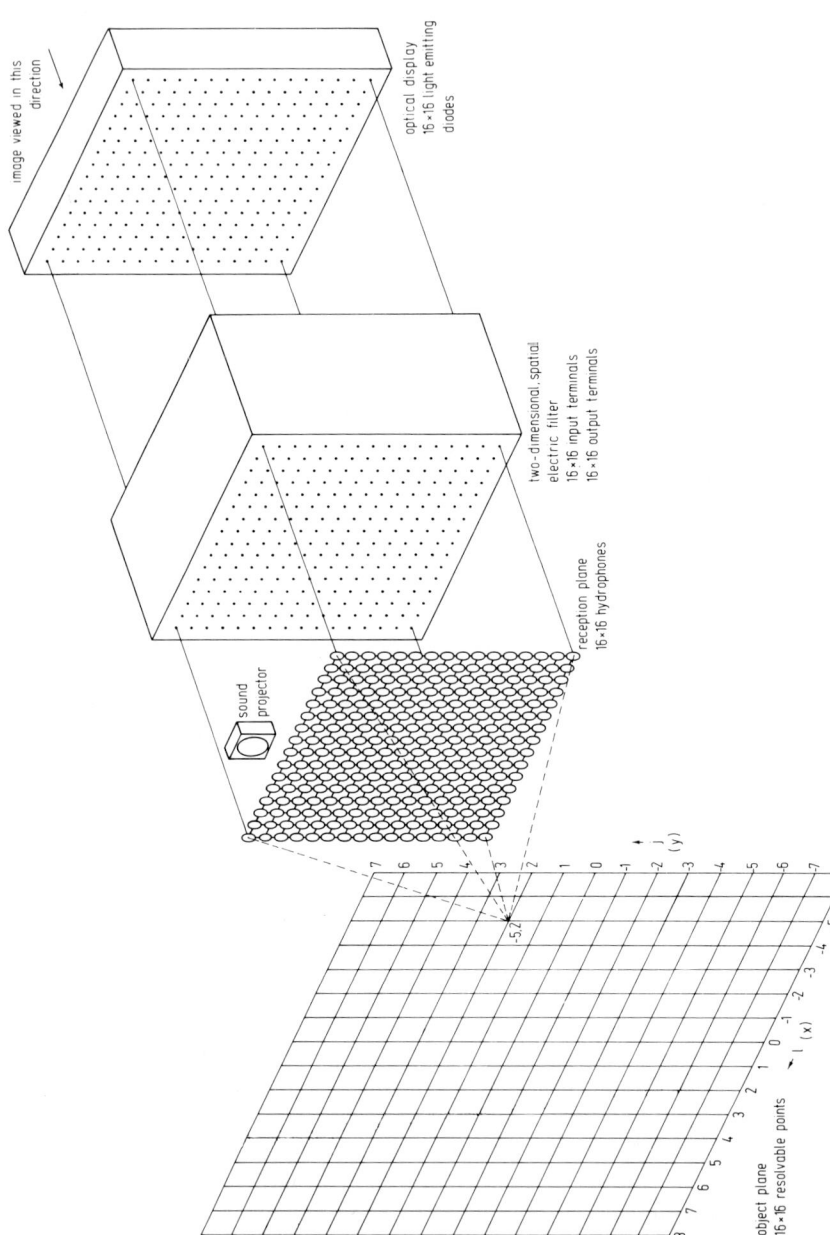

FIG. 1. The principle of image generation by two-dimensional, spatial electric filters.

that produces the same undesirable effects as reflected light waves in photography; the reflected wave will be ignored for the time being. The sum of all scattered, spherical waves produces the wavefront of the scattered waves in the object plane. This wavefront propagates to the reception plane. There will be a linear relationship between the wavefront in the object plane and the wavefront in the reception plane. If one knows this linear transformation and can perform the inverse transformation, one regains the wavefront in the object plane. The regenerated wavefront is the image (of the object plane).

For the practical implementation of the inverse transformation one converts the acoustic wavefront into a two-dimensional spatial array of voltages. This is done by the two-dimensional hydrophone array in Fig. 1. For practical reasons, the sound *pressure* is converted into proportional voltages.

The output voltages of the hydrophones are fed to the input terminals of the two-dimensional, spatial electric filter. It performs the inverse transformation. At the output terminals of the filter one obtains voltages that are proportional to the sound pressures in the object plane. In other words, one obtains a representation of the acoustic image by electric voltages.

Since only our eyes and not our ears can receive two-dimensional spatial signals, we do not want an acoustic image but rather an optical one. This is readily done by feeding the output voltages of the filter to an optical display, such as an array of light-emitting diodes. The brightness of the diodes then represents the sound pressures at the corresponding points in the object plane.

The conversion of sound pressures into electric voltages by means of hydrophones and the conversion of electric voltages into light by means of light-emitting diodes are standard operations. What needs to be discussed is the conversion of input voltages, representing sound pressures in the reception plane, into output voltages, representing sound pressures in the object plane, by means of a two-dimensional, spatial electric filter. We will first discuss the mathematical operations for an infinite distance between object and reception plane, then we will discuss the circuits, switch back to theory and investigate focusing for a finite distance between object and reception plane, and finally discuss more sophisticated features that provide some scope of the potential of electric filters for imaging. The discussion will be in terms of underwater acoustics, but it is obvious that the method works for acoustic waves in air too, and beyond that for electromagnetic waves that vary sufficiently slowly with time to permit the practical implementation of the equipment. The hydrophones would have to be replaced by microphones or receiving antennas, and the sound projector by a loudspeaker or a radiating antenna, but the two-dimensional filter and the electrooptic display would not be affected.

II. Image Generation by Linear Transformations

A. Transformation in the Medium and Inverse Transformation by Tapped Delay Circuits

Let a wave with sinusoidal time variation originate at one point of the object plane:

$$u(t) = U \sin (2\pi vt/\lambda + \alpha). \tag{1}$$

$u(t)$ may be considered to represent the sound pressure, but the investigation applies to any other acoustic or electromagnetic quantity represented by $u(t)$. The wave produces output voltages at all hydrophones in the reception plane, as indicated for the point $l = -5, j = +2$ in Fig. 1. Let us first calculate the output voltages for one dimension, assuming a plane wave coming from a point at infinity. Figure 2 shows an array with $2^n = 16$ hydrophones equally spaced with distance d. A hydrophone k has the distance $(k - 2^n/2 + 1)d = id$ from the center of the array. Note that the center of the array is somewhat awkward to define for an even number 2^n of hydrophones. An odd number of hydrophones would avoid this difficulty, but powers of 2 lead to simple circuits.

Let the plane wave $W(\beta)$ arrive from the direction β and let it have the time variation* of $u(t)$ of Eq. (1). Let it produce the voltage $V \sin 2\pi vt/\lambda$ at the terminals of the hydrophone $k = 2^n/2 - 1 = 7$ in the center of the array. This voltage is the arbitrary reference for the output voltage of all other hydrophones. One may readily recognize from Fig. 2 that the output voltage $v(k)$ of the hydrophone k is defined by the relation

$$v(k) = V \sin \{2\pi[vt + (k - 2^n/2 + 1)d \sin \beta]/\lambda\}. \tag{2}$$

The change from Eq. (1) to Eq. (2) is the transformation of the sound wave by the medium between object plane and reception plane.† The inverse transformation will produce the image of the point radiating the sound wave of Eq. (1). To perform this inverse transformation we need a circuit with 2^n input terminals connected to the hydrophones, and producing 2^n output voltages. The 2^n input voltages $v(k)$ of Eq. (2) should produce an output voltage at only one terminal, or at least very nearly so.

---

* We will discuss three types of filters, one based on time-invariant delay circuits, one on sampling, and one on the Fourier transform. The filters based on the Fourier transform work for sinusoidal waves only, while the others do not suffer from this restriction.

† The ratio of the amplitudes V and U depends on the attenuation in the medium and the characteristics of the hydrophones. It is of no interest here.

We will discuss three types of circuits that can perform this inverse transformation. The first is based on time-invariant delay circuits. It is the most understandable one, but it is not practical for acoustics. The second is based on the Fourier transform. It is at the present the most practical circuit for acoustic arrays with up to 64 × 64 hydrophones. The third type is based on sampled delay circuits. It is an attractive alternative to the Fourier transform for arrays with more than 64 × 64 hydrophones since it lends itself to integrated circuit construction.

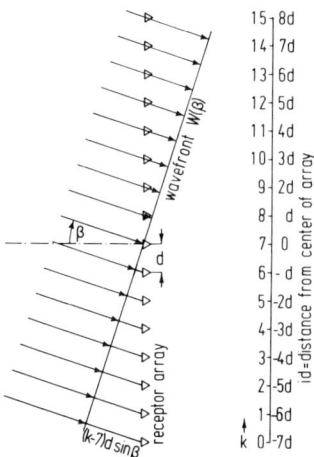

FIG. 2. Calculation of the output voltages of the receptors of an array produced by a plane wavefront coming from the point β at infinity.

Let us assume we have a delay line that delays the voltage $v(k)$ of Eq. (2) by the time

$$t_\gamma = (k - 2^n/2 + 1)(d/v) \sin \gamma, \qquad (3)$$

where γ is the *observation angle*.* This is a nominal angle for which the circuit is designed to receive a wave, while the wave actually comes from direction β. An array of 2^n hydrophones permits one to design the circuit for 2^n observation angles. Hence, γ can have 2^n different values.

---

* The negative delay times are no problem since one can add a constant delay t_0 that makes $t_0 + t_\gamma$ always positive. The same result can be achieved by using the output voltage of the hydrophone $k = 15$ in Fig. 2 as reference for $\beta > 0$, and that of the hydrophone $k = 0$ for $\beta < 0$.

The sum of all delayed voltages $v(k)$ is denoted $v(\gamma, \beta)$:

$$v(\gamma, \beta) = \sum_{k=0}^{2^n-1} V \sin\{2\pi[vt + (k - 2^n/2 + 1)d\delta]/\lambda\}$$

$$= 2^n V \left\{ \sin(2\pi vt/\lambda) \sum_{k=0}^{2^n-1} 2^{-n} \cos[2\pi d\delta(k - 2^n/2 + 1)/\lambda] \right.$$

$$\left. + \cos(2\pi vt/\lambda) \sum_{k=0}^{2^n-1} 2^{-n} \sin[2\pi d\delta(k - 2^n/2 + 1)/\lambda] \right\}, \quad (4)$$

$$\delta = \sin\beta - \sin\gamma = 2\sin\tfrac{1}{2}(\beta - \gamma)\cos\tfrac{1}{2}(\beta + \gamma)$$

$$\doteq (\beta - \gamma)\cos\tfrac{1}{2}(\beta + \gamma) \quad \text{for } \beta - \gamma \ll 1$$

$$\doteq \beta - \gamma \quad \text{for } \beta \ll 1 \text{ and } \gamma \ll 1,$$

$\beta - \gamma =$ reception angle referred to the observation angle γ rather than to the array axis.

Each term in the last two sums in Eq. (4) represents the area of a rectangle with the base equal to 2^{-n} and the height equal to $\cos[2\pi d\delta(k - 2^n/2 + 1)/\lambda]$ or $\sin[2\pi d\delta(k - 2^n/2 + 1)/\lambda]$. The sums themselves represent the areas under step functions with steps of width 2^{-n}. For large values of 2^n one may approximate these sums by integrals:

$$\sum_{k=0}^{2^n-1} 2^{-n} \cos[2\pi d\delta(k - 2^n/2 + 1)/\lambda] \doteq \int_0^1 \cos\left[\frac{2\pi A\delta}{\lambda}(2^{-n}k - \tfrac{1}{2})d(2^{-n}k)\right]$$

$$= \int_{-1/2}^{1/2} \cos(2\pi A\delta x/\lambda)\, dx = \frac{\sin \pi A\delta/\lambda}{\pi A\delta/\lambda}, \quad (5)$$

$$\sum_{k=0}^{2^n-1} 2^{-n} \sin[2\pi d\delta(k - 2^n/2 + 1)/\lambda] \doteq \int_{-1/2}^{1/2} \sin(2\pi A\delta x/\lambda)\, dx = 0,$$

$$A = 2^n d = \text{aperture}.$$

The substitution of Eq. (5) into Eq. (4) yields:

$$v(\gamma, \beta) = V \frac{A}{d} \frac{\sin[\pi A(\sin\beta - \sin\gamma)/\lambda]}{\pi A(\sin\beta - \sin\gamma)/\lambda} \sin(2\pi vt/\lambda)$$

$$\doteq V \frac{A}{d} \frac{\sin[\pi A(\beta - \gamma)/\lambda]}{\pi A(\beta - \gamma)/\lambda} \sin(2\pi vt/\lambda) \quad \text{for } \beta, \gamma \ll 1. \quad (6)$$

We are interested here in the principle of imaging. Hence, the simplified function $[\sin \pi A(\beta - \gamma)/\lambda]/[\pi A(\beta - \gamma)/\lambda]$ holding for small values of β and γ is plotted in Fig. 3 for $\gamma = 0, \pm\lambda/A, \pm 2\lambda/A$. These curves give the variation of the amplitude $v(\gamma, \beta)$ as a function of β for these particular choices of the observation angle γ.

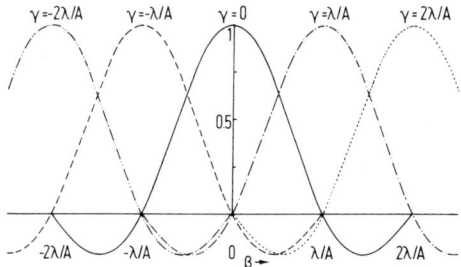

FIG. 3. The voltages $v(\gamma, \beta)$ according to Eq. (6) for several values of γ for $\beta \ll 1$, $\gamma \ll 1$. For larger values of β and γ one has to substitute $\sin \beta$ and $\sin \gamma$ for β and γ; the values $\pm \lambda/A$, $\pm 2\lambda/A, \ldots$ of the scale are replaced by $\pm \sin(\lambda/A)$, $\pm \sin(2\lambda/A), \ldots$.

The principle of a circuit producing $v(\gamma, \beta)$ from 16 input voltages for 16 values of γ is shown by Fig. 4. Tapped delay circuits are connected to the output terminals of the hydrophones. The delay times at the taps are $t_0 + (k - 2^4/2 + 1)(d/v) \sin \gamma$. These delay times are shown in detail for the taps $k = 4$, $\gamma = -5\lambda/A$ and $k = 11$, $\gamma = 5\lambda/A$. The voltages for a fixed value of γ and all values $k = 0 \cdots 15$ are summed to yield $v(\gamma, \beta)$ of Eq. (6). For the values $\gamma = -7\lambda/A, \ldots, +8\lambda/A$ shown in Fig. 4 one obtains the 16 voltages $v(-7\lambda/A, \beta)$ to $v(+8\lambda/A, \beta)$.

Let a sound source be located at infinity. The amplitude of the voltage $v(0, \beta)$ obtained by summing the voltages at the taps along the line $\gamma = 0$ in Fig. 4 will vary like the curve $\gamma = 0$ in Fig. 3 as function of β; the voltage $v(\lambda/A, \beta)$ obtained by summing the voltages at the taps along the line $\gamma = \lambda/A$ in Fig. 4 will vary like the curve $\gamma = \lambda/A$ in Fig. 3, and so on. Let

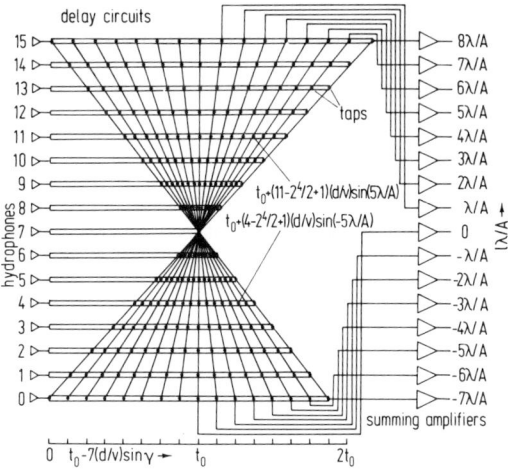

FIG. 4. Producing the inverse transform by means of delay circuits.

two sound sources located in the direction $\sin \beta = 0$ and $\sin \beta = \lambda/A$ at infinity generate waves with amplitude V_0 and $V_{\lambda/A}$ according to Eq. (1). They will produce the amplitudes $v(0, 0) = AV_0/d$ and $v(\lambda/A, \lambda/A) = AV_{\lambda/A}/d$, with $V_0/V_{\lambda/A} = U_0/U_{\lambda/A}$, while all the other amplitudes $v(\gamma, 0)$ and $v(\gamma, \lambda/A)$, with $\gamma = -7\lambda/A, \ldots, -\lambda/A, +2\lambda/A, \ldots, +8\lambda/A$, are zero. Hence, the two sound sources will be resolved without mutual interference. The ratio $\lambda/A = \sin \varepsilon$ defines the resolution angle ε.

The delay principle discussed here is the principle of image forming by a lens. The lens and the space between the lens and the focal plane provide paths with different delays, so that a plane wave coming from infinity requires the same time to reach its image point in the focal plane, regardless of where the plane wave reaches the surface of the lens. The summing of all the waves with various delays is done by the eye or a photographic film. Hence, the lens can perform the linear operations of delay and summation. Linearity of the operations is a necessity if we want to produce the images of many points simultaneously. But there are many more linear operations than delay and summation. For instance, multiplication is a very useful operation that the lens cannot perform; time-variable linear operations are a whole class of operations beyond the capabilities of the lens. Electronic circuits can readily perform multiplications and time-variable operations. The time-invariant delay circuits of Fig. 4, on the other hand, are not practical for acoustics due to the vastly different velocities of acoustic and electromagnetic waves.

Before we look for more practical principles for electrical circuits let us generalize the array of Fig. 2 and the delay circuits of Fig. 4 from one to two dimensions. Figure 5 shows on the left a planar wave $W(\beta_x, \beta_y)$ represented by either its wavefront or the vector normal to it. It is decomposed into a wave $W(\beta_x)$ with its vector in the xz plane and a wave $W(\beta_y)$ with its vector in the yz plane. Let the wavefront arrive at the hydrophone $k = 2^n/2 - 1 = 7$ and $m = 2^n/2 - 1 = 7$. The wavefront still has to travel the distance

$$(k - 2^n/2 + 1)d \sin \beta_x \tag{7}$$

to reach the hydrophone k in the row $m = 2^n/2 - 1 = 7$. This is the same distance as in Eq. (2), except that β is replaced by β_x. In addition, the wave has to travel the distance

$$(m - 2^n/2 + 1)d \sin \beta_y \tag{8}$$

to reach the hydrophone m. The generalization of the voltage $v(k)$ of Eq. (2) to two dimensions is thus the voltage $v(k, m)$:

$$v(k, m) = V \sin \{2\pi[vt + (k - 2^n/2 + 1)d \sin \beta_x \\ + (m - 2^n/2 + 1)d \sin \beta_y]/\lambda\}. \tag{9}$$

IMAGE GENERATION BY ELECTRICAL FILTERS

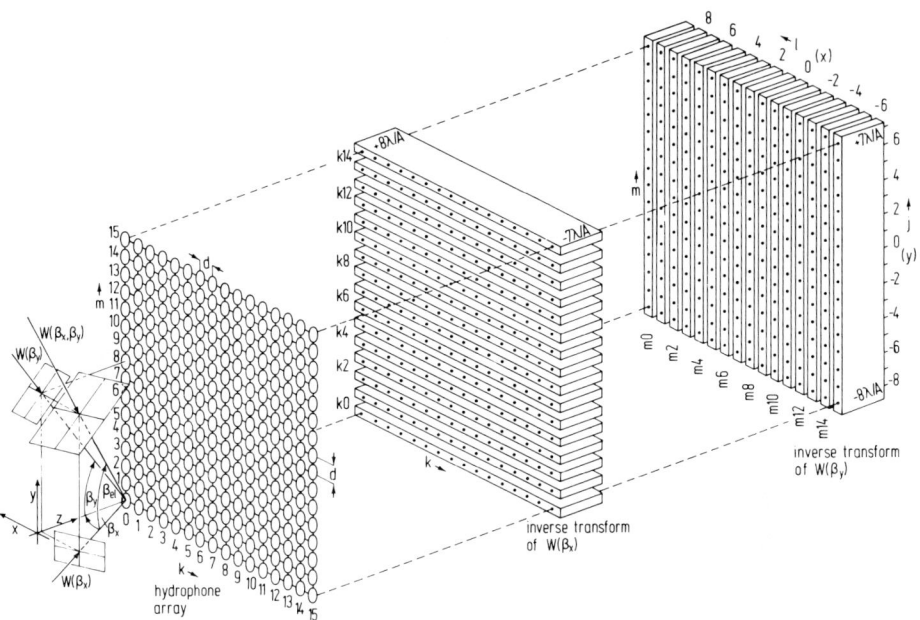

FIG. 5. Producing the inverse transform in two dimensions using Cartesian coordinates.

β_x is usually called azimuth, but β_y is not the elevation. The elevation is denoted β_{el} in Fig. 5. Azimuth and elevation refer to spherical coordinates, while we are using Cartesian coordinates. Filters based on spherical coordinates will be discussed later on.

The delay time t_y of Eq. (3) is replaced by $t_{y,xy}$:

$$t_{y,xy} = (k - 2^n/2 + 1)(d/v)\sin \gamma_x + (m - 2^n/2 + 1)(d/v)\sin \gamma_y; \quad (10)$$

γ_x is the observation angle in the xz plane and γ_y the observation angle in the yz plane.

$v(\gamma, \beta)$ in Eq. (4) is replaced by $v(\gamma_x, \gamma_y, \beta_x, \beta_y)$ and the single sum by a double sum:

$$v(\gamma_x, \gamma_y, \beta_x, \beta_y) = \sum_{m=0}^{2^n-1} \sum_{k=0}^{2^n-1} V \sin \{2\pi[vt + (k - 2^n/2 + 1)d\delta_x$$
$$+ (m - 2^n/2 + 1)d\delta_y]/\lambda\}, \quad (11)$$

$$\delta_x = \sin \beta_x - \sin \gamma_x, \quad \delta_y = \sin \beta_y - \sin \gamma_y.$$

Using the approximation of Eq. (5), which shows that the sum of terms

$\sin\left[2\pi d\delta_x(k - 2^n/2 + 1)/\lambda\right]$ and $\sin\left[2\pi d\delta_y(m - 2^n/2 + 1)/\lambda\right]$ equals zero, one obtains the following double sum:

$$\begin{aligned}
v(\gamma_x, \gamma_y, \beta_x, \beta_y) &\doteq 2^{2n} V \sin(2\pi vt/\lambda) \\
&\times \sum_{k=0}^{2^n-1} 2^{-n} \cos\left[2\pi d\delta_x(k - 2^n/2 + 1)/\lambda\right] \\
&\times \sum_{m=0}^{2^n-1} 2^{-n} \cos\left[2\pi d\delta_y(m - 2^n/2 + 1)/\lambda\right] \\
&\doteq V\left(\frac{A}{d}\right)^2 \frac{\sin\left[\pi A(\sin\beta_x - \sin\gamma_x)/\lambda\right]}{\pi A(\sin\beta_x - \sin\gamma_x)/\lambda} \\
&\times \frac{\sin\left[\pi A(\sin\beta_y - \sin\gamma_y)/\lambda\right]}{\pi A(\sin\beta_y - \sin\gamma_y)/\lambda} \sin(2\pi vt/\lambda).
\end{aligned} \quad (12)$$

For small values of β_x, β_y, γ_x, and γ_y one obtains the functions of Fig. 3 in two dimensions. A computer plot of the function for $\gamma_x = \gamma_y = 0$ and $\sin\beta_x \doteq \beta_x$, $\sin\gamma_y \doteq \gamma_y$ is shown in Fig. 6.

FIG. 6. The function $[(\sin \pi A\beta_x/\lambda)/\pi A\beta_x/\lambda][(\sin \pi A\beta_y/\lambda)/\pi A\beta_y/\lambda]$ representing $v(0, 0, \beta_x, \beta_y)$ for small values of β_x and β_y.

A practical circuit to obtain $v(\gamma_x, \gamma_y, \beta_x, \beta_y)$ may be derived from Eq. (11). The received $2^n \times 2^n$ voltages are first summed over k, then over m. This can be done by producing printed circuit cards with the circuit of Fig. 4, stacking 2^n such cards parallel to the xz plane (horizontally), another 2^n parallel to the yz plane (vertically), and interconnecting them according to

Fig. 5. The voltages $v(\gamma_x, \gamma_y, \beta_x, \beta_y)$ are obtained at the output terminals of the second stack of cards. These terminals are labeled horizontally from $l = -7$ to $l = +8$, corresponding to the values $\gamma_x = -7\lambda/A$ to $\gamma_x = +8\lambda/A$; vertically they are labeled from $j = -8$ to $j = +7$ corresponding to the values $\gamma_y = -8\lambda/A$ to $\gamma_y = +7\lambda/A$. The different labeling of the horizontally stacked and the vertically stacked cards is caused by the even number of cards and one's desire to have a wave arriving with the angles $\beta_x = 0, \beta_y = 0$ produce an image as close as possible to the center of the display, which means that the voltage $v(0, 0, 0, 0)$ should be produced at an output terminal of the vertical stack of cards in Fig. 5 as close as possible to the center. The different labeling causes no practical problem, since the cards of Fig. 4 only have to be flipped around so that the current input terminal 15 becomes the new input terminal 0. The output terminals must then be labeled from $-8\lambda/A$ at the bottom to $+7\lambda/A$ at the top, which corresponds to the labeling of the vertically stacked cards in Fig. 5.

B. *Inverse Transformation by Means of the Fourier Transform*

The use of the tapped delay circuits for image generation is handicapped by practical problems. Let us try a completely different approach to produce the output voltage $v(\gamma, \beta)$ of Eq. (6) from the input voltages $v(k)$ of Eq. (2). We will first discuss this approach qualitatively.

It is well known that correlation with sample functions is a good way to detect a signal. Figure 7 shows the principle. There are two possible signals denoted "expected function" on top of Fig. 7a and b. These signals are sampled sinusoidal pulses with either one or two cycles. Note that "sampled function" means that only the voltages shown by heavy lines are received.

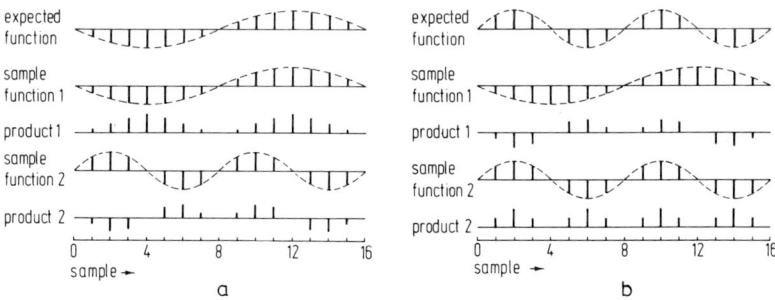

FIG. 7. Signal detection by cross-correlation with sample functions. The expected functions on top are multiplied with the two sample functions 1 and 2. Product 1 has only positive samples if the expected function equaled the sample function 1. Product 2 has only positive samples if the expected function equaled the sample function 2.

The dashed envelope of these samples is plotted only to emphasize that the samples represent two sinusoidal functions.

In order to decide which of the two expected functions has been received we multiply the expected function in Figs. 7a and 7b with the sample functions 1 and 2. These multiplications are represented by the products 1 and 2.

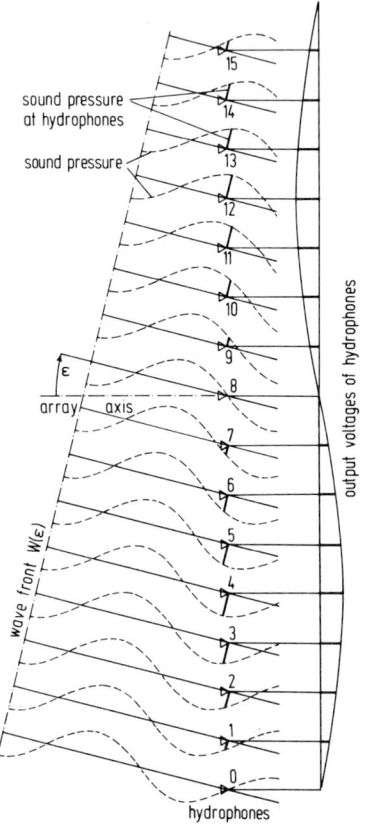

FIG. 8. Planar wave with sinusoidal time variation arriving at an angle ε with the axis of a line array of hydrophones.

Let the products be summed. The sum of product 2 in Fig. 7a and the sum of product 1 in Fig. 7b equal zero since for any positive sample there is a negative one with equal magnitude. The sum of product 1 in Fig. 7a and the sum of product 2 in Fig. 7b, on the other hand, are not zero. Hence, "sum of product 1 unequal zero" means that the sinusoidal function with one cycle was received, while "sum of product 2 unequal zero" means that the sinusoidal function with two cycles was received. This is the principle of signal detection by cross-correlation with sample functions.

There are two things one must keep in mind for the application of this principle to image generation:

(a) The signals or functions do not have to be continuous, but can be represented by samples.

(b) The signals or functions do not have to be *time* signals or *time* functions. For this reason, the abscissa in Fig. 7a and b is marked "sample" and not "time."

Consider now 16 hydrophones in a straight line with equal distances as shown in Fig. 8. A planar wave with sinusoidal time variation arrives at an angle ε with the array axis. The pressure of the wave at a certain time is indicated by the 16 sinusoidal functions. The pressure at the hydrophones is represented by the samples of these functions shown by heavy lines. The pressure at the hydrophone 0 equals zero, it decreases for the hydrophones 1, 2, 3, 4, increases for the hydrophones 5, 6, 7, and becomes zero for hydrophone 8. The pressure then becomes positive for the hydrophones 9 to 15.

The hydrophones produce output voltages proportional to the pressure. These voltages are represented by the samples 0 to 15 on the right of Fig. 8. One may readily see that these output voltages are nothing else than the "expected function" in Fig. 7a.

Let us turn to Fig. 9. It shows the same hydrophone array as before, but the sinusoidal wave now arrives at an angle 2ε with the array axis. Again, the pressure at a certain time is represented by 16 sinusoidal functions, and the pressure at the hydrophones by the samples of these functions shown by heavy lines. The pressure is zero at the hydrophones 0, 4, 8, and 12; it is positive at the hydrophones 1, 2, 3, 9, 10, 11, and it is negative at the remaining hydrophones. The output voltages are represented by the samples on the right, and they are equal to the samples of the "expected function" in Fig. 7b.

Let us now elaborate this principle quantitatively. We start from the voltage $v(k)$ in Eq. (2). Instead of delaying it by t_y according to Eq. (3) before summation, we multiply $v(k)$ by the following factors:

$$\tfrac{1}{2}\sqrt{2}, \quad \sin[2\pi k(d/\lambda)\sin\gamma], \quad \cos[2\pi k(d/\lambda)\sin\gamma]. \tag{13}$$

The calculation is simplified by using the hydrophone $k = 0$ rather than $k = 2^n/2 - 1$ as reference for the phase. Hence, $v(k)$ of Eq. (2) is replaced by

$$v(k) = V\sin[2\pi(vt + kd\sin\beta)/\lambda]. \tag{14}$$

One obtains the following three formulas by multiplying Eqs. (13) and (14),

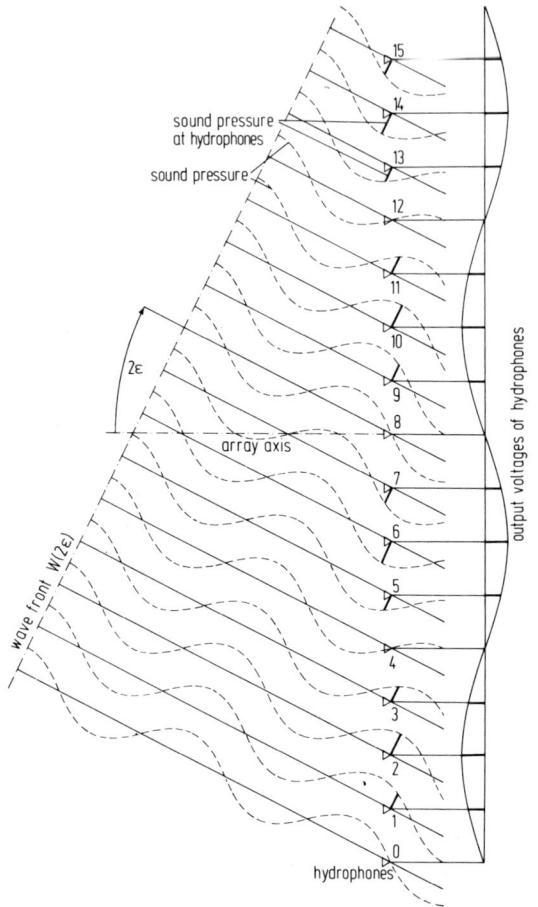

FIG. 9. Planar wave with sinusoidal time variation arriving at an angle 2ε with the axis of a line array of hydrophones.

and summing over k:

$$\begin{aligned}
v(0, \beta) &= \tfrac{1}{2}\sqrt{2} \sum_{k=0}^{2^n-1} V \sin\left[2\pi(vt + kd \sin \beta)/\lambda\right] \\
&= \tfrac{1}{2}\sqrt{2}\, 2^n V \Big\{ \sin(2\pi vt/\lambda) \sum_{k=0}^{2^n-1} 2^{-n} \cos\left[2\pi k(d/\lambda) \sin \beta\right] \\
&\quad + \cos(2\pi vt/\lambda) \sum_{k=0}^{2^n-1} 2^{-n} \sin\left[2\pi k(d/\lambda) \sin \beta\right] \Big\};
\end{aligned} \qquad (15)$$

$$v(\gamma, \beta, s) = \sum_{k=0}^{2^n-1} V \sin\left[2\pi(vt + kd \sin \beta)/\lambda\right] \sin\left[2\pi k(d/\lambda) \sin \gamma\right]$$

$$= 2^n V \left\{ \sin(2\pi vt/\lambda) \sum_{k=0}^{2^n-1} 2^{-n} \cos\left[2\pi k(d/\lambda) \sin \beta\right] \right.$$

$$\times \sin\left[2\pi k(d/\lambda) \sin \gamma\right] + \cos(2\pi vt/\lambda)$$

$$\left. \times \sum_{k=0}^{2^n-1} 2^{-n} \sin\left[2\pi k(d/\lambda) \sin \beta\right] \sin\left[2\pi k(d/\lambda) \sin \gamma\right] \right\}; \quad (16)$$

$$v(\gamma, \beta, c) = \sum_{k=0}^{2^n-1} V \sin\left[2\pi(vt + kd \sin \beta)/\lambda\right] \cos\left[2\pi k(d/\lambda) \sin \gamma\right]$$

$$= 2^n V \left\{ \sin(2\pi vt/\lambda) \sum_{k=0}^{2^n-1} 2^{-n} \cos\left[2\pi k(d/\lambda) \sin \beta\right] \right.$$

$$\times \cos\left[2\pi k(d/\lambda) \sin \gamma\right] + \cos(2\pi vt/\lambda)$$

$$\left. \times \sum_{k=0}^{2^n-1} 2^{-n} \sin\left[2\pi k(d/\lambda) \sin \beta\right] \cos\left[2\pi k(d/\lambda) \sin \gamma\right] \right\}. \quad (17)$$

One may readily recognize that the functions $v(0, \beta)$, $v(\gamma, \beta, s)$ and $v(\gamma, \beta, c)$ are the discrete Fourier transform of the space–time function $v(k)$,

$$v(k) = V \sin\left[2\pi(vt + kd \sin \beta)/\lambda\right],$$

for the discrete spatial variable k, if $(d/\lambda) \sin \gamma$ satisfies the following condition:

$$\begin{aligned} 2^n(d/\lambda) \sin \gamma = (A/\lambda) \sin \gamma &= l & \text{for } \gamma > 0 \\ &= -l & \text{for } \gamma < 0 \quad (18) \\ &= 0 & \text{for } \gamma = 0 \end{aligned}$$

$$l = 1, 2, \ldots.$$

These are the same values $\sin \gamma = 0, \pm \lambda/A, \pm 2\lambda/A, \ldots$ for which Fig. 3 was plotted.

For large values of 2^n one may approximate the sums in Eqs. (15)–(17) by integrals as shown in the Appendix:

$$v(0, \beta) \doteq \frac{1}{2}\sqrt{2} V(A/d) \frac{\sin\left[\pi(A/\lambda) \sin \beta\right]}{\pi(A/\lambda) \sin \beta} \sin(2\pi vt/\lambda); \quad (19)$$

$$v(\gamma, \beta, s) \doteq \frac{VA}{2d} \left(\frac{\sin [\pi(A/\lambda)(\sin \beta - \sin \gamma)]}{\pi(A/\lambda)(\sin \beta - \sin \gamma)} \right.$$
$$\left. - \frac{\sin [\pi(A/\lambda)(\sin \beta + \sin \gamma)]}{\pi(A/\lambda)(\sin \beta + \sin \gamma)} \right) \cos (2\pi v t/\lambda); \quad (20)$$

$$v(\gamma, \beta, c) \doteq \frac{VA}{2d} \left(\frac{\sin [\pi(A/\lambda)(\sin \beta - \sin \gamma)]}{\pi(A/\lambda)(\sin \beta - \sin \gamma)} \right.$$
$$\left. + \frac{\sin [\pi(A/\lambda)(\sin \beta + \sin \gamma)]}{\pi(A/\lambda)(\sin \beta + \sin \gamma)} \right) \sin (2\pi v t/\lambda). \quad (21)$$

Let $v(\gamma, \beta, s)$ be integrated in time to change the time variation from $\cos (2\pi v t/\lambda)$ to $\sin (2\pi v t/\lambda)$:

$$v^\dagger(\gamma, \beta, s) = -(2\pi v/\lambda) \int v(\gamma, \beta, s) dt. \quad (22)$$

Let us substitute l or $-l$ for $(A/\lambda) \sin \gamma$ from Eq. (18) and form the following sums and differences:

$$v(l, \beta) = \tfrac{1}{2}[v(\gamma, \beta, c) + v^\dagger(\gamma, \beta, s)]$$
$$= \frac{VA}{2d} \frac{\sin \pi[(A/\lambda) \sin \beta - l]}{\pi[(A/\lambda) \sin \beta - l]} \sin (2\pi v t/\lambda); \quad (23)$$

$$v(-l, \beta) = \tfrac{1}{2}[v(\gamma, \beta, c) - v^\dagger(\gamma, \beta, s)]$$
$$= \frac{VA}{2d} \frac{\sin \pi[(A/\lambda) \sin \beta + l]}{\pi[(A/\lambda) \sin \beta + l]} \sin (2\pi v t/\lambda). \quad (24)$$

The voltages $v(0, \beta)$, $v(l, \beta)$ and $v(-l, \beta)$ in Eqs. (19), (23), and (24) are identical with $v(\gamma, \beta)$ in Eq. (6) if one substitutes 0, $+l$, or $-l$ for $(A/\lambda) \sin \gamma$. Hence, one may replace the delay circuits in Fig. 4 by circuits that perform a discrete Fourier transform of the sinusoidal input voltages according to Eqs. (15)–(17), a phase shift according to Eq. (22) and summations according to Eqs. (23) and (24). Note that the delay circuits work for any time variation of the input voltage, while the phase shift according to Eq. (22) restricts the circuits based on the Fourier transform to sinusoidal input voltages.

For the implementation of practical circuits, refer to Fig. 10. It shows the functions required for the discrete Fourier transform for $2^n = 16$. Note that only the amplitude samples at $k = 0, 1, 2, \ldots$ are needed. The dashed lines are intended to help visualize the sine–cosine functions. Note further that the function $\cos 16\pi x = \cos \pi k$ occurs, but not the function $\sin 16\pi x$. The sampled amplitudes have only the magnitudes 0, $p = \sin \pi/8$, $q = \sin \pi/4$, $r = \sin 3\pi/8$, 1 for the sine–cosine functions and $q = \tfrac{1}{2}\sqrt{2}$ for the constant function $f(0, x)$.

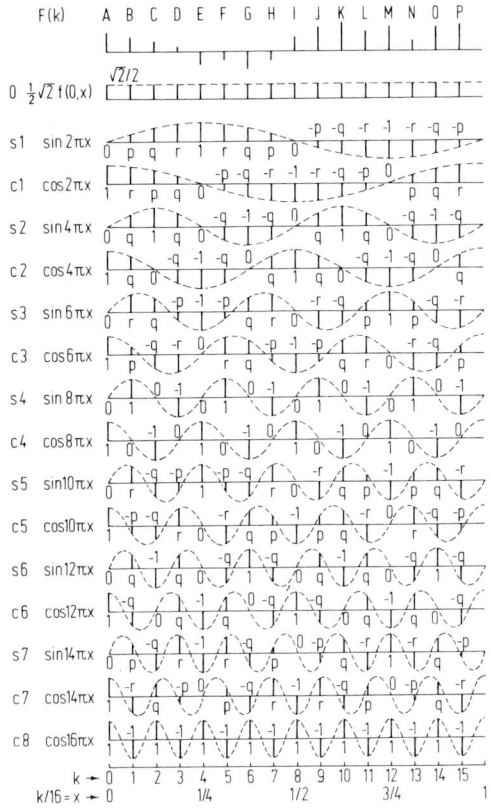

FIG. 10. Discrete Fourier transform of a sampled spatial function $F(k) = A, B, \ldots, P$. $p = \sin \pi/8$, $q = \sin \pi/4$, $r = \sin 3\pi/8$.

Consider the Fourier transform of the function $F(k)$ with the samples $A \cdots P$ shown on top of Fig. 10. For $\sin 2\pi x = \sin 2\pi k/16$ one obtains

$$\sum_{k=0}^{15} F(k) \sin 2\pi k/16 = 0A + pB + qC + rD + 1E + rF + qG + pH$$

$$+ 0I - pJ - qK - rL - 1M - rN - q0 - pP.$$

A circuit performing these 16 multiplications and one summation is shown in Fig. 11. The values of its resistors are determined by the magnitude 0, p, q, r, 1 of the samples of $\sin 2\pi k/16$, and their connection to the adding or subtracting input terminal of the operational amplifier OP by the sign at the samples.

A total of 16 circuits according to Fig. 11 are needed to transform $F(k)$ in Fig. 10 with all 16 functions $\frac{1}{2}\sqrt{2} f(0, x)$ to $\cos 16\pi x$. These circuits have a

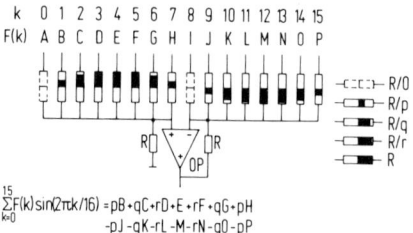

$$\sum_{k=0}^{15} F(k)\sin(2\pi k/16) = pB + qC + rD + E + rF + qG + pH$$
$$- pJ - qK - rL - M - rN - qO - pP$$

FIG. 11. Circuit based on the usual Fourier transform producing one Fourier component of the sampled function $F(k)$ in Fig. 10. (OP) operational amplifier. $p = \sin \pi/8$, $q = \sin \pi/4$, $r = \sin 3\pi/8$.

practical drawback. An operational amplifier does not work well if up to 16 resistors are connected to its input terminals. Furthermore, the hydrophones supplying the voltages $A, B, \ldots$ of $F(k)$ are loaded by up to 16 resistors. Both problems can be overcome by using circuits derived from the fast Fourier

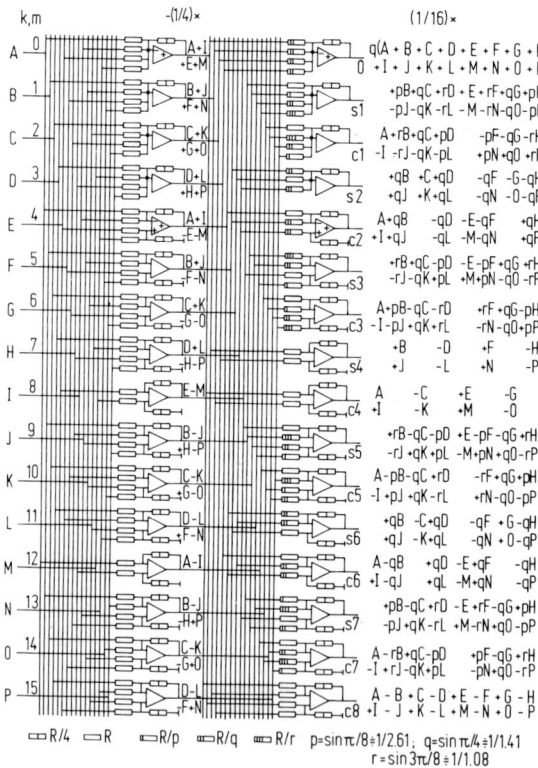

FIG. 12. Circuit performing a Fourier transform of the input voltages $A \cdots P$ using a variation of the fast Fourier transform that produces the sum of four voltages at a time.

transform. Figure 12 shows such a circuit. It would be too difficult to discuss how it was derived, but one may readily verify that the output voltages have the values required by Fig. 10. This particular circuit is based on the "radex 4 fast Fourier transform," hence every output terminal is loaded by at most four resistors, and there are never more than four resistors connected to any input terminal. The usual fast Fourier transform is a "radex 2 transform" in this terminology. At most two resistors would be connected to any input or output terminal for this transform, but twice as many operational amplifiers would be needed. Circuits for the radex 2 transform can be built for 2^n terminals.

A circuit performing the integration according to Eq. (22) and the summations of Eqs. (23) and (24) is shown in Fig. 13. Note that $v(0, \beta)$ in Eq. (19) has $\sqrt{2}$ times the amplitude of $v(l, \beta)$ and $v(-l, \beta)$ in Eqs. (23) and (24). The resistor $\sqrt{2} R$ of the multiplying circuit on top of Fig. 13 corrects this difference. The output terminals 0, s1, c1, ..., c8 of the circuit of Fig. 12 feed right into the input terminals 0, s1, c1, ... of the circuit in Fig. 13.

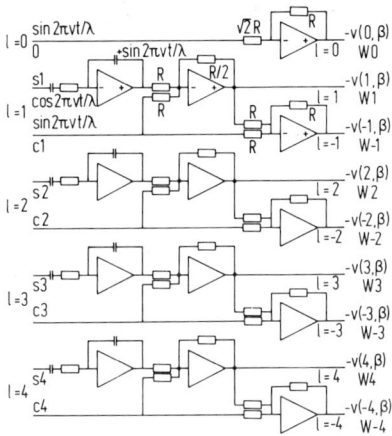

FIG. 13. Circuit to be connected to the output terminals of the circuit in Fig. 12 to distinguish between waves from the upper and lower half-planes in Fig. 2. This circuit works for a continuously variable angle of incidence β, but requires a fixed frequency.

Going from one to two dimensions one may arrange cards having the circuits of Figs. 12 and 13 on them according to Fig. 5. The printed circuit cards become rather crowded with components in this case. It was found more practical to perform a two-dimensional Fourier transform with the circuit of Fig. 12 first and do the phase shifting, adding or subtracting according to Fig. 13, afterward. Some elaboration on the circuit of Fig. 13 is required for this case.

One may see from Eqs. (23) and (24) that $v(l, \beta)$ has its largest amplitude for $\sin \beta = +l\lambda/A$, and $v(-l, \beta)$ for $\sin \beta = -l\lambda/A$. The wavefronts $W(\beta)$ corresponding to these values of β arrive either with the angle of incidence $|\beta|$ or $-|\beta|$, that is either from the upper or the lower half-plane. One may readily see this from the definitions of β and $W(\beta)$ in Fig. 2.

If one does a Fourier transform first for the variable k, then for the variable m according to Fig. 5, one obtains output voltages at four terminals. This is shown more clearly by Fig. 14. The four points or wavefronts

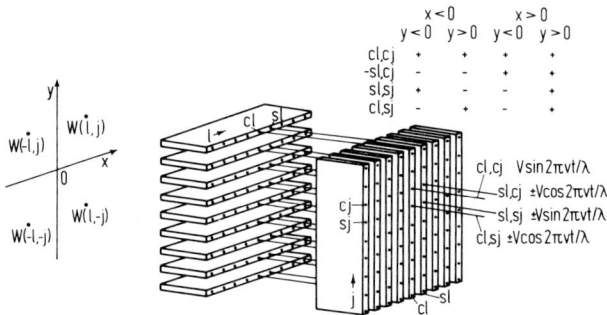

FIG. 14. Signs of the output voltages of a two-dimensional spatial filter caused by waves originating at the points $-l$, $-j$ to $+l$, $+j$ in the four quadrants of the object plane.

$W(+|l|, +|j|)$ to $W(-|l|, -|j|)$ located in the four quadrants at infinite distance produce the voltages $V \sin 2\pi ct/\lambda$ or $V \cos 2\pi ct/\lambda$ at the output terminals cl, cj to sl, sj. The signs of these four voltages determine the quadrant as shown in the table in the upper right corner of Fig. 14. The circuit of Fig. 15 connected to these four terminals will produce a voltage at only the one output terminal that corresponds to the correct quadrant.

The complete circuit for a two-dimensional processor using the Fourier transform is shown in Fig. 16. Two stacks of cards perform the two-

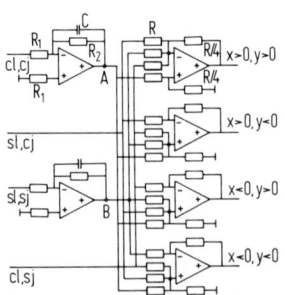

FIG. 15. Circuit to be connected to the output terminals cl, cj to sl, sj in Fig. 14 to distinguish among waves from the four quadrants.

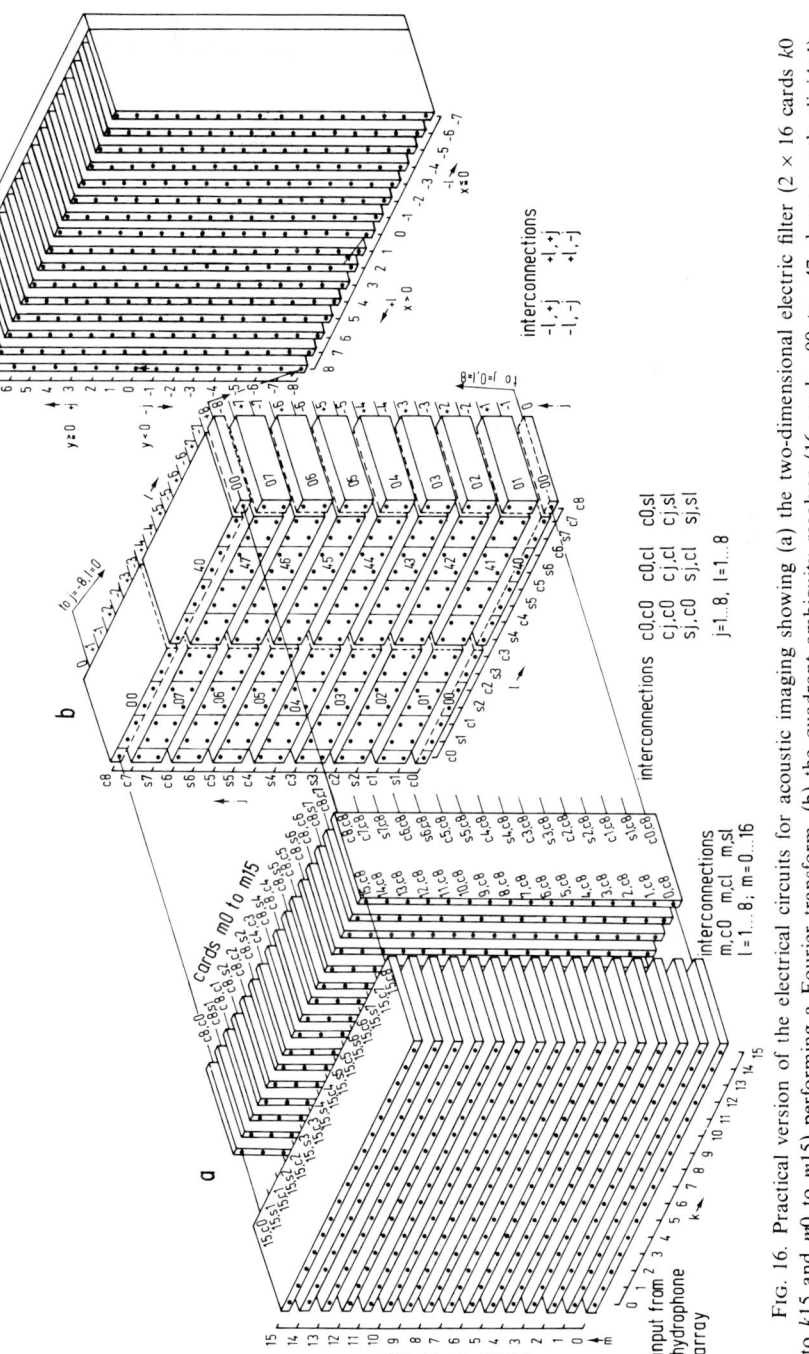

FIG. 16. Practical version of the electrical circuits for acoustic imaging showing (a) the two-dimensional electric filter (2 × 16 cards $k0$ to $k15$ and $m0$ to $m15$) performing a Fourier transform, (b) the quadrant ambiguity resolver (16 cards $q00$ to $q47$ shown partly divided), and (c) a light-emitting diode display (16 cards dl, $l = -7$ to $+8$) with the associated driving circuits. This illustration shows how the various printed circuit cards are interconnected. The actual cards are mounted in the usual way and interconnected by cable trees.

dimensional Fourier transform. It is followed by 16 cards that resolve the quadrant ambiguity according to Figs. 14 and 15. The output voltages of the quadrant ambiguity resolver feed driving circuits for a light-emitting diode display. Note that the cards in Fig. 16 are shown symbolically to illustrate their interconnection. In reality the cards are housed in the usual way in printed circuit card hangers, but the wiring between the hangers is done according to Fig. 16.

Let us observe that the operations discussed in this section can be performed by digital as well as analog circuits. Digital circuits are too expensive at the present time, but the development of semiconductor technology is so rapid that one can hardly predict what will be more practical ten years from now. The circuit of Fig. 16 would not be affected by a change from analog to digital; only the circuitry on the individual cards would be changed to digital.

C. *Inverse Transformation by Means of Sampled Storage Circuits*

The delay circuits in Fig. 4 may be implemented by shift registers. The input signal must be transformed from analog to digital for the use of digital shift registers, but one could also use analog shift registers implemented, e.g., by charge-coupled devices. The output voltages of the summing amplifiers in Fig. 4 would be step functions if analog shift registers were used, and binary numbers if digital shift registers were used. In both cases one obtains sampled functions rather than continuous ones.

The use of sampled functions leads to the implementation of the inverse transform by sampling filters. The principle will be explained with the help of Fig. 17. A plane wave producing the voltage $f(t)$ at the center receptor $i = 0$ produces the voltage $f[t + i(d/v) \sin \beta]$ at the receptor i. Let the switch $s(i, l)$ be closed momentarily at the time $t_0 - i(d/v) \sin \gamma$. The sampled vol-

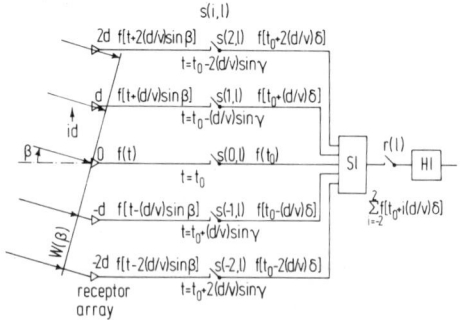

FIG. 17. Principle of the inverse transformation by means of sampled storage circuits. $\delta = \sin \beta - \sin \gamma$; (H$l$) analog hold circuit; (Sl) analog sum-and-hold circuit.

tage $f[t_0 + i(d/v)(\sin \beta - \sin \gamma)] = f[t_0 + i(d/v)\delta]$ is obtained and fed to the analog sum-and-hold circuit S*l*. After samples have been summed from all switches $s(i, l)$, $i = -2 \cdots +2$, one obtains the sum

$$\sum_{i=-2}^{+2} f[t_0 + i(d/v)\delta] = \sum_{k=0}^{5} f\{[vt_0 + (k-2)d\delta]/\lambda\}. \tag{25}$$

This is the same sum as that of Eq. (4), except that $V \sin(2\pi vt/\lambda)$ has been replaced by $f(vt_0/\lambda)$ and $2^n - 1$ by 5. Hence, a sample, sum, and hold circuit according to Fig. 17 can be substituted for the delay and summing circuits of Fig. 4, if one is satisfied with a sampled output voltage.

The sum-and-hold circuit in Fig. 17 is connected via the readout switch

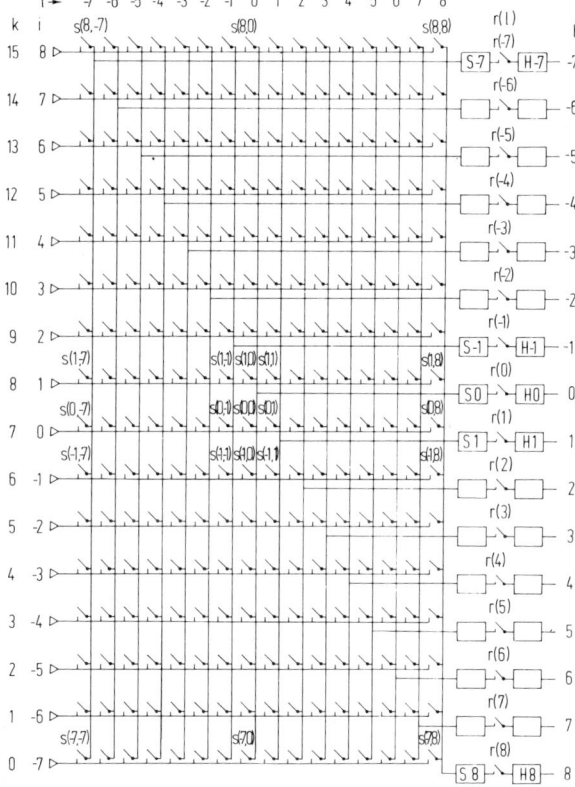

FIG. 18. Circuit for the resolution of 16 points using 16 receptors and sampled storage circuits. The location of the hydrophones relative to the center $i = 0$ is id. The observation angles are $\sin \gamma = l\lambda/A$. The selection switches are denoted $s(i, l)$ and the readout switches $r(l)$. (S*l*) sum-and-hold circuit; (H*l*) hold circuit. The output voltage at any terminal l in the notation of Fig. 17 is given by the sum $\sum_{i=-7}^{8} f[t_0 - i(d/v)(\sin \beta - l\lambda/A)]$.

r(l) to a hold circuit Hl. This switch transfers a completed sum $\sum f[t_0 + i(d/v)\delta]$ from the summing circuit Sl to the hold circuit Hl and permits a new cycle of sampling and summing to start.

The circuit of Fig. 17 is shown in Fig. 18 for 16 input terminals $k = 0 \cdots 15$ or $i = -7 \cdots +8$. The 16 input terminals imply 16 possible observation angles γ determined by the relation $\sin \gamma = l\lambda/A, l = -7 \cdots +8$, according to Eq. (18). Hence, there are 16^2 switches s(i, l), $i = k - (2^n/2 - 1) = k - (16/2 - 1) = -7 \cdots +8$ and $l = -7 \cdots +8$. Furthermore, there are 16 sum-and-hold circuits Sl, read-out switches r(l) and hold circuits Hl.

The timing diagram for the switches s(i, l) and r(l) of Fig. 18 is shown in Fig. 19. The number i is plotted vertically on the left, the number l denotes the slanting lines. The intersection of a line i with a line l, marked by a dot, indicates the closing time of switch s(i, l). The time is plotted horizontally; t_0 is an arbitrary reference time. Note that all switches s(0, -7) to s(0, 8) are closed at the same time $t = t_0$; this time is represented by the intersection of the 15 lines $i = 0, l = -7 \cdots +8$.

The extension of the sampled storage principle from one to two dimensions is rather difficult. The 16 input terminals in Fig. 18 call for 16^2 switches. A two-dimensional array with 16^2 input terminals would require 16^4 switches. The timing circuits associated with those switches and the wiring would make the method undesirable. For practical equipment one must be able to replace one two-dimensional transform by two one-dimensional transforms. Figure 20 shows a possible way to accomplish this.

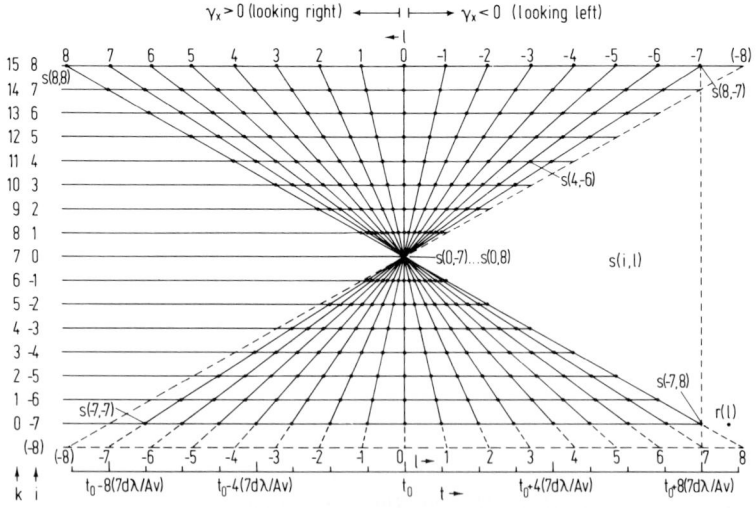

FIG. 19. Closing times of the selection switches s(i, l) and the readout switches r(l). The observation angles are $\sin \gamma = l\lambda/A$.

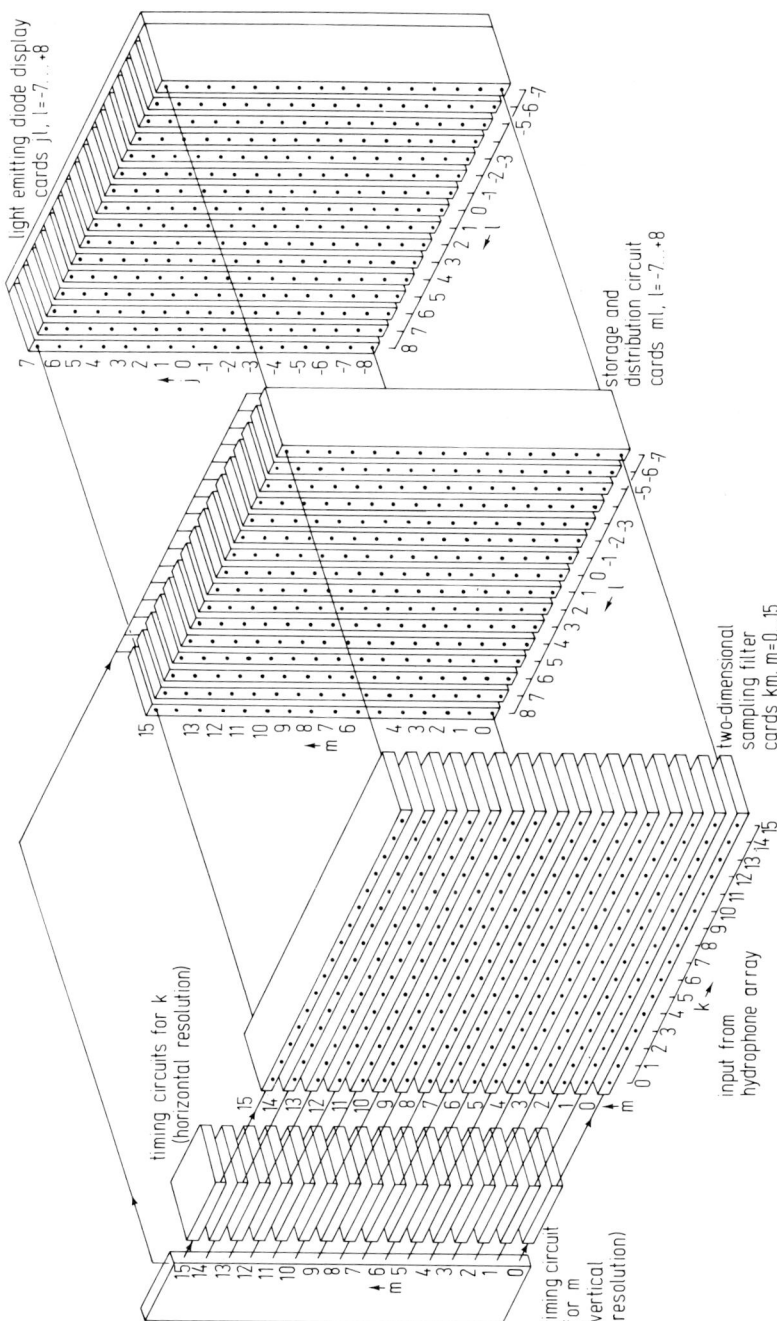

FIG. 20. Image generation by means of a two-dimensional sampling filter.

The 16^2 receptors feed into the two-dimensional sampling filter consisting of a stack of 16 cards with the circuit of Fig. 18. The cards are denoted km with $m = 0 \cdots 15$. The output terminals of these cards feed to a storage and distribution circuit with 16 cards ml, $l = -7 \cdots +8$. The circuit of these cards is shown in Fig. 21. The cards ml sum the output voltages of the cards

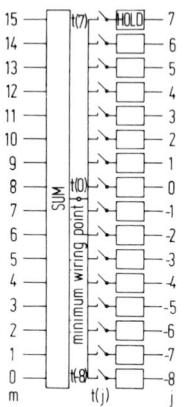

FIG. 21. Storage and distribution circuit for sampling in two-dimensional Cartesian coordinates. (SUM) summer; (HOLD) hold circuit.

km for all values $m = 0 \cdots 15$ and a certain value of l. Only four of the 16^2 interconnecting wires are shown in Fig. 20 to avoid obscuring the picture. Let the switches on all cards km, $m = 0 \cdots 15$, of the two-dimensional sampling filter be operated according to the time diagram of Fig. 19. Each card km will then produce the same 16 output voltages, and the summation in one of the cards of the storage and distribution circuit will yield the voltage of one output terminal of a card km multiplied by 16. Let now the switch t(0) in Fig. 21 be closed. Each card ml in Fig. 20 will then feed a voltage to the input terminal $j = 0$ of the cards jl, $l = -7 \cdots +8$, of the light-emitting diode display. The 16 light-emitting diodes of the row $j = 0$ will show the *line image* for the Cartesian elevation angle $\gamma_y = 0$, or $(A/\lambda) \sin \gamma_y = j = 0$, and all azimuth angles $(A/\lambda) \sin \gamma_x = l = -7 \cdots +8$.

The center column with the heading $j = 0$ in Fig. 22 shows the time diagram for the generation of the line image for the Cartesian elevation angle $\gamma_y = 0$. The 16 timing diagrams for $m = 0 \cdots 15$ are all identical, and a smaller scale version of the time diagram of Fig. 19, except that the operating time of the switch t(0) of Fig. 21 is shown on top of Fig. 22 in addition to the operating times of the switches $s(i, l)$ and $r(l)$.

If we want to generate a line image for a Cartesian elevation angle $(A/\lambda) \sin \gamma_y = j = -8 \cdots +7$ one must operate the switches of the cards km

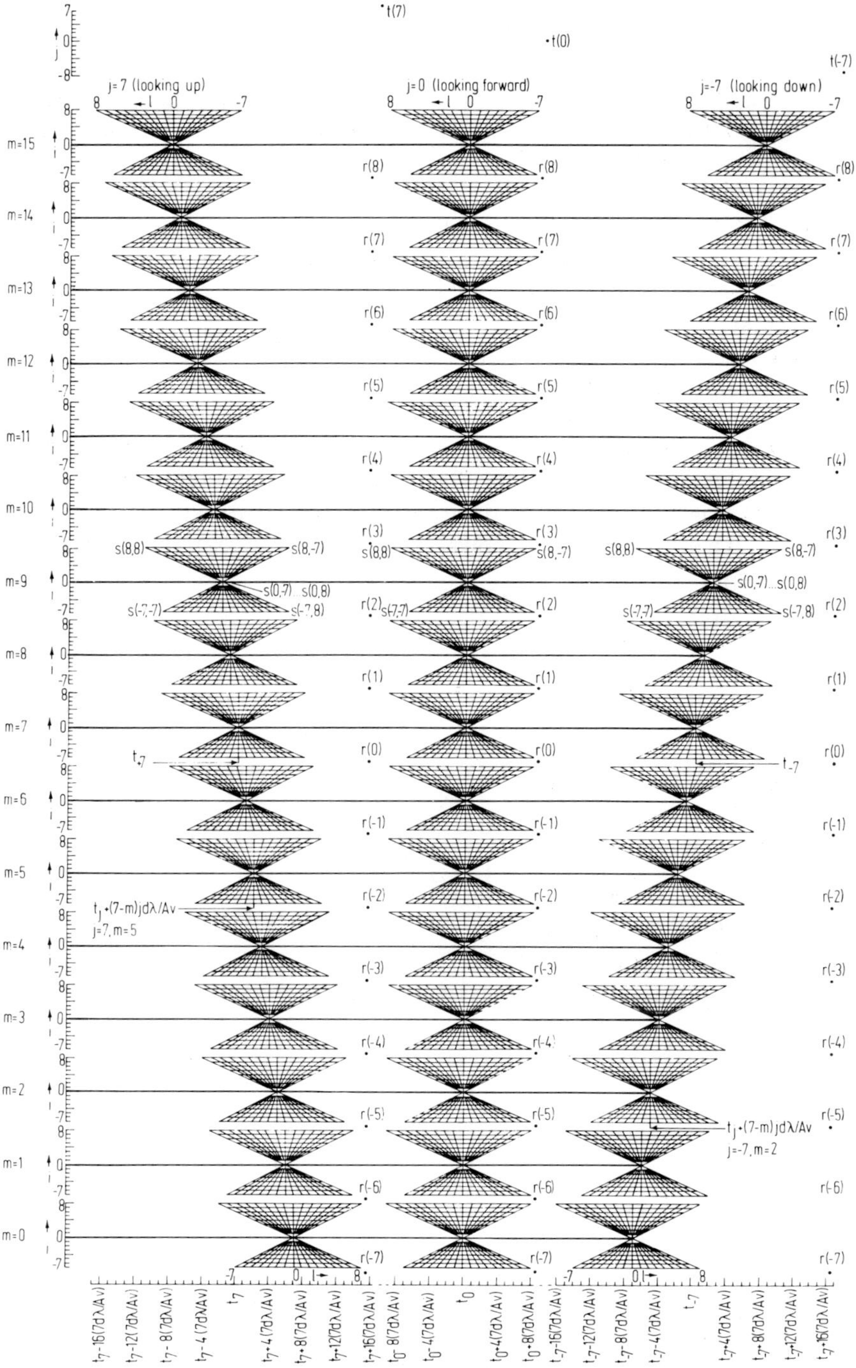

FIG. 22. Time diagram for the two-dimensional sampling filter of Fig. 20 for the horizontal angles $(A/\lambda)\sin\gamma_x = i$, $i = -7 \cdots +8$, and the vertical angles $(A/\lambda)\sin\gamma_y = j$, $j = -7, 0, +7$. The times $t_j + (7-m)jd\lambda/Av$ give the operating times for the switches $s(-7, 0) \cdots s(8, 0)$; the number 7 must be replaced by $2^n/2 - 1$ for arrays with 2^n input terminals.

with certain delays for $m = 0, \ldots, 15$. The resulting time diagrams are shown in Fig. 22 for $(A/\lambda) \sin \gamma_y = j = +7$ and $(A/\lambda) \sin \gamma_y = j = -7$. The relative delays of the "center points" of the time diagrams for $m = 0 \cdots 15$ are the same as the relative delays for $l = +7$ or $l = -7$ in Fig. 19; the "center points" show the operating times of the switches $s(0, -7) \cdots s(0, 8)$. Note that the operating times in Fig. 22 are not indicated as in Fig. 19 by dots at the intersections of the horizontal lines i and the slanting lines l, but by the intersections only.

Let us discuss the practical implementation of the timing circuits for the switches $s(i, l)$, $r(l)$ and $t(j)$. For each card km, $m = 0 \cdots 15$, in Fig. 20 one needs a timing circuit that operates the switches $s(i, l)$ and $r(l)$ according to Fig. 19. These 16 *timing circuits for k* are shown in Fig. 20 on the left. It is assumed that they are driven by clock pulses provided by the *timing circuit for m* in Fig. 20. The time diagram for the pulses at the terminals $m = 0 \cdots 15$ of this circuit is shown in Fig. 23. Let us assume that the line

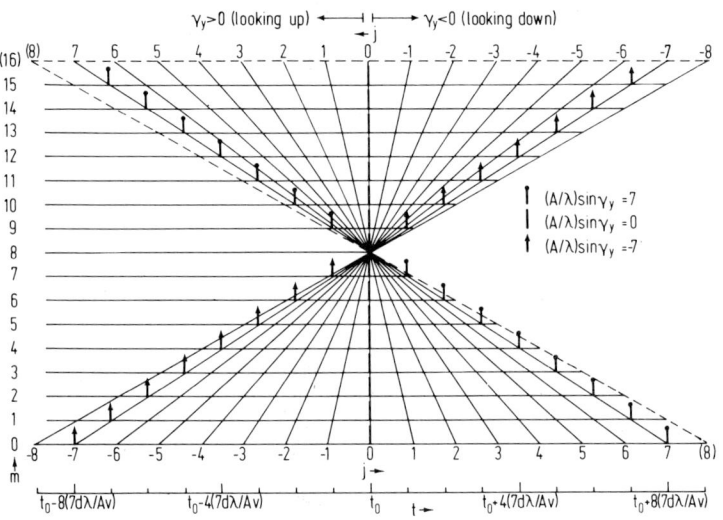

FIG. 23. Time diagram for the synchronizing pulses from the "timing circuit for m" to the "timing circuit for k" in Fig. 20. The synchronizing pulses for $j = -7, 0, +7$ are shown explicitly.

image for the Cartesian elevation angle $(A/\lambda) \sin \gamma_y = 0$ is to be produced. The pulses at the terminals $m = 0 \cdots 15$ have to be simultaneous as shown by the pulses for $j = 0$, $m = 0 \cdots 15$ in Fig. 23. An elevation angle $(A/\lambda) \sin \gamma_y = +7$ calls for a timing of the pulses as shown for $j = +7$, $m = 0 \cdots 15$. Similarly, an elevation angle $(A/\lambda) \sin \gamma_y = -7$ requires the timing shown for $j = -7$, $m = 0 \cdots 15$. Note that the timing diagrams of

Figs. 19 and 23 differ in two points only: the dots representing the closing times of switches in Fig. 19 are replaced by pulses in Fig. 23, occurring at the terminals of the *timing circuit for m* in Fig. 20, and the lines $l = -8$ or 8 in Fig. 19 are interchanged with the lines $j = 8$ or -8 in Fig. 23.

III. Focusing for Spherical Wavefronts

Plane waves represent points at infinity. A point at a finite distance R creates a spherical wavefront at the array. Figure 24 shows such a wavefront coming from the point W0 located at the distance R on the axis of the hydrophone array. Let the output voltage of the hydrophone at the center be $V \sin 2\pi v t/\lambda$. The distance from W0 to the hydrophone i located at id is

$$[R^2 + (id)^2]^{1/2} \doteq R + i^2 d^2/2R, \qquad i^2 d^2 \ll R^2. \tag{26}$$

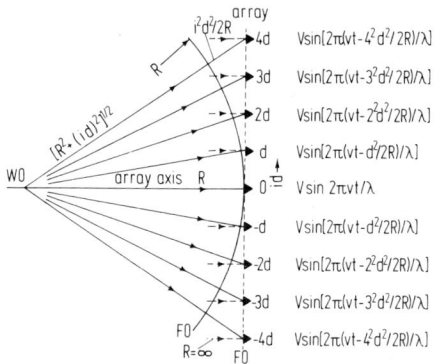

Fig. 24. Output voltages of a hydrophone array due to a spherical wave originating in the point W0 at the distance R on the axis of the array.

The wave has to travel the additional distance $i^2 d^2/2R$ to reach this hydrophone and its output voltage will thus be $V \sin [2\pi(vt - i^2d^2/2R)/\lambda]$. These output voltages are shown in Fig. 24.

Consider now a point Wβ located at the distance R from the center of the array but at an angle β from the array axis as shown by Fig. 25. All possible points Wβ are located on a circle or a sphere with radius R around the center of the array. The sphere is called the *sphere of focus*. The distance from Wβ to the hydrophone i is

$$[R^2 + (id)^2 - 2iR \cos(\pi/2 - \beta)]^{1/2} \doteq R - id \sin \beta$$
$$+ (i^2d^2/2R)(1 - \tfrac{1}{4} \sin^2 \beta), \tag{27}$$
$$R^2 \gg i^2 d^2.$$

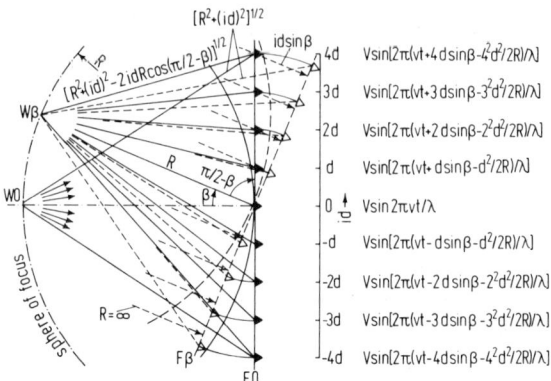

FIG. 25. Output voltages of a hydrophone array due to a spherical wave originating in the point $W\beta$ at the distance R on a ray forming the angle β with the axis of the array.

For small values of β one may ignore the last term in Eq. (27). The wave has then to travel the additional distance $-id\sin\beta + i^2d^2/2R$ to reach the hydrophone i rather than the center $i = 0$ of the array. The output voltage of hydrophone i will thus be

$$V \sin[2\pi(vt + id\sin\beta - i^2d^2/2R)/\lambda].$$

For $R \to \infty$ one obtains the output voltage due to a plane wave. The correction term $i^2d^2/2R$ for finite values of R is independent of the angle β if the conditions

$$R^2 \gg A^2 \geq i^2d^2, \quad \tfrac{1}{4}\sin^2\beta \ll 1 \tag{28}$$

are satisfied, where $A = 2nd$ is the aperture of the array having $2n + 1$ hydrophones spaced at multiples of d.

Focusing is achieved by advancing the output voltage of a hydrophone i in Fig. 24 by the time $i^2d^2/2R$ relative to the output voltage of the hydrophone $i = 0$. If the array has $2n + 1$ hydrophones, the voltage for the hydrophone $i = n$ has to be advanced $n^2d^2/2Rv$, or the voltage of the hydrophone $i = 0$ delayed by $n^2d^2/2Rv$. Generally, the output voltage of a hydrophone i has to be delayed by the focusing time t_{n-i}:

$$t_{n-i} = (n^2 - i^2)d^2/2Rv. \tag{29}$$

To obtain some idea about the magnitude of t_{n-i} let us divide the largest delay time t_0 by the period $T = \lambda/v$ of the wave:

$$t_0/T = \tfrac{1}{2}(nd/R)(nd/\lambda); \tag{30}$$

nd/R must be small compared with 1 according to Eq. (26). The wavelength λ

is typically of the order of d. Hence, the ratio t_0/T ranges from 0 for $R = \infty$ to about $t_0/T = 1$. At a frequency of 100 kHz a delay by one period yields $t_0 = T = 10$ μsec. A coaxial delay line would have to be more than one kilometer long to yield such a delay. A more practical approach to the delay problem is to use sample-and-hold circuits.

Figure 26 shows sampling switches si at the output terminals of the hydrophones. The switches are closed at the time indicated in the time diagram on the right. The holding circuits HO store the sampled voltages. Shortly after the time $t_4 = 16d^2/2Rv$ the holding circuits contain voltage samples corresponding to a certain focusing distance R. The readout switches ri are then closed simultaneously and the voltages are transferred to the input of the two-dimensional spatial filter performing the Fourier transform.

Figure 27 shows the determination of the sampling times t_{n-i} for a

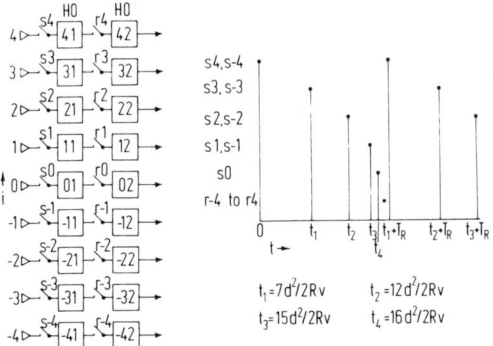

FIG. 26. Sample-and-hold circuits for focusing an array for a finite distance R, and the time diagram for the sampling switches si and the readout switches ri. (HO) hold circuit.

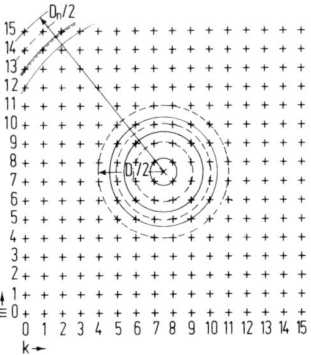

FIG. 27. Determination of the sampling times t_{n-i} for a two-dimensional Cartesian array.

two-dimensional square array with 16 × 16 hydrophones. Equation (29) must be replaced by the equation

$$t_{n-i} = (D_n/2)^2 - (D_i/2)^2/2Rv,$$

where D_i is the diameter of the circle i and D_n the diameter of the circle n.

It is obvious from Fig. 27 that the timing of the switches for focusing will be greatly simplified if the two-dimensional filter is not based on Cartesian coordinates but on polar coordinates. Such filters have been discussed (Harmuth et al., 1974). The reason for this distinction of polar coordinates is that we investigated the image of a point source, which emits spherical waves. A line source would emit cylindrical waves, and favor Cartesian coordinates.

We will return to the focusing problem in Section VI,E and show that a more satisfying solution can be obtained for Cartesian coordinates. Some additional material must be developed first to make this solution more understandable.

In optical photography we have two methods of focusing. One is to change the distance between the lens and the photographic film. This is the equivalent of the focusing process discussed here. The other method is the reduction of the area of the lens used by means of a circular mask with smaller diameter, marked as f stops on a camera. This second process achieves focusing indirectly by increasing the depth of field at the expense of resolution. Trading resolution for depth of field is acceptable for optical lenses due to the large size of the lenses compared with the wavelength used. The diameter of a small optical lens equals about 10^4 wavelengths, while the diameter of a large acoustic array equals about 10^2 wavelengths. A reduction of the lens area by masks corresponds to using the hydrophones close to the center of the array only. The high cost of an acoustic array and its poor resolution compared with an optical lens make it mandatory to use all the hydrophones. The loss of resolution caused by not using all hydrophones to increase the depth of field has been discussed by Thorn et al. (1974). Fortunately, there is a method to reduce the array size and thus to increase the depth of field without a reduction of the number of hydrophones used. This method is based on the introduction of time-variable elements, and it can thus not be applied to the optical lens. The method is discussed in the following section.

IV. Reduction of the Array Size

A. Array Size for the Classical Limit of Resolution

The observation angle of Eq. (18) may be written in the form

$$\sin \gamma = l\lambda/A, \qquad l = 0, \pm 1, \pm 2, \ldots \qquad (31)$$

l may assume 2^n values if there are $2^n \times 2^n$ receptors in a square array. A typical choice for l is $-2^n/2, \ldots, 0, \ldots, +2^n/2 - 1$ or $-2^n/2 + 1, \ldots, 0, \ldots, +2^n/2$. The *viewing angle* or *field of view angle* is defined by inserting the smallest and largest value of l into Eq. (31):

$$\sin \alpha_- = -2^n \lambda / 2A \quad \text{or} \quad \sin \alpha_- = (-2^n/2 + 1)\lambda/A;$$
$$\sin \alpha_+ = (2^n/2 - 1)\lambda/A \quad \text{or} \quad \sin \alpha_+ = 2^n \lambda / 2A. \tag{32}$$

For large values of n one does not need to distinguish between $2^n/2$ and $2^n/2 - 1$. One obtains the following simpler formula:

$$\sin \frac{\alpha}{2} = 2^n \lambda / 2A = \lambda / 2d;$$
$$A = 2^n d, \quad \alpha \doteq 2\alpha_- \doteq 2\alpha_+ = \text{viewing angle.} \tag{33}$$

The *field of view* at a distance L from the array equals $D \times D$, where D is defined by the equation

$$D = 2L \tan (\alpha/2). \tag{34}$$

The distance of resolvable points at the distance L is defined by $D/2^n$. Let us point out that one could space the curves in Fig. 3 differently. For instance, the maximum of the curve $\gamma = \lambda/A$ could be shifted from the zero crossing of the curve $\gamma = 0$ at $\beta = \lambda/A$ to the minimum of the curve $\gamma = 0$ further to the right. One would obtain different values for the viewing angle α and the field of view $D \times D$. The spacing used in Fig. 3 has the advantage that all curves are zero when one has its maximum.

Table I shows some representative values derived from Eqs. (33) and (34). The three operating frequencies of 10 kHz, 100 kHz, and 1 MHz are assumed. The corresponding wavelengths for sound in water are 15 cm, 1.5 cm, and 1.5 mm. Furthermore, hydrophone arrays with 16×16, 64×64, and 256×256 hydrophones are assumed. The two viewing angles $\alpha = 30°$ and $\alpha = 10°$ are considered. The distance d between adjacent hydrophones is defined by Eq. (33), the size of the array by $2^n d \times 2^n d$, the field of view $D \times D$ by Eq. (34), the distance of resolvable points by $D/2^n$.

Let us note that hydrophone arrays with thousands of hydrophones are well within the state of the art. The fabrication of such arrays was described by Prokhorov (1972).

For a frequency of 10 kHz one obtains dimensions that are too large for a movable array. A frequency of 1 MHz, on the other hand, implies absorption losses that are too high. The frequency of 100 kHz is representative for practical equipment. An array with a viewing angle of $\alpha = 10°$ and 16×16 hydrophones has been built. The size of 1.4×1.4 m appears to be about the

TABLE 1: REPRESENTATIVE FIGURES FOR ACOUSTIC IMAGING IN WATER BY MEANS OF TWO-DIMENSIONAL ELECTRIC FILTERS

Frequency of insonifying wave Wavelength	10 kHz 15 cm		100 kHz 1.5 cm		1 MHz 1.5 mm	
Hydrophone array	16×16	64×64	256×256	16×16	64×64	256×256	16×16	64×64	256×256
Number of hydrophones	256	4096	65,536	256	4096	65,536	256	4096	65,536
$\alpha = 30°$ $d = \lambda/2 \sin(\alpha/2)$		29 cm			2.9 cm			2.9 mm	
Size of hydrophone array $D = 2L \tan(\alpha/2)$ $D \times D$ = field of view at $L = 200$ m	4.6^2 m^2	18.5^2 m^2	74^2 m^2	0.46^2 m^2	1.85^2 m^2	7.4^2 m^2	4.6^2 cm^2	18.5^2 cm^2	74^2 cm^2
Distance of resolvable points at $L = 200$ m	6.6 m	1.66 m	41.4 cm	6.6 m	1.66 m	41.4 cm	6.6 m	1.66 m	41.4 cm
Field of view for $L = 20$ m					10.6^2 m^2				
Distance of resolvable points at $L = 20$ m	66 cm	—	—	66 cm	16.6 cm	4.2 cm	66 cm	16.6 cm	4.2 cm
$\alpha = 10°$ $d = \lambda/2 \sin(\alpha/2)$		86 cm			8.6 cm			8.6 mm	
Size of hydrophone array $D = 2L \tan(\alpha/2)$ $D \times D$ = field of view at $L = 200$ m	13.8^2 m^2	55^2 m^2	220^2 m^2	1.4^2 m^2	5.5^2 m^2	22^2 m^2	14^2 cm^2	55^2 cm^2	2.2^2 m^2
Distance of resolvable points at $L = 200$ m					35^2 m^2				
Field of view for $L = 20$ m	2.2 m	55 cm	—	2.2 m	55 cm	13.7 cm	2.2 m	55 cm	13.7 cm
Distance of resolvable points at $L = 20$ m					3.5^2 m^2				
	22 cm	—	—	22 cm	5.5 cm	1.37 cm	22 cm	5.5 cm	1.37 cm

upper limit for a practical movable array. An array with 64 × 64 hydrophones and a viewing angle $\alpha = 10°$ calls for an array of 5.5 × 5.5 m according to Table I. This is too large. An increase of α to 30° yields an array size of 1.85 × 1.85 m. This is practical, particularly if one increases the frequency from 100 to about 150 kHz. For an array of 256 × 256 hydrophones one obtains the array size of 7.4 × 7.4 m even for $\alpha = 30°$. This is too large. An increase of the frequency to about 750 kHz would be required to reduce the array size to a practical value, but the absorption losses would increase from about 27 dB per kilometer to about 200 dB per kilometer. Hence, there is no realistic way to resolve more than about 64 × 64 points in underwater acoustic imaging, unless one finds a method to improve on the classical limit of resolution defined by Eq. (18) and the curves of Fig. 3. Using the notation

$$\sin \gamma_l = l\lambda/A,$$

one may derive from Eq. (18) the classical resolution* angle ε:

$$\sin \gamma_l - \sin \gamma_{l-1} \doteq \lambda/A = \varepsilon, \qquad \gamma_l \doteq \gamma_{l-1}. \tag{35}$$

A considerable amount of work has gone into overcoming this limit. A good survey of this work is provided by the papers of Capon, Goodman, Greenfield, Kolker, Lee, Lewis, Lo, Schultheiss, Seligson and Vanderkulk.

It is generally understood now that the signal-to-noise ratio is the limit, but the question remains how to make practical use of this knowledge. A typical difficulty is that very high accuracy is required for the receptors as well as the processing equipment. In the case of seismic waves one can overcome the problem of processing equipment by the use of digital computers with essentially unlimited accuracy, but the digital computer is too slow for other applications (Woods and Lintz, 1973).

The limit set by the classical resolution angle was overcome successfully in synthetic aperture radar (Harger, 1970). This type of radar uses transmitting and receiving antennas made rapidly time-variable by mounting them aboard an airplane. We do not want any movement of the hydrophone array. Hence, we must find a way to make it rapidly time-variable without moving it physically.

B. *Super-Resolution by Means of Sampled Arrays*

Figure 28a shows a receptor or antenna array consisting of 13 individual receptors. The receptors 1–8 are connected to a processor. In Figs. 28b–d the receptors 2–9, 3–10, and 4–11 are connected to the processor. As far as the

---

* A factor C may be written in front of λ/A which allows for somewhat different definitions of the resolution angle ε, and for the shape of the array in two dimensions, e.g., circular or square array.

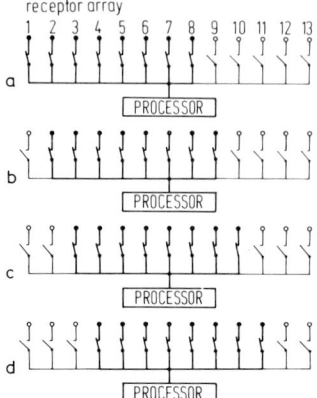

FIG. 28. Simple electric movement of a receptor array by means of switches. (a) Receptors 1–8 connected to a processor; (b–d) receptors 2–9, 3–10, 4–11, respectively, connected to the processor.

processor is concerned, one might as well have moved eight receptors physically from left to right. Hence, the electric movement by switches can replace the physical movement.

An airplane can move receptors only in the way corresponding to Fig. 28. Switches can simulate much more complicated movements. We can operate the switches in Fig. 28 in any way we want. This makes the electric movement more general than the physical movement, and one will expect results that go beyond synthetic aperture radar.

For an array made time-variable by means of sampling switches the resolution depends on the maximum rate of change of the wave (Harmuth, 1976). For instance, a sinusoidal wave $V \sin 2\pi ft$ has the maximum rate of change at the zero crossings and its magnitude is $2\pi Vf$. An increase of either the frequency f or the amplitude V brings an equal improvement of the resolution. The signal-to-noise ratio will obviously determine how close to the zero crossings one can sample a sinusoidal function, and the signal-to-noise ratio thus determines the possible resolution.

As a standard of comparison we use the factor $A(\gamma, \beta)$ of $v(\gamma, \beta)$ of Eq. (6) that contains the angular variation. For simplification we consider only the case $\gamma = 0$:

$$A(0, \beta) = A(\beta) = \frac{\sin\left[(\pi A \sin \beta)/vT\right]}{(\pi A \sin \beta)/vT} \qquad vT = \lambda. \qquad (36)$$

$A(\beta)$ is called the amplitude diagram. A plot of $A(\beta)$ is shown by the solid line in Fig. 29.

Equation (36) applies if the output voltages of the sensors in Fig. 2 are summed, either with or without a constant delay for all voltages. Let now the

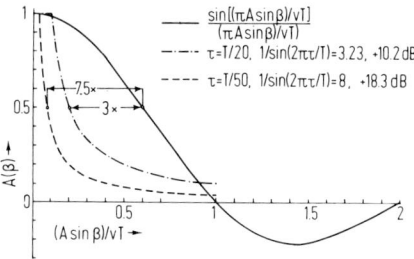

FIG. 29. Amplitude (reception) diagrams for the sensor array of Fig. 2. Solid curve, $A(\beta)$ according to Eq. (36); dashed and dashed-dotted curves, $A(\beta)$ according to Eq. (42).

output voltages be fed to the input terminals of the circuit of Fig. 30. This circuit has $2n + 1$ input terminals for greater generality, while Fig. 2 has 16 receptors. The use of an odd number $2n + 1$ makes the calculations simpler than the use of an even number $2n$. This difference is not important since we must assume n to be large in order to carry through the analytical investigation. Computer plots for even numbers $2n$ may readily be obtained by eliminating either the terminal n or $-n$ in Fig. 30 and changing the limits of summation in the following equations accordingly.

The voltages coming from the sensor array pass in Fig. 30 through narrow-band filters FI tuned to the frequency $f = 1/T$ of the received wave. The filtered voltages are fed to the upper terminals of the selection switches SE. The voltage $V \sin (2\pi t/T)$ from the terminal $j = 0$ is also fed to the selection switches, directly to their lower terminals and via an amplitude-reversing amplifier AR to their center terminals. The selection switches SE are operated by the amplitude comparators AC, which compare the voltages

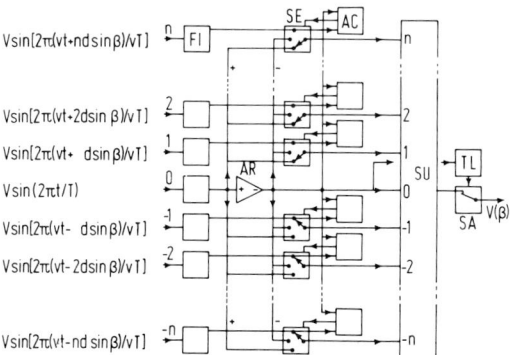

FIG. 30. Simplified processing circuit for a sampled array receiving a sinusoidal wave. (FI) filter tuned to the frequency of the wave; (AR) amplitude reversing amplifier; (SE) selection switch; (AC) amplitude comparator; (SU) summing circuit; (TL) tracking loop for the voltage $V \sin (2\pi t/T)$ and timing circuit for the sampling switch; (SA) sampling switch.

$V \sin [2\pi(v t + i d \sin \beta)/v T]$ and $-V \sin (2\pi t/T)$, $i = \pm 1 \cdots \pm n$. If the magnitude of a voltage $V \sin [2\pi(v t + i d \sin \beta)/v T]$ is smaller than the magnitude of $-V \sin (2\pi t/T)$ it passes through the selection switch SE to the summing circuit SU. Otherwise, either the voltage $+V \sin (2\pi t/T)$ or $-V \sin (2\pi t/T)$ is passed through to the summing circuit; the positive or negative sign is chosen to yield the same polarity as that of the voltage $V \sin [2\pi(v t + i d \sin \beta)/v T]$.

Figure 31 shows the selection process more clearly. The voltage at the center terminal $i = 0$ has the magnitude $V \sin (2\pi\tau/T)$ at the time $t = \tau$. The voltages for $i \geq 1$ are larger. Hence, the switches SE must be in the lower position as shown in Fig. 30 for $i = 1 \cdots n$. The magnitude of the voltages for $i = -1$ and $i = -2$, on the other hand, is less than $V \sin (2\pi\tau/T)$ in Fig. 31. Hence, the switches for $i = -1 \cdots -n$ in Fig. 30 are in the upper position.

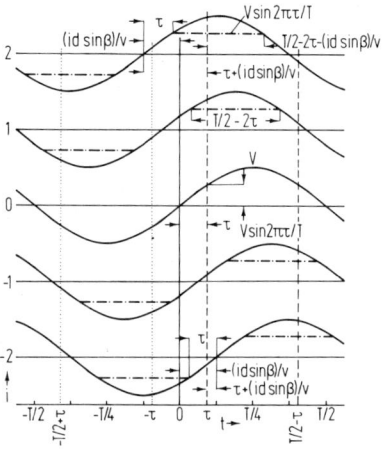

FIG. 31. Sinusoidal voltages arriving with various time shifts at the input terminals of the circuit of Fig. 30. The amplitude deformations due to the selection switches SE are shown by the dashed-dotted lines. The sampling points are indicated by dashed lines, and the sampling points requiring amplitude reversal by dotted lines.

The output voltage of the summing circuit in Fig. 30 is sampled by the sampling switch SA at a certain time τ, as indicated by the dashed line at $t = \tau$ in Fig. 31. The sampling switch SA is controlled by a tracking loop TL which is locked to the sinusoidal voltage $V \sin (2\pi t/T)$ at the center terminal $i = 0$. The output voltage $V(\beta)$ of the sampling switch SA is the sum of the $2n + 1$ input voltages at the time τ. According to Fig. 31, this sum consists of voltages $V \sin [2\pi(v\tau + i d \sin \beta)/v T]$ and $V \sin (2\pi\tau/T)$, whichever has the smaller magnitude, but the polarity is always that of $V \sin [2\pi(v\tau +$

$id \sin \beta)/vT]$. In the absence of noise the effect of the selection switches SE is the same as if the sinusoidal functions in Fig. 31 were amplitude-limited as $\pm V \sin (2\pi\tau/T)$, as shown by the dashed-dotted lines. However, the action of the selection switches SE in the presence of noise is not identical to amplitude limiting. Superimposed noise would be clipped along the dashed-dotted lines in Fig. 31 by an amplitude limiter, while the selection switches SE have the effect of clipping the signal but not the noise.

Let us observe that the effective limitation of the signal amplitudes to $\pm V \sin (2\pi\tau/T)$ represents a power penalty that would not have to be paid if square waves with amplitude $\pm V \sin (2\pi\tau/T)$ and a linear transient of duration 2τ between $+V \sin (2\pi\tau/T)$ and $-V \sin (2\pi\tau/T)$ were used.

The amplitude selection process as discussed here yields a very poor signal-to-noise ratio since the input voltages of only $n + 1$ sensors are used for small values of β, while the voltage $V \sin (2\pi\tau/T)$ of the center receptor $i = 0$ is substituted for the voltages of the n remaining sensors. Still fewer input voltages are used for increasing values of β; in an extreme case only the voltage $V \sin (2\pi\tau/T)$ for $i = 0$ is used. However, the circuit of Fig. 30 may readily be modified to use all the $2n + 1$ input voltages all the time. We will discuss this improved but mathematically more complicated circuit later on.

For small positive values of β the output voltage $V(\beta)$ in Fig. 30 follows readily from Fig. 31:

$$V(\beta) = \sum_{i=-n}^{0} V \sin [2\pi(v\tau + id \sin \beta)/vT] + nV \sin (2\pi\tau/T); \quad (37)$$

$$0 < \tau < T/4, \qquad 0 \le (nd \sin \beta)/v < \tau.$$

Note that all voltages from terminals $i = +1 \cdots +n$ are equal to $V \sin (2\pi\tau/T)$. For larger values of β some of the voltages for $i = -n, -(n-1), \ldots$ become equal to $-V \sin (2\pi\tau/T)$:

$$V(\beta) = \sum_{i=-n'}^{0} V \sin [2\pi(v\tau + id \sin \beta)/vT] + nV \sin (2\pi\tau/T)$$
$$- (n - n')V \sin (2\pi\tau/T); \quad (38)$$

$$0 < \tau < T/6, \qquad \tau \le (nd \sin \beta)/v < T/2 - 2\tau;$$

n' is the largest value of $|i|$ for which the magnitude of $\sin [2\pi(v\tau - |i|d \sin \beta)/vT]$ is still smaller than that of $V \sin (2\pi\tau/T)$:

$$-[(v\tau - |i|d \sin \beta)/vT] \le \tau/T;$$

$$n' \doteq \frac{2\tau/T}{(d \sin \beta)/vT}. \quad (39)$$

The sums in Eqs. (37) and (38) may be approximated by integrals:

$$\sum_{i=-n}^{0} \frac{1}{n} \sin[2\pi(v\tau + id\sin\beta)/vT]$$

$$\doteq \sin(2\pi\tau/T) \int_{-1}^{1} \cos\left(\frac{2\pi nd\sin\beta}{vT}\frac{i}{n}\right) d\left(\frac{i}{n}\right)$$

$$+ \cos(2\pi\tau/T) \int_{-1}^{1} \sin\left(\frac{2\pi nd\sin\beta}{vT}\frac{i}{n}\right) d\left(\frac{i}{n}\right)$$

$$\doteq \sin(2\pi\tau/T) \frac{\sin[(2\pi nd\sin\beta)/vT]}{(2\pi nd\sin\beta)/vT}$$

$$- \cos(2\pi\tau/T) \frac{\sin^2[(\pi nd\sin\beta)/vT]}{(\pi nd\sin\beta)/vT};$$

$$\sum_{i=-n'}^{0} \frac{1}{n} \sin[2\pi(v\tau + id\sin\beta)/vT] \qquad (40)$$

$$\doteq \sin(2\pi\tau/T) \int_{-n'/n}^{0} \cos\left(\frac{2\pi nd\sin\beta}{vT}\frac{i}{n}\right) d\left(\frac{i}{n}\right)$$

$$+ \cos(2\pi\tau/T) \int_{-n'/n}^{0} \sin\left(\frac{2\pi nd\sin\beta}{vT}\frac{i}{n}\right) d\left(\frac{i}{n}\right)$$

$$\doteq \sin(2\pi\tau/T) \frac{\sin(4\pi\tau/T)}{(2\pi nd\sin\beta)/vT}$$

$$- \cos(2\pi\tau/T) \frac{\sin^2(2\pi\tau/T)}{(\pi nd\sin\beta)/vT};$$

$$n, n' \gg 1.$$

One obtains from Eqs. (37) to (40)

$$V(\beta) \doteq nV\left\{\left[1 + \frac{\sin[(2\pi nd\sin\beta)/vT]}{(2\pi nd\sin\beta)/vT}\right] \sin(2\pi\tau/T)\right.$$

$$\left. - \frac{\sin^2[(\pi nd\sin\beta)/vT]}{(\pi nd\sin\beta)/vT} \cos(2\pi\tau/T)\right\} \qquad (41)$$

for $0 \le (nd\sin\beta)/v < \tau, \quad 0 < \tau < T/4;$

$$V(\beta) \doteq nV\left\{\left[\frac{2\tau/T}{(nd\sin\beta)vT} + \frac{\sin(4\pi\tau/T)}{(2\pi nd\sin\beta)/vT}\right]\sin(2\pi\tau/T)\right.$$

$$\left. - \frac{\sin^2(2\pi\tau/T)}{(\pi nd\sin\beta)/vT}\cos(2\pi\tau/T)\right\}$$

for $\tau \leq (nd\sin\beta)/v < T/2 - 2\tau$,

$0 < \tau < T/6$, $2\tau/T \gg (d\sin\beta)/vT$;

$$V(\beta) \doteq nV\frac{2\tau/T}{(nd\sin\beta)/vT}\sin(2\pi\tau/T)$$

for $\tau \leq (nd\sin\beta)/v < T/2 - 2\tau$, $0 < \tau \ll T/4$.

For $\beta = 0$ one obtains

$$V(0) = 2nV\sin(2\pi\tau/T).$$

The amplitude diagram $V(\beta)/V(0) = A(\beta)$ becomes

$$A(\beta) \doteq \frac{1}{2}\left\{1 + \frac{\sin[(2\pi nd\sin\beta)/vT]}{(2\pi nd\sin\beta)/vT}\right.$$

$$\left. - \frac{\sin^2[(\pi nd\sin\beta)/vT]}{(\pi nd\sin\beta)/vT}\frac{\cos(2\pi\tau/T)}{\sin(2\pi\tau/T)}\right\}$$

for $0 \leq (2nd\sin\beta)/vT < 2\tau/T$, $0 < \tau/T < 1/4$;

$$A(\beta) \doteq \frac{2\tau/T}{(2nd\sin\beta)/vT}\left\{1 + \frac{\sin(4\pi\tau/T)}{4\pi\tau/T}\right.$$

$$\left. - \frac{\sin^2(2\pi\tau/T)}{2\pi\tau/T}\frac{\cos(2\pi\tau/T)}{\sin(2\pi\tau/T)}\right\} \quad (42)$$

for $2\tau/T \leq (2nd\sin\beta)/vT < 1 - 4\tau/T$,

$0 < \tau/T < 1/6$, $2\tau/T \gg (d\sin\beta)/vT$;

$$A(\beta) \doteq \frac{2\tau/T}{(2nd\sin\beta)/vT}$$

for $2\tau/T \leq (2nd\sin\beta)/vT < 1 - 4\tau/T$, $0 < \tau/T \ll 1/4$.

Let us observe that the process yielding the output voltage $V(\beta)$ in Fig. 30 is linear. The input voltages have the arbitrary amplitude V, and this arbitrary amplitude is retained in Eq. (41). Hence, the proportionality law is satisfied. The superposition law is trivially satisfied since the filters FI in Fig. 30 suppress any sinusoidal voltage orthogonal to $V\sin(2\pi t/T)$ due to a different frequency $f' = 1/T' \neq 1/T$. The function $V'\cos(2\pi t/T)$ would not be suppressed, but the sum of $V \sin(2\pi t/T)$ and $V' \cos(2\pi t/T)$ can be written in the form $V'' \sin(2\pi t'/T)$ and the superposition law is thus satisfied as a result of the proportionality law being satisfied.*

$A(\beta)$ is given by Eq. (42) for the range $0 \leq (2nd \sin \beta)/vT < 1 - 4\tau/T$. It can be extended to negative values of β from symmetry considerations to be $A(-\beta) = A(\beta)$. The extension to larger values of $(2nd \sin \beta)/vT$ is analytically too cumbersome, and one has to resort to computer plotting.

Curves for $\tau = T/20$ and $\tau = T/50$ according to Eq. (42) are shown in Fig. 29. They are substantially better than the solid curve. What is the price for this improvement?

The sampled amplitude of an input voltage $V \sin(2\pi t/T)$ at the time $t = T/4$ equals V, while it equals $V \sin(2\pi\tau/T)$ at the smaller time $t = \tau$. Hence, the amplitude V of the input signal must be increased to $V/\sin(2\pi\tau/T)$ in order to obtain the same signal-to-noise ratio in Fig. 29. The necessary increase in signal power equals $[1/\sin(2\pi\tau/T)]^2$; this factor is shown in Fig. 29 in decibels ($+10.2$ dB for $\tau = T/20$ and $+18.3$ dB for $\tau = T/50$).

The resolution represented by the dashed and dashed-dotted curves in Fig. 29 is obviously much better than the classical one represented by the solid curve, but the different shapes of the curves make a characterization of the improvement by one number very difficult. A possible way is to compare the values of $(A \sin \beta)/vT$ for which the three curves drop to one-half. This value is 3 times as large for the solid curve as for the dashed-dotted curve, and 7.5 times as large as for the dashed curve. Putting it differently, either the resolution angle or the aperture for the classical curve would be 3 or 7.5 times larger than for the other curves. This is the return for the required increase in signal power.

The calculation of $A(\beta)$ for larger values of $(A \sin \beta)/vT$ in Fig. 29 is too cumbersome, but one may readily simulate the circuit of Fig. 30 by computer. The results of computer simulation for an array of 64 hydrophones are shown in Figs. 32–33. The shape of the main lobe in Fig. 32 agrees with the calculated main lobes in Fig. 29 for small angles. The first two side lobes are much improved in Fig. 32. The zero crossings are not changed by the sampling process according to Fig. 32. However, the zero crossings for the curves $\tau/T = 1/16$ and $\tau/T = 1/64$ may readily be shifted to

---

* It is somewhat surprising that the circuit of Fig. 30 is linear even though amplitude comparators AC are used to operate the selection switches SE. As a bridge to understanding let us note that a ring modulator used for amplitude modulation is also a linear device, even though the diodes are definitely nonlinear. Hence, nonlinear components or circuits can be used to produce linear, time-variable effects. The difficulty of comprehending this fact is well documented by the many textbooks that call amplitude modulators nonlinear devices, and go on to discuss their use for the transmission of voice and music.

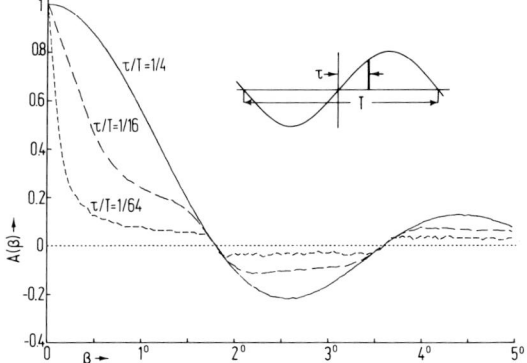

FIG. 32. Computer plots for three values of the sampling time for small angles β. The plot for $\tau/T = 1/4$ is identical with the classical curve shown by a solid line in Fig. 29. (Courtesy T. Frank of Catholic University.)

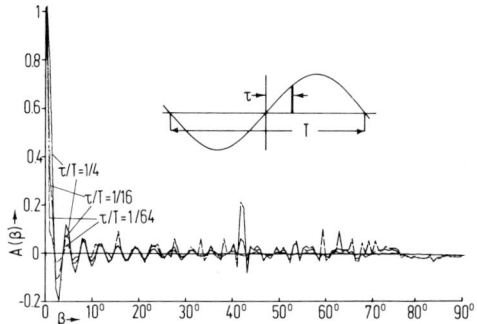

FIG. 33. Computer plots for three values of the sampling time for $0 \leq \beta \leq 90°$. (Courtesy T. Frank of Catholic University.)

smaller values of β by an additional linear transformation (Harmuth, 1976). Figure 33 shows the computer plot for the whole range $0 \leq \beta \leq 90°$. The only conspicuous feature of this plot is the side lobe at about 42° for $\tau/T = 1/16$.

C. *Improving the Signal-to-Noise Ratio*

We had mentioned that the weak point of the circuit of Fig. 30 is the multiple use of the voltage $V \sin(2\pi t/T)$ of the receptor $i = 0$, since the signal-to-noise ratio of this voltage is that of a single receptor rather than that of the array of $2n + 1$ receptors. A summing circuit is used in Fig. 34 to produce the average voltage of all the input voltages. This average voltage is the sum of the $2n + 1$ input voltages divided by $2n + 1$. For large values of n one obtains $v(0, \beta)$ divided by the number of receptors, where $v(0, \beta)$ is

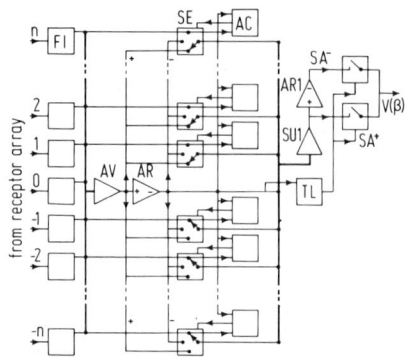

FIG. 34. More sophisticated version of the circuit of Fig. 30. An averager AV and a sampling switch SA$^-$ are added to improve the signal-to-noise ratio. (FI) filter; (AV) averager; (SE) selection switch; (AC) amplitude comparator; (SU1) summing circuit; (SA$^+$, SA$^-$) sampling switches; (TL) tracking loop and timing circuit for the sampling switches.

defined by Eq. (6). The number of receptors is A/d, and $v(0, \beta)/(A/d)$ thus equals $V \sin 2\pi vt/\lambda = V \sin 2\pi t/T$ for $\beta = 0$, but the signal-to-noise ratio is that of the $2n + 1$ receptors rather than that of the single receptor $i = 0$. This solves the signal-to-noise problem.* One also gets a fringe benefit. $v(0, \beta)$ in Eq. (6) drops according to the amplitude diagram $A(\beta)$ of Eq. (36) for larger values of β. Hence, the sinusoidal functions in Fig. 31 are "limited" at $\pm A(\beta)V \sin (2\pi \tau/T)$ rather than at $\pm V \sin (2\pi \tau/T)$. The effect is the same as a reduction of τ for larger values of β, and it leads to amplitude diagrams that are better for larger values of β than the ones shown in Figs. 32 and 33.

The summer SU of Fig. 30 is replaced by the summer SU1 in Fig. 34, which sums the same $2n + 1$ voltages. Similarly, the sampling switch SA is replaced by SA$^+$. The amplitude reversing amplifier AR1 and the sampling switch SA$^-$ are used to sample the sinusoidal functions in Fig. 31 at the time $t = -\tau$, which improves the signal-to-noise ratio of the output voltage $V(\beta)$. At the same time, the switches SA$^+$ and SA$^-$ may be used to solve the sampling problem at the zero crossings of $V \sin (2\pi t/T)$ with negative slope. According to Fig. 31 the switch SA$^+$ should be closed at the times $\tau + sT$ and $T/2 - \tau + sT$, where $s = 1, 2, \ldots$, as indicated by the dashed lines. The switch SA$^-$ should be closed at the times $-\tau + sT$ and $-T/2 + \tau + sT$ as indicated by the dotted lines.

* The weakest point of the circuit in the presence of noise is the amplitude comparator AC. It compares one signal with high signal-to-noise ratio with one with low signal-to-noise ratio. An improvement is possible on the basis of statistical signal theory by using the fact that the delay $(id \sin \beta)/v$ of a wave at the receptor i is proportional to i.

D. Practical Use of Super-Resolution for Image Generation

The advantages of super-resolution for two-dimensional arrays are dramatic since a linear reduction of an array by a factor 1/3 or 1/7.5 means a reduction of the area of a two-dimensional array by factors $(1/3)^2$ or $(1/7.5)^2$. The weight and cost of the array decrease even more. Table I shows that for a frequency of 100 kHz, a viewing angle $\alpha = 30°$, and 256 × 256 hydrophones, the array size is 7.4^2 m². Using super-resolution this area is reduced to about $(7.4/3)^2 \doteq (2.5)^2$ m² or $(7.4/7.5)^2 \doteq 1$ m². These are practical values.

Let us investigate the cost of electronic components in Fig. 34 that has to be paid for the mechanical simplification of the array. The tracking loop TL is the most complicated circuit, but it is required only once and is thus of little consequence. One filter FI is required for each hydrophone. The rest of the circuit—from the averaging amplifier AV to the sampling switches SA^+ and SA^-—is also required once for every hydrophone. Let us look at examples for the implementation of some of these circuits to obtain an idea of the complexity.

A typical filter FI is shown in Fig. 35. Its costliest and bulkiest component is the inductance. The progress in mechanical resonators used for filters in telephony multiplexing equipment might provide a superior way to implement these filters.

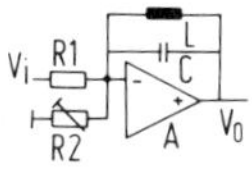

FIG. 35. Example of a circuit for the filters FI in Fig. 34.

Figure 36 shows circuits for the selection switches SE and the amplitude comparators AC in Fig. 34. The voltages from the filter FI and the averager AV pass through absolute-value circuits, which are high-precision full-wave rectifiers, to a comparator COM1 and a single-pole, double-throw analog switch SPDT1. The comparator is in essence a differential operational amplifier, the switch contains two field-effect transistors with driving circuits. In the shown position of the switch SPDT1, the voltage coming from the filter F1 is absolutely smaller than the one coming from the averager AV. If this voltage is absolutely larger, the switch SPDT1 must be in the lower position. The switch SPDT2 feeds either the voltage from the averager AV or from the amplitude reverser AR to SPDT1, depending on which voltage has the same polarity as the one coming from the filter FI. Let the voltages from FI and from AR have the same polarity, either positive or negative. The output voltages of the comparators COM2 and COM3 will then both be

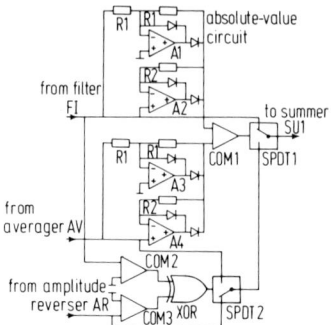

FIG. 36. Example of a circuit for the selection switches SE and the amplitude comparators AC in Fig. 34. The following components are commercially available as integrated circuits: (A) operational amplifier; (COM) voltage comparator; (XOR) exclusive OR gate; (SPDT) analog switch.

either +1 (positive saturation) or 0 (negative saturation). The output of the exclusive OR gate XOR will be 0 in both cases and the switch SPDT2 will remain in the lower position. If the voltages from FI and AR have opposite polarity, the exclusive OR gate XOR will receive one input 0 and one input 1 from the comparators COM2 and COM3. The output of XOR will be 1, and the switch SPDT2 will be turned to the upper position.

Figure 37 shows a simple block diagram for the super-resolution of a one-dimensional array with 16 hydrophones. The output voltages of the hydrophones pass through filters FI as in Fig. 34. The averager AV in Fig. 34 produced the beam pattern for the observation angle $\gamma = 0$ only. Hence, it is replaced in Fig. 37 by the beam former BF which produces 16 observation angles. This beam former contains either the delay circuit of Fig. 4 or the Fourier transform circuits of Figs. 12 and 13. The 16 output voltages of the beam former BF are fed directly and via amplitude reversers AR0 ··· AR15 to 16 *super-resolvers* SR0 ··· SR15. Each one of these super-resolvers contains 16 selection switches SE and amplitude comparators AC according to Fig. 36. In addition, each super-resolver contains one summer SU, amplitude reverser AR1, sampling switch SA$^-$ and sampling switch SA$^+$, as shown in Fig. 34. Each super-resolver in Fig. 37 produces one output voltage $v(\beta)$ according to Fig. 34; these voltages are denoted $V_0(\beta)$ to $V_{15}(\beta)$.

Let us now try to transform Fig. 37 into two dimensions. An array of 16 × 16 hydrophones would be connected to 16 × 16 filters FI. The beam former BF for two dimensions would typically consist of the two-dimensional sequency filter and the quadrant ambiguity resolver of Fig. 16. The 16 × 16 output voltages would be fed directly and via 16 × 16 amplitude reversers to 16 × 16 super-resolvers. Each one of the super-resolvers

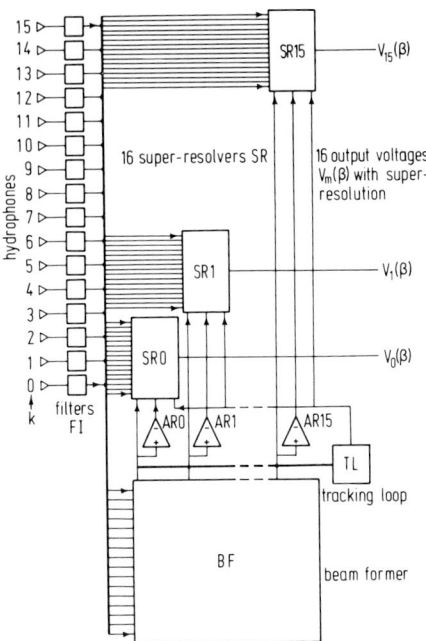

FIG. 37. Super-resolution for a one-dimensional array of 16 hydrophones. The beam former BF contains either the circuit of Fig. 4 or the circuits of Figs. 12 and 13. Each super-resolver contains 16 circuits SE and AC according to Fig. 36.

would contain 16^2 times the circuit of Fig. 36. In addition, each super-resolver would contain one summer SU, amplitude reverser AR1, sampling switch SA^- and sampling switch SA^+ according to Fig. 34. The 16 × 16 output voltages of the super-resolvers would be fed to the light-emitting diode display in Fig. 16.

One may readily conclude from this discussion that super-resolution by means of sampled arrays is presently well within our technological capabilities for one-dimensional arrays. Such arrays with a need for super-resolution exist in underwater acoustics and in over-the-horizon radar. The use in radar calls for circuits that operate at about 10 MHz. This is a very high frequency for a circuit as shown in Fig. 36. On the other hand, the beam forming in radar by a delay circuit as shown in Fig. 4 is standard practice.

Applying super-resolution as discussed here to a two-dimensional array with 256 × 256 hydrophones appears at first sight a task far beyond our present technological capability. One seems to need 256^2 super-resolvers with 256^2 circuits each, according to Fig. 37. However, we do not need to produce the whole image at once. We can produce one line at a time, as discussed in Section II,C, or we can produce one point at a time as a television set does. One needs 256 super-resolvers with 256^2 circuits each, according to Fig. 36, to produce one line at a time, and only one such super-resolver to produce one point at a time. These numbers are well within our capabilities if the super-resolvers are implemented as integrated circuits. Let us observe that the amplifiers, comparators, switches, and the exclusive OR gate in Fig. 36 are already available as integrated circuits. Hence, it is a problem of sufficient demand and not of technology to produce the whole circuit in integrated form.

It was stated without proof that only one tracking loop TL is needed in Fig. 37, even if one extends the circuit to two dimensions. One may see from Eq. (12) that this statement is correct. The output voltage of every beam former will have the time variation of $\sin 2\pi vt/\lambda$, only the amplitude will depend on γ_x, γ_y, β_x and β_y. Hence, the sum of the output voltages of many beam formers must have the time variation $\sin 2\pi vt/\lambda$ too. It might be theoretically possible that the sign of the amplitude changes, which would appear to the tracking loop like a phase shift of 180°, but this would be of no consequence if samples are taken in the neighborhood of the zero crossings of $\sin 2\pi vt/\lambda$ with positive and negative slope.

V. Implementation of Filters by Digital Circuits

A. Digital Delay Circuits

Let the output voltages of the hydrophones in Fig. 1 be fed to 16×16 analog-to-digital converters that convert at a sufficiently high sampling rate. We may then use digital circuits for processing instead of the analog circuits used so far. The delay circuits may be replaced by digital shift registers. Each delay circuit has to be replaced by r shift registers if the analog-to-digital converter produces r binary digits per sample. The analog circuits had a dynamic range for the voltage of about 1000 : 1. Digital circuits would require $r = 10$ binary digits to obtain the same dynamic range ($2^{10} = 1024$). Hence, the circuit of Fig. 4 calls for $10 \times 16 = 160$ shift registers, while the two-dimensional circuit of Fig. 5 requires $160 \times 16 \times 2 = 5120$ shift registers.

To get a rough estimate of the number of stages in the shift registers let us calculate the delay time in Fig. 4 between the taps of the delay circuit connected to hydrophone 6, and leading to the summing amplifiers denoted $-7\lambda/A$ and $-6\lambda/A$:

$$t_0 + (6 - 2^4/2 + 1)(d/v) \sin(-7\lambda/A)$$
$$- [t_0 + (6 - 2^4/2 + 1) \sin(-6\lambda/A)]$$
$$= (d/v)[\sin(7\lambda/A) - \sin(6\lambda/A)] \doteq d\lambda/vA.$$

This is the smallest incremental delay required, which means that each stage of the shift registers should produce that much delay. The total delay of a delay circuit divided by $d\lambda/vA$ yields the number of stages required. The total delay of the circuit connected to hydrophone zero can be expressed in two ways. Counting the number of taps one sees that it must be equal to $15(t_0/8)$. The delay can also be expressed by $t_0 + (0 - 4^4/2 + 1)(d/v) \sin(-7\lambda/A)$. Hence, one can calculate t_0:

$$15(t_0/8) = t_0 + (0 - 4^4/2 + 1)(d/v) \sin(-7\lambda/A);$$

$$t_0 = 8(d/v) \sin(7\lambda/A) \doteq 56d\lambda/vA.$$

All delay lines in Fig. 4 require a delay between t_0 and little more than $2t_0$. The average delay is thus $1.5t_0$, and the average number of stages per shift register equals $1.5 \times 56 = 84$. The total number of shift register stages in Fig. 4 is thus about $160 \times 84 = 13{,}440$, and about $5120 \times 84 = 430{,}080$ in Fig. 5. These numbers are very large. The problem of cost becomes even more impressive when one considers the $16 \times 16 = 256$ analog-to-digital converters in front of the delay circuits in Fig. 5. As a result, digital delay circuits are currently important for one-dimensional arrays with a low dynamic range only. Typically, the value $r = 1$ is used, which means a positive voltage is converted into the digit 1 and a negative voltage into the digit 0. The analog-to-digital converter becomes in this case a simple threshold detector.

Let us see whether the implementation of the inverse Fourier transform by digital circuits is more promising.

B. Digital Circuits for the Inverse Fourier Transform

Figure 38 shows the digital version of the circuit of Fig. 12. The major difference between the two circuits is the summers. Analog circuits may readily sum four voltages at a time, but digital circuits typically add only two numbers at a time, hence the need for almost twice as many adders $(+\,+)$ or subtractors $(+\,-)$ as there are operational amplifiers in Fig. 12. The multiplications performed by resistors in Fig. 12 require digital multipliers, represented by the blocks with the notation $\times p$, $\times q$, and $\times r$. Digital multipliers are rather expensive, and the cost of the circuit of Fig. 38 is determined by the few multipliers, not by the many adders and subtractors.

At the present, the circuit of Fig. 38 is too expensive to be acceptable. However, it is perfectly suited for large-scale integration and it is quite possible that it will be one day more practical than the analog circuit of Fig. 12.

A digital implementation of one of the circuits for $l = 1, 2, \ldots$ in Fig. 13 is shown in Fig. 39. The integration is performed by adding an incoming

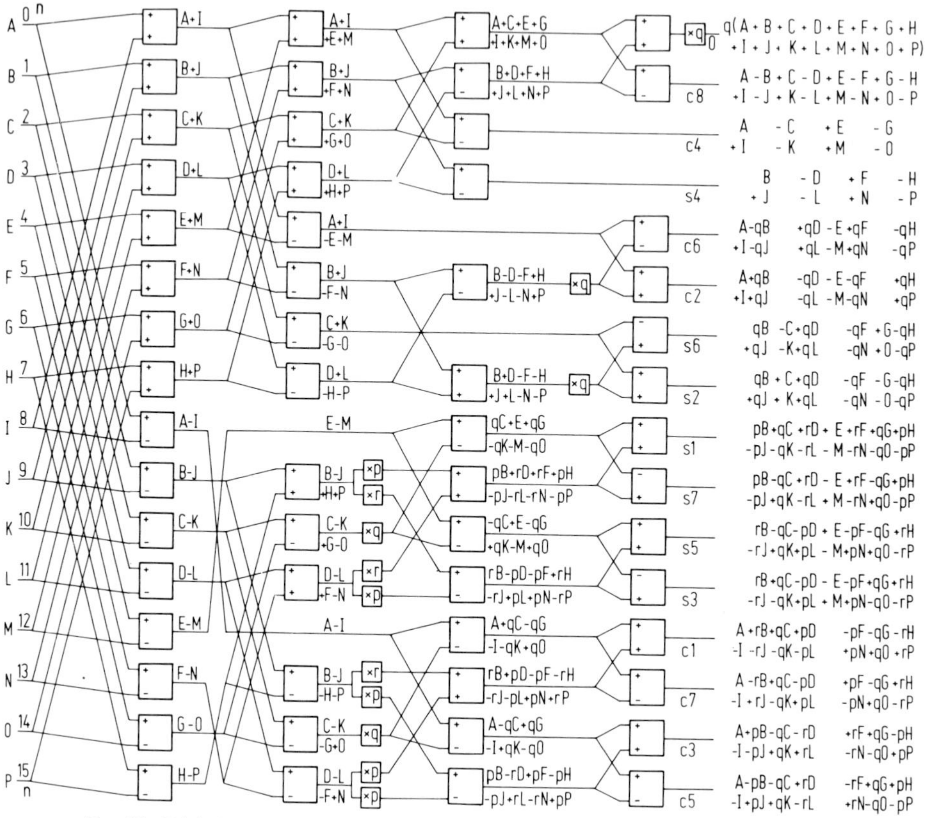

FIG. 38. Digital circuit performing the fast Fourier transform of 16 input voltages. The circuit blocks with two positive signs (+ +) are adders, those with a positive and a negative sign (+ −) subtractors, those with a multiplication sign (× p, × q, × r) multipliers. $p = \sin \pi/8$, $q = \sin \pi/4$, $r = \sin 3\pi/8$.

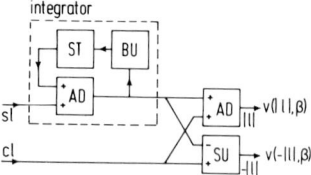

FIG. 39. Digital version of one of the circuits for $l = 1, 2, \ldots$ in Fig. 13. (AD) adder; (SU) subtractor; (ST) storage; (BU) buffer storage.

number and a number stored in storage ST by the adder AD. The sum of the two numbers is fed back via a buffer storage BU to the storage ST. Hence, the storage ST contains the sum of all received numbers.

One can use time sharing to make the digital circuits of Figs. 38 and 39 less expensive. This is what is done if the inverse Fourier transform is performed by computer processing. A new problem is encountered. If the acoustic wave used has a frequency of 100 kHz one should take upwards of 200,000 samples per second and convert them into digital numbers. This is a very fast rate for a digital circuit using multipliers, and the rate is prohibitive if one wants to time share the multipliers. Synchronous demodulation permits operation at much slower rates.

C. Synchronous Demodulation as the Key to Practical Circuits

The output voltage $v(k)$ of a hydrophone k according to Eq. (14),

$$v(k) = V \sin [2\pi(vt + kd \sin \beta)/\lambda] = V \sin 2\pi(ft + kd\lambda^{-1} \sin \beta), \quad (43)$$

is multiplied by a voltage $e(t)$ with amplitude 2:

$$e(t) = 2 \sin 2\pi f_1(t - t_0). \quad (44)$$

One obtains:

$$v(k)e(t) = V\{\cos 2\pi[(f - f_1)t + f_1 t_0 + kd\lambda^{-1} \sin \beta]$$
$$- \cos 2\pi[(f + f_1)t - f_1 t_0 + kd\lambda^{-1} \sin \beta]\}. \quad (45)$$

Let the frequencies f and f_1 be about equal. The difference frequency $f - f_1 = f'$ is then very small, while the sum frequency $f + f_1$ equals about $2f$. One may suppress the term with the sum frequency in Eq. (45) by a simple filter. Stray capacitances of the circuits will usually do this filtering. The remaining low-frequency term is denoted $v'(k)$:

$$v'(k) = V \cos 2\pi[f'(t + t'_0) + kd\lambda^{-1} \sin \beta]$$
$$= V \sin [2\pi(v't' + kd \sin \beta)/\lambda]; \quad (46)$$
$$f' = f - f_1, \qquad t'_0 = t_0 f_1/f', \qquad v' = \lambda f' = v(1 - \lambda/\lambda_1),$$
$$t' = t + t'_0 + 1/4f', \qquad \lambda f = v, \qquad \lambda_1 = v/f_1.$$

A comparison of Eq. (46) with Eq. (43) shows that we have the same voltage except that the frequency f is replaced by f', the velocity v by v' and the time variable t by t'. The results of Eqs. (14)–(24) remain unchanged except for the replacement of the time-variable terms $\sin 2\pi vt/\lambda$ and $\cos 2\pi vt/\lambda$ by $\sin 2\pi v't'/\lambda$ and $\cos 2\pi v't'/\lambda$.

The reduction of the operating frequency f in the water to the lower processing frequency f' in the equipment has obvious advantages. One can choose f to be 100 kHz to keep the hydrophone array small, but convert f to a frequency f' of the order of a few hundred or thousand hertz to avoid the need for shielded wiring, and to improve the performance of the operational amplifiers in the circuits of Figs. 12 and 13. However, one must be somewhat careful not to choose the processing frequency f' too low, lest the integrators in Fig. 13, as well as certain filters to be discussed later on, become too bulky. If super-resolution is to be used, one cannot convert to the operating frequency f' ahead of the super-resolution circuits, since this would reduce the slope of the sinusoidal voltages at the zero crossings from $\pm 2\pi f V$ to $\pm 2\pi f' V$.

When one uses digital circuits one does not have to be concerned about bulkiness caused by a low processing frequency. Hence, one may reasonably choose $f_1 = f$ and convert the sinusoidal output voltages of the hydrophones into dc voltages. Let us consider $v'(k)$ in the form shown by Eq. (45) for $f_1 \to f$:

$$v'(k) = V[\cos 2\pi(f - f_1)t \cdot \cos 2\pi(f_1 t_0 + kd\lambda^{-1} \sin \beta) \\ - \sin 2\pi(f - f_1)t \cdot \sin 2\pi(f_1 t_0 + kd\lambda^{-1} \sin \beta)]. \tag{47}$$

For $f - f_1 = 0$, the first term becomes

$$v'(k) = V \cos 2\pi(f_1 t_0 + kd\lambda^{-1} \sin \beta), \tag{48}$$

but the second term vanishes, and with it the value of $\sin 2\pi(f_1 t_0 + kd\lambda^{-1} \sin \beta)$. One cannot reconstruct $v'(k)$ of Eq. (47) without the knowledge of $\sin 2\pi(f_1 t_0 + kd\lambda^{-1} \sin \beta)$. Hence, Eq. (48) contains less information than Eq. (47). To regain the lost term we multiply Eq. (43) with the voltage $e'(t)$

$$e'(t) = 2 \cos 2\pi f_1(t - t_0). \tag{49}$$

We obtain

$$v(k)e'(t) = V\{\sin 2\pi[(f - f_1)t + f_1 t_0 + kd\lambda^{-1} \sin \beta] \\ + \sin 2\pi[(f + f_1)t - f_1 t_0 + kd\lambda^{-1} \sin \beta]\}. \tag{50}$$

The low-frequency term is denoted $v''(k)$

$$v''(k) = V[\sin 2\pi(f - f_1)t \cdot \cos 2\pi(f_1 t_0 + kd\lambda^{-1} \sin \beta) \\ + \cos 2\pi(f - f_1)t \cdot \sin 2\pi(f_1 t_0 + kd\lambda^{-1} \sin \beta)]. \tag{51}$$

Now the first term vanishes for $f - f_1 = 0$ and we obtain the term lost by Eq. (48)

$$v''(k) = V \sin 2\pi(f_1 t_0 + kd\lambda^{-1} \sin \beta). \tag{52}$$

In the present context the term $f_1 t_0$ in Eqs. (48) and (52) is undesirable. One may eliminate it by using synchronized voltages $e(t)$ and $e'(t)$ in Eqs. (44) and (49). For instance, one may use $v(k)$ for $k = 0$ in Eq. (43) for $e(t)$, and produce $e'(t)$ by a phase shift of $v(0)$. It was discussed in Sections IV,B–D how one can obtain voltages with the same phase as $v(0)$ but with a better signal-to-noise ratio. Hence, we will accept without further discussion that $v'(k)$ and $v''(k)$ can be produced without the term $f_1 t_0$:

$$v'(k) = V \cos (2\pi k d \lambda^{-1} \sin \beta), \qquad v''(k) = V \sin (2\pi k d \lambda^{-1} \sin \beta). \qquad (53)$$

We may now multiply $v'(k)$ by $\sqrt{2}/2$ and $\cos [2\pi k(d/\lambda) \sin \gamma]$ from Eq. (13), while $v''(k)$ is multiplied by $\sin [2\pi k(d/\lambda) \sin \gamma]$, and produce the sums of Eqs. (15)–(17). However, one may just as well add $v'(k)$ and $v''(k)$, multiply and sum, since the sums of the terms

$$V \cos [2\pi k(d/\lambda) \sin \beta] \sin [2\pi k(d/\lambda) \sin \gamma],$$

$$V \sin [2\pi k(d/\lambda) \sin \beta] \cos [2\pi k(d/\lambda) \sin \gamma]$$

and $(\sqrt{2}/2)V \sin [2\pi k(d/\lambda) \sin \gamma]$ vanish. One obtains

$$v'(0, \beta) = \tfrac{1}{2}\sqrt{2} \sum_{k=0}^{2^n-1} [v'(k) + v''(k)]$$

$$= \tfrac{1}{2}\sqrt{2} \sum_{k=0}^{2^n-1} \cos (2kd\lambda^{-1} \sin \beta);$$

$$v'(\gamma, \beta, s) = \sum_{k=0}^{2^n-1} [v'(k) + v''(k)] \sin (2\pi k d\lambda^{-1} \sin \gamma) \qquad (54)$$

$$= V \sum_{k=0}^{2^n-1} \sin (2\pi k d\lambda^{-1} \sin \beta) \sin (2\pi k d\lambda^{-1} \sin \gamma);$$

$$v'(\gamma, \beta, c) = \sum_{k=0}^{2^n-1} [v'(k) + v''(k)] \cos (2\pi k d\lambda^{-1} \sin \beta)$$

$$= V \sum_{k=0}^{2^n-1} \cos (2\pi k d\lambda^{-1} \sin \beta) \cos (2\pi k d\lambda^{-1} \sin \gamma).$$

A comparison with Eqs. (15)–(17) and the formulas in the Appendix shows that $v'(0, \beta)$, $v'(\gamma, \beta, s)$, and $v'(\gamma, \beta, c)$ are identical with $v(0, \beta)$, $v(\gamma, \beta, s)$, and $v(\gamma, \beta, c)$, except that the time-variable terms $\sin (2\pi v t/\lambda)$ and $\cos (2\pi v t/\lambda)$ have to be replaced by 1. As a result, the integration of Eq. (22) is not needed to obtain the voltages $v(l, \beta)$ and $v(-l, \beta)$ of Eqs. (23) and (24), but without the time-variable term $\sin(2\pi vt/\lambda)$:

$$v'(0, \beta) = \frac{1}{2}\sqrt{2}\, V \frac{A \sin(\pi A \lambda^{-1} \sin \beta)}{d \quad \pi A \lambda^{-1} \sin \beta};$$

$$v'(l, \beta) = v'(\gamma, \beta, c) + v'(\gamma, \beta, s) = V \frac{A}{d} \frac{\sin \pi(A\lambda^{-1} \sin \beta - l)}{\pi(A\lambda^{-1} \sin \beta - l)} \quad (55)$$

$$v'(-l, \beta) = v'(\gamma, \beta, c) - v'(\gamma, \beta, s) = V \frac{A}{d} \frac{\sin \pi(A\lambda^{-1} \sin \beta + l)}{\pi(A\lambda^{-1} \sin \beta + l)}$$

$$l = A\lambda^{-1} \sin \gamma \quad \text{for} \quad \gamma > 0,$$
$$-l = A\lambda^{-1} \sin \gamma \quad \text{for} \quad \gamma < 0.$$

Let us discuss the practical difference between the conversion of the operating frequency f to a lower processing frequency $f - f_1$ and the synchronous demodulation of f to dc. The one multiplier per hydrophone required for frequency conversion is shown in Fig. 40a. No filter is needed to suppress the term with the frequency $f + f_1$. The output voltage $v'(k)$ is fed to the input terminals in Fig. 16 or—after analog-to-digital conversion—to the input terminals in Fig. 38. The integrators in Figs. 13, 15, and 39 are required.

Synchronous demodulation requires the two multipliers in Fig. 40b and the summer. In return, the integrators in Figs. 13, 15, and 39 are eliminated.

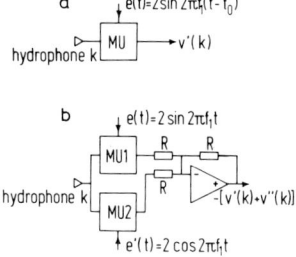

FIG. 40. (a) Conversion of the operating frequency to a lower processing frequency; (b) synchronous demodulation to dc. (MU) multiplier.

In essence one trades a multiplier for an integrating capacitor. The main objection to synchronous demodulation is the need to provide synchronized voltages $e(t) = 2 \sin 2\pi f_1 t$ and $e'(t) = 2 \cos 2\pi f_1 t$. There is little incentive to use dc instead of ac with a frequency of a few hundred to a few thousand hertz, if the Fourier transform is done with an analog circuit as in Fig. 12. The same holds true if a parallel digital circuit as in Fig. 38 is used. However, if this circuit is used with time-shared multipliers, adders, and subtractors, a factor between a few hundred and a few thousand translates directly into an increase in processing time by the same factor. In this case one will definitely prefer synchronous demodulation to frequency conversion.

The term synchronous demodulation has been used in electrical engineering for upwards of half a century. Optical holography is in essence an implementation of synchronous demodulation at the frequencies of coherent light. The word holography was then transferred from optics to acoustics. Holography in acoustics may be implemented by electric voltages as discussed here under the name synchronous demodulation. This is done, e.g., in the equipment developed by Booth and Sutton (1974) or by Marom et al. (1971). Let us observe that square waves can and often are used instead of the sinusoidal voltages with the same period in Eqs. (44) and (49), since the harmonics are readily suppressed by stray capacitances. There are, however, other ways to implement acoustical holography for which the term synchronous demodulation would not mean much. These methods typically use interference patterns produced by the sound waves on the interface between two materials, and a laser beam. Examples of this technique are discussed in papers by Mueller and Keating (1969) and Fritzler et al. (1969).

VI. Beyond the Capabilities of Photography and Holography

A. Range Images and Three-Dimensional Images

We have seen in the chapter on the reduction of the array size that electric filters with a time-variable element can achieve a resolution limited only by the signal-to-noise ratio rather than by the classical wavelength-to-aperture ratio. This was an example in which electric filters went beyond the capabilities of photography and holography.

Photography by means of a lens is mainly useful for light waves. The physical dimensions of lenses have restricted their use in acoustics or radar. The lens produces transformations that are almost exclusively based on delay and summation. Multiplication is used only in the simple forms provided by masks and wavelength-selective filter glasses. More general linear and nonlinear operations are currently being developed under the name optical processing and should eventually become applicable for image generation and processing. Time-variable devices that can be changed in times short compared with the period of the used light are beyond the foreseeable development of our technology.

Holography is a means to obtain and store the amplitudes and phases of wavefronts. In the case of light waves one may use this stored information to generate an image or one may use it as input for optical processing. There is no equivalent to optical processing in acoustics since acoustic lenses are so cumbersome compared with optical lenses.

The main feature of electrical processing of acoustic waves is the possibility of using time-variable devices that can be changed in a time that is short compared with the period of the wave. A secondary feature, compared with optical processing, is the advanced technology of electronics. Amplification, variation of the frequency, time filtering, amplitude clipping, and so forth can be done more readily and more accurately in electronics than in optics. Of course, electronic components do not work as fast as optical devices, but this is of little concern in acoustics.

Super-resolution is one application exploiting rapid time variation. A second one is the generation of range images or three-dimensional images. According to Fig. 41 a projector radiates a sinusoidal pulse with n periods of

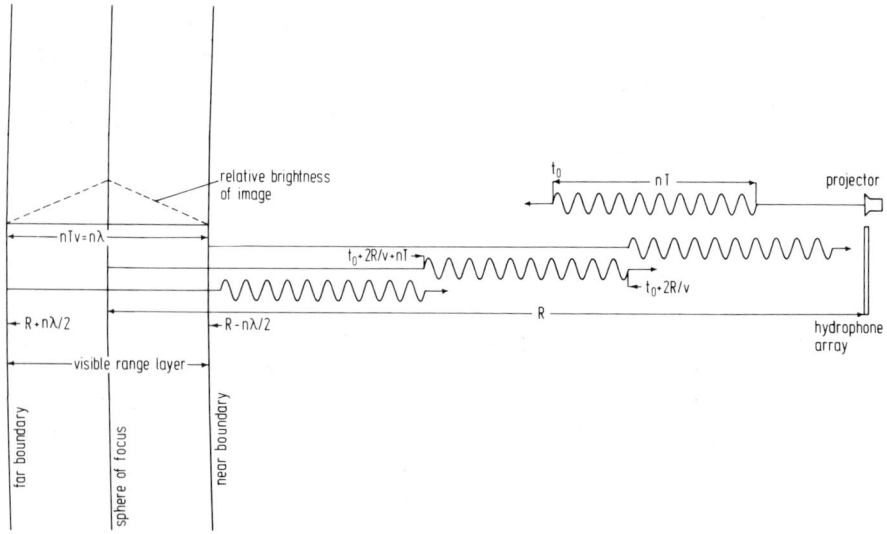

FIG. 41. Generation of the image of a spherical layer of thickness $n\lambda$ by means of pulsed sinusoidal waves with n cycles per pulse.

duration T rather than a continuous wave. The projector is connected via a switch to the power oscillator at the time t_0 and disconnected at the time $t_0 + nT$. The pulse travels to a scatterer at a distance R and returns. The front of the pulse arrives at the time $t_0 + 2R/v$ at the hydrophone array, the end of the pulse at the time $t_0 + 2R/v + nT$. Let the hydrophones be connected during the time $t_0 + 2R/v < t < t_0 + 2R/v + nT$ to the processing circuit. The waves scattered on a surface of a sphere of radius R will be processed as if there had been no switches. Waves scattered at a distance $r < R$ will have their "heads" cut off by the switches, and pulses scattered at a distance $r > R$ will have their "tails" cut off. Waves scattered at distances

$r < R - n\lambda/2$ or $r > R + n\lambda/2$ will be suppressed completely. These distances are denoted *near boundary* and *far boundary* in Fig. 41. The fraction of the wave returned from the region between these boundaries will vary like the triangle denoted *relative brightness*.

Let the pulse have $n = 100$ periods and let the period be $T = 10$ μsec, corresponding to a frequency of 100 kHz for the continuous sinusoidal wave. The thickness of the *visible range layer* will then be $nTv = n\lambda = 1.5$ m. The advantage of producing only an image of this layer is the suppression of returns from other scatterers such as air bubbles at different distances that would make the image appear "foggy." The optical equivalent of a range image would permit us to see through fog. For instance, shutters in front of the headlights of a car synchronized with shutters in front of the eyes of the driver would permit the driver to see only the light scattered at a certain distance, but suppress the light scattered by the fog at closer or larger distances.

A slight modification permits one to obtain three-dimensional images from range images. Let the projector in Fig. 41 again send out pulses with n periods, but let no switches be used at the output terminals of the hydrophone array. Instead, the optical display—shown as a light-emitting diode display in Fig. 42—is projected onto a moving ground-glass plate. The

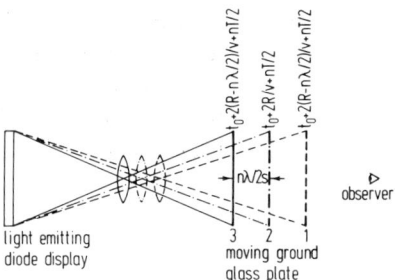

FIG. 42. Generation of a genuine three-dimensional image by means of pulsed sinusoidal waves and a display on a moving ground-glass plate.

return from the near boundary in Fig. 41 produces an image during the time interval $t_0 + 2(R - n\lambda/2)/v < t < t_0 + 2(R - n\lambda/2) + nT$. The peak brightness will occur at the time $t_0 + 2(R - n\lambda/2)/v + nT/2$. The moving ground-glass plate in Fig. 42 is at that time in the position 1. At the time $t_0 + 2R/v + nT/2$ the image of the sphere of focus in Fig. 41 will appear with peak brightness; the moving ground-glass plate in Fig. 42 will be in position 2. At the time $t_0 + 2(R + n\lambda/2)/v + nT/2$ the image of the "far boundary" in Fig. 41 will appear with peak brightness, and the moving ground-glass plate in Fig. 42 will be in position 3. Hence, images of objects at different distances from the hydrophone array will be seen on the ground-glass plate at different distances from the observer.

The position of the ground-glass plate in Fig. 42 must be properly scaled to the position of the imaged layers. A linear reduction of $1/s$ means that the position of the ground-glass plate in Fig. 42 must change $n\lambda/2s$ if the distance of the imaged layer in Fig. 41 is changed by $n\lambda/2$. If the ground-glass plate is moved physically, its velocity must be v/s if v is the velocity of sound in water.

The displayed image in Fig. 42 is a genuine three-dimensional image. The observer can move relative to the display to see the three-dimensional image from various angles. This is different from the stereoscopic pictures in photography that have a fixed viewing angle, and which require that one eye see one two-dimensional image and the other eye another two-dimensional image. Optical holography, on the other hand, also yields genuine three-dimensional images, but by a different process.

The movement of the ground-glass plate in Fig. 42 does not have to be made mechanically. One can substitute many thin layers of a nematic liquid crystal. One layer at a time can be made translucent by applying a voltage across it; the other layers can be made transparent by applying no voltage or only a very small voltage. This layer then acts like a ground-glass plate. The mechanical movement is thus replaced by an electrical movement. The movement of the lens can be avoided by making the ground-glass plate much larger than the light-emitting diode display and by using a mask with a small focal opening. This is basically the same problem as increasing the depth of field of a camera. A fixed lens would make an image on the ground-glass plate in position 1 in Fig. 42 appear larger than in position 2 or 3 and thus produce a perspective image.

B. *Doppler Images*

Let an object approach with the velocity v_r relative to a projector and a hydrophone array. A sinusoidal wave with frequency f produced by the projector will be returned to the hydrophones with a frequency $f(1 + 2v_r/v)$, where v is the velocity of sound in water. For $f = 100$ kHz the frequency shift $\Delta f = 2fv_r/v$ equals 1 kHz for $v_r = 7.5$ m/sec $= 14.6$ knots. A resonance filter for 100 kHz with a bandwidth of about 1 kHz can be built fairly easily. Let 16×16 filters be inserted in Fig. 16 between the quadrant ambiguity resolver and the light-emitting diode display. Phase distortions caused by the filters are of no concern if they are inserted at this place. Let the projector radiate a wave with frequency 99 kHz. The image produced by a stationary object $v_r = 0$ will be blocked by the filters and will not be displayed. An object approaching with a velocity $v_r = 7.5$ m/sec will return a wave with frequency 100 kHz. The image of this object will not be blocked by the filters. Hence, by changing the frequency of the insonifying wave one can make objects with certain velocities visible and suppress others.

What is the limit for the Doppler resolution? We have seen in the section on synchronous demodulation that a sinusoidal voltage with a frequency of 100 kHz can readily be converted to a sinusoidal voltage with a much lower frequency. The frequency Δf of the Doppler shift remains unchanged during such a conversion. Hence, instead of building a resonance filter at 100 kHz with an approximate bandwidth of 1 kHz, one can convert from 100 kHz to 10 kHz. A resonance filter with a bandwidth of 100 Hz can be built at 10 kHz with about as much effort as one with a bandwidth of 1 kHz at 100 kHz. A Doppler shift of 100 Hz is produced by a relative velocity $v_r = 0.75$ m/sec $= 1.46$ knots. There is no theoretical limit for the Doppler resolution, since one can convert the frequency of 100 kHz to 1 kHz, 100 Hz, etc. However, building good, small, and inexpensive resonance filters at such low frequencies is a major problem.

Range and Doppler imaging can be combined to yield images of objects at a certain distance with a certain velocity. The practical significance of such images is the elimination of the images of air bubbles, of boundary layers in water or of the sea bottom, which obscure the image of a wanted object. We have explained previously range images in terms of optics by means of shutters that permit one to look through fog. An optical equivalent of a range-Doppler image is to use in addition the Doppler effect to, e.g., make visible only cars that move with excessive velocity through fog. To do so with light waves is beyond our current technology, but there is no technological barrier to the use of range-Doppler images to look through fog or tropical foliage by means of sound waves.

C. Telelens Effect

The resolution angle $\varepsilon = \lambda/A$ according to Eq. (35) is a function of the wavelength λ for an array with aperture A. Figure 43 shows five representative beam patterns for the wavelength λ in the direction of integer multiples of $\pm \varepsilon$. Let a wave with the wavelength $\lambda/2$ be used, but let the array remain unchanged. The resolution angle is now $\varepsilon = \lambda/2A$ and the beam patterns shown by dashed lines apply. The change of the wavelength has the same effect as a change of the focal length of a lens in photography.

Let us see what influence a change of wavelength has on the equipment. Hydrophones can work over a fairly wide frequency range, particularly since the sensitivity does not have to be constant. The two-dimensional sequency filter in Fig. 16 contains cards with the circuit shown in Fig. 12. This circuit works from dc to about 100 kHz. The frequency sensitive circuit in Fig. 16 is

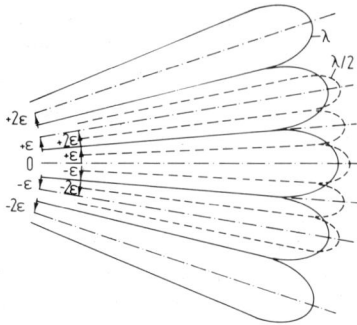

FIG. 43. Telelens effect for a fixed array of hydrophones and a variable wavelength of the insonificating waves.

the quadrant ambiguity resolver, which is shown in more detail in Fig. 15. By using multipliers for frequency conversion at the outputs of the hydrophone array as discussed in the section on synchronous demodulation, one can transform a wave with any wavelength into oscillations with fixed frequency. Hence, the circuit of Fig. 16 combined with frequency conversion is suitable to make practical use of the telelens effect.

One may derive from Fig. 43 a curious effect of imaging of active sound sources. Two sources at essentially the same location radiating waves with wavelengths λ and $\lambda/2$ will appear at different locations on the light-emitting diode display of Fig. 16. Furthermore, since the quadrant ambiguity resolver operates at one frequency only, the image of at least one of the active sources will appear in all four quadrants. Of course, both phenomena disappear as soon as one inserts narrow-band filters between the quadrant ambiguity resolver and the light-emitting diode display, as discussed in the section on Doppler images.

D. Pseudo-Color Images

We have so far assumed insonification by a sine wave with a certain frequency. In terms of photography this would mean illumination with monochromatic light. Such illumination would produce black-and-white pictures, although the shades of gray would differ from the ones obtained with the usual, nonmonochromatic light sources. To obtain color pictures in photography, we can illuminate either with white light or we can use three monochromatic light sources emitting red, green, and blue light, respectively. Substituting three monochromatic light sources for one source of white light is not practical in photography, but it is the proper way to produce color images by means of optical holography or spatial electric filters.

The equivalent of three monochromatic light sources in underwater acoustics is three projectors radiating sound waves with three different frequencies. One can choose these frequencies within the technical limitations of the equipment. This gives one more freedom than one has in photography or optical holography, where the light sources must emit the three colors for which the eye is sensitive. Only infrared photography provides some degree of independence between the frequency of the light used to produce a picture and the frequencies to which the eye is sensitive.

Using three acoustic waves with different frequencies will provide us with information about the frequency dependence of the scattering of the observed object and the frequency dependence of the absorption in the water. The absorption in seawater changes drastically with frequency and one will primarily get a range effect. The wave with the highest frequency will produce images primarily of near objects, while the waves with lower frequency will show this effect less strongly. We have discussed previously how images of objects at a certain distance can be produced by gating the projector and the hydrophones. Combining the use of three frequencies with gating eliminates the influence of the frequency dependence of absorption in the water on the final color of the image. The brightness of the image will still depend on the frequency, but this can be compensated by proper amplification since the distance to the object and thus the absorption in the water is known.

Let us now see how color can be used to discriminate between different materials of scatterers. Rayleigh derived the following expression for the relative amplitude* σ_i of a wave returned from an incompressible spherical scatterer of diameter D, if D is small compared with the wavelength λ:

$$\sigma_i = (D/\lambda)(4\pi/9)(4C_0^2 + 3C_1^2). \tag{56}$$

The constants C_0 and C_1 depend on the density and elasticity of the material of the sphere. The relative amplitude σ_c of a wave returned from a highly compressible air bubble of diameter D was also derived by Rayleigh:†

$$\sigma_c = (D/\lambda)^4 4\pi^4 v^4 (\rho/3\gamma p_0)^{1/2}; \tag{57}$$

v = sound velocity in water; ρ = density of water; p_0 = pressure of the incident wave; γ = ratio of specific heat at constant pressure to specific heat at constant volume (Albers, 1965).

Let us assume that a wavelength λ is used that yields the same value for σ_i and σ_c. An incompressible and a highly compressible scatterer yield in

---

* The term "scattering cross section" or "effective target area" is used for $(D^2\pi/4)\sigma_i$. This terminology would be misleading in image generation, since a point-like scatterer with large effective target area is represented on the display by a very bright point, not by a large bright area.

† Rayleigh's law of scattering uses $(D^2\pi/4)\sigma_c$ instead of σ_c.

this case the same brightness and cannot be distinguished. Let a second wave with shorter wavelength $\lambda' < \lambda$ be used. σ_c will increase much faster than σ_i with decreasing wavelength. Let the image produced by the wave with length λ be displayed in red and the image produced by the wave with length λ' in green. The image of the incompressible scatterer will then appear yellow, while that of the compressible scatterer will be green-yellow. Hence, pseudo-color can be used to discriminate between scatterers of different compressibility. As a practical example one may think of a skindiver with a foam-rubber suit of high compressibility against a background of clay, sand, or rock with low compressibility.

In addition to the intrinsic color of objects we have in optics the color of thin films. What is an equivalent for acoustic waves? Consider a ship bottom with a layer of marine growth 1 cm thick. The ship bottom will appear rough to a wave with a wavelength significantly shorter than 1 cm, but smooth to one with a wavelength significantly longer than 1 cm. Hence, pseudo-color can be used to show the surface structure of scatterers.

How can one produce pseudo-color images practically? To have three projectors, three hydrophone arrays, three processors, and one color TV tube would not be practical. Let us instead have three projectors radiating waves with frequencies f_r, f_g and f_b, but time-share the hydrophone array and the processor. To do so one may insert multipliers between the output terminals of the hydrophone array and the input terminals of the processor. Somewhere in the processor one needs narrow-band filters with center frequency f_0. During a first time interval of about 100-msec duration one feeds the voltage $V_r \sin 2\pi(f_r - f_0)t$ to the multipliers. The resulting voltages with frequency $f_r - (f_r - f_0) = f_0$ will pass through the filters, while the voltages $f_g \pm (f_r - f_0)$, $f_b \pm (f_r - f_0)$ and $f_r + (f_r - f_0)$ will be blocked. These voltages will be represented by a red image. One then applies the voltages $V_g \sin 2\pi(f_g - f_0)t$ for 100 msec to obtain a green image, and $V_b \sin 2\pi(f_b - f_0)t$ to obtain a blue image. The amplitudes V_r, V_g, and V_b are to be chosen so as to compensate the different attenuations of the three waves.

If the processor is using the delay principle according to Fig. 4 or the sampling principle according to Fig. 17 one will see all scatterers at the same location in the red, green, and blue images. This is not so if one uses the Fourier transform according to Figs. 12–16. We have seen from Fig. 43 that a telelens effect is produced by this circuit if the frequency of the wave is changed. However, this telelens effect can be eliminated by using super-resolution according to Figs. 30–37 to reduce the beams for the wavelength λ in Fig. 43 to those for the wavelength $\lambda/2$. This application of super-resolution gives us a fringe benefit. Let the frequencies f_r, f_g, and f_b be used, where f_b is a higher frequency than f_g, and f_g a higher frequency than f_r. Let the sampling time $\tau = T/4$ be used for the frequency f_b. This means, according to Fig. 31, that the sinusoidal voltage for $i = 0$ is sampled at its peak. No super-resolution, but the best signal-to-noise ratio, will be obtained. The voltages with frequency f_g must be sampled at a smaller value of τ, and those with frequency f_r at a still smaller value of τ. Hence, the signal-to-noise ratio is reduced for f_g and f_r, but the waves with lower frequency are less attenuated in water and thus have an inherently better signal-to-noise ratio.

E. Focusing for Spherical Wavefronts Revisited

Let us return to Fig. 25. The output voltages of the hydrophones are rewritten:

$$v(i) = V \sin [2\pi(vt + id \sin \beta - i^2 d^2/2R)/\lambda \\ = V \sin 2\pi(ft + id\lambda^{-1} \sin \beta - i^2 d^2/2R\lambda). \tag{58}$$

In analogy to Eq. (44) we multiply with the voltage $e(i, t)$:

$$e(i, t) = 2 \sin 2\pi[f_1(t - t_0) - i^2 d^2/2R\lambda]. \tag{59}$$

These voltages can be produced from a voltage $2 \sin 2\pi f_1(t - t_0)$ by adding phase shifts $2\pi i^2(d^2/2R\lambda)$ with $i = 1, 2, \ldots$.

Let $v(i)$ be multiplied by $e(i, t)$. The term with the sum frequency $f + f_1$ shall be suppressed. The term with the difference frequency $f' = f - f_1$ is denoted $v'(i)$:

$$v'(i) = V \cos 2\pi(f't' + id\lambda^{-1} \sin \beta); \\ f' = f - f_1, \quad t' = t + t_0 f_1/f'. \tag{60}$$

The term $i^2(d^2/2R\lambda)$ is eliminated, and Eq. (60) is identical with Eq. (46). Hence, if synchronous demodulation or frequency conversion is used one may obtain focusing by adding the phase shifts $2\pi i^2(d^2/2R\lambda)$ to the multiplying voltages.

Let us elaborate this method for two-dimensional Cartesian coordinates. In polar coordinates the circle denoted sphere of focus in Fig. 25 becomes a sphere, but in Cartesian coordinates it becomes a cylinder. This cylinder is denoted *cylinder of focus for x* in Fig. 44. For the vertical coordinate one obtains the *cylinder of focus for y*. The image of any straight line on either cylinder will be in focus. Two such lines are shown in Fig. 39 as *horizontal line in focus* and *vertical line in focus*. The image of the point where these two lines intersect will be in focus. The two curves denoted *locus of focused points* represent all the points that can be in focus. The surfaces of the horizontal and the vertical cylinder represent all the horizontal and the vertical lines that can be in focus.

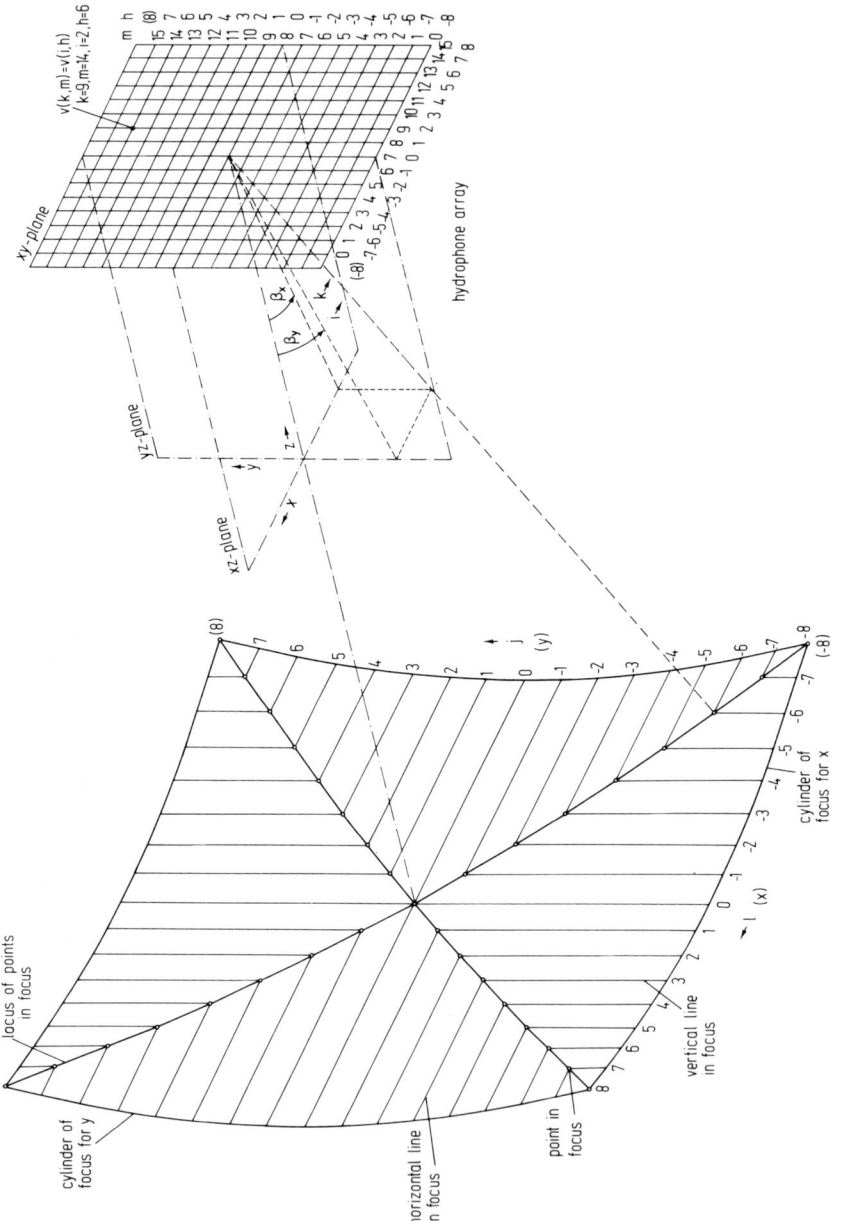

FIG. 44. Focusing in Cartesian coordinates by means of the multiplication principle.

These principles of optics in Cartesian coordinates are somewhat confusing when first encountered, since optics is universally based on polar coordinates. The reason is, of course, that a lens with rotational symmetry is the easiest one to grind, and this method of production suggests the use of a circular rather than a square piece of glass. The optical equivalent of our image generation in Cartesian coordinates would be to use a cylindrical lens with horizontal axis followed by a cylindrical lens with vertical axis instead of one spherical lens.

Equation (58) is readily generalized for two dimensions by following the transition from $v(k)$ in Eq. (2) to $v(k, m)$ in Eq. (9):

$$v(i, h) = V \sin [2\pi(vt + id \sin \beta_x$$
$$+ hd \sin \beta_y - i^2 d^2/2R - h^2 d^2/2R)/\lambda]$$
$$= V \sin 2\pi [ft + id\lambda^{-1} \sin \beta_x + hd\lambda^{-1} \sin \beta_y$$
$$- i^2 d^2/2R\lambda - h^2 d^2/2R\lambda]. \tag{61}$$

We multiply first with a sinusoidal voltage $e_x(i, t)$ having the frequency

$$f_0 = (f - f')/2 \tag{62}$$

and the proper phase shift for the x direction:

$$e_x(i, t) = 2 \sin 2\pi [f_0(t - t_0) - i^2 d^2/2R\lambda]. \tag{63}$$

Then we multiply with a voltage $e_y(h, t)$ of the same frequency but the proper phase shift for the y direction:

$$e_y(h, t) = 2 \cos 2\pi [f_0(t - t_0) - h^2 d^2/2R\lambda]. \tag{64}$$

The product $v(i, h)e_x(i, t)$ contains a term $v_x(i, h)$ with the difference frequency $f - f_0 = \tfrac{1}{2}(f + f')$:

$$v_x(i, h) = V \cos 2\pi [\tfrac{1}{2}(f + f')t_x + id\lambda^{-1} \sin \beta_x$$
$$+ hd\lambda^{-1} \sin \beta_y - h^2 d^2/2R\lambda]; \tag{65}$$
$$t_x = t + t_0 f_0/(f - f_0) = t + t_0(f - f')/(f + f').$$

The product $v(i, h)e_x(i, t)$ also contains a term with the sum frequency

$$f + f_0 = \tfrac{1}{2}(3f - f'). \tag{66}$$

Let $v_x(i, h)$ be multiplied now by $e_y(h, t)$. The product contains a term $v_{xy}(i, h) = v'(i, h)$ with the difference frequency f':

$$v'(i, h) = V \cos 2\pi [f't' + id\lambda^{-1} \sin \beta_x + hd\lambda^{-1} \sin \beta_y];$$
$$f' = \tfrac{1}{2}(f + f') - \tfrac{1}{2}(f - f'), \qquad t' = t + t_0(f - f')/f'. \tag{67}$$

In addition to the term with the difference frequency f' one also obtains a term with the sum frequency

$$\tfrac{1}{2}(f+f') + \tfrac{1}{2}(f-f') = f. \tag{68}$$

Furthermore, one also obtains two terms with the frequency of Eq. (65) plus and minus the frequency f_0:

$$(3f-f')/2 + (f-f')/2 = 2f - f'; \tag{69}$$

$$(3f-f')/2 - (f-f')/2 = f. \tag{70}$$

The frequency f' is zero for synchronous demodulation and of the order of 1 kHz for frequency conversion, while the frequency f is of the order of 100 kHz. The terms with frequency f and $2f - f'$ are thus readily suppressed by the stray capacitances of the circuit.

Equation (67) is identical with Eq. (9), except that the variables i and h are used in place of k and m. Hence, we have succeeded in solving the focusing problem for Cartesian coordinates. In addition, we have succeeded in doing so first for the variable x, then for the variable y. The practical importance of this focusing in two steps is readily comprehended. $i = k - 7$ and $h = m - 8$ in Fig. 16 assume only the nine absolute values 0, 1, 2, 3, 4, 5, 6, 7, 8. This means that only eight phase shifts $2\pi(1^2 d^2/2R\lambda)$ to $2\pi(8^2 d^2/2R\lambda)$ have to be produced. In addition, conversions from $\sin \omega t$ to $\cos \omega t$ may be required, which are easily performed by integrators. In contrast, Fig. 27 shows 11 circles, each one calling for a different timing of the respective switches, and only the smallest and largest circles have been drawn to avoid obscuring the picture. For a processor with 256 × 256 input terminals we need only $256/2 = 128$ different phase shifts by focusing first for x and then for y, while one would need of the order of $(256/2)^2$ phase shifts without this separation of variables. The price paid for the simplification is the replacement of a spherical surface of points in focus by the cylindrical surfaces in Fig. 44 with lines in focus, and the loci with points in focus.

We do not go into a more detailed discussion of focusing since the field opened here for Cartesian coordinates is as large as and more complicated than the field of high-quality lens design for polar coordinates. To obtain the depth of field, one must calculate the elongation of a point at the distance R and the two angles of incidence β_x, β_y for the directions x and y. The point will be displayed as a point on a light-emitting diode display if the elongations are smaller than the distance between adjacent light-emitting diodes. In optics based on polar coordinates one obtains the depth of field as a function of the distance R and only the one angle of incidence between the point and the optical axis. Let us also observe that horizontal and vertical lines in focus are quite acceptable since most man-made objects are characterized by horizontal and vertical lines.

F. Focusing without Approximation

Focusing as discussed so far contained the condition that R was large compared with the aperture A of the array, and the angle of incidence β, or β_x and β_y, was small. These are the usual assumptions of image generation by means of a lens. Spatial electric filters make it possible to go beyond these conditions and solve the focusing problem without any approximations.

Let Fig. 25 be replaced by Fig. 45. The important difference is that the sphere of focus is replaced by a plane of focus. The distance between this plane and the hydrophone array is R. The point $W\beta$ on the plane has the distance

$$(R^2 + R^2 \tan^2 \beta)^{1/2} = R(1 + \tan^2 \beta)^{1/2} \tag{71}$$

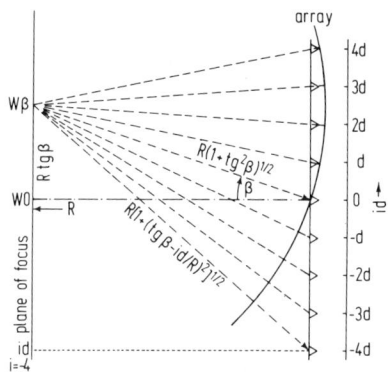

FIG. 45. Exact values of the output voltages of a hydrophone array due to a spherical wavefront originating in the point $W\beta$ at the distance $R(1 + \tan^2 \beta)$ on a ray forming the angle β with the axis of the array.

from the hydrophone $i = 0$ in the center of the array. The distance between the point $W\beta$ and an arbitrary hydrophone i is defined by the relation

$$[R^2 + (R \tan \beta - id)^2]^{1/2} = R[1 + (\tan \beta - id/R)^2]^{1/2}. \tag{72}$$

The difference of the distances of Eqs. (72) and (71) divided by the wavelength λ is called the relative phase or the relative propagation time $\psi(i, R, \beta)$:

$$\psi(i, R, \beta) = R\lambda^{-1}\{[1 + (\tan \beta - id/R)^2]^{1/2} - (1 + \tan^2 \beta)^{1/2}\}. \tag{73}$$

For small values of β one may replace $(1 + \tan^2 \beta)^{1/2}$ by $1 + \tfrac{1}{2} \tan^2 \beta$. If in addition id/R is small for the largest absolute value of i, which we denote by m, one may use the following approximation:

$$[1 + (\tan \beta - id/R)^2]^{1/2} \doteq 1 + \tfrac{1}{2}(\tan \beta - id/R)^2; \quad |i| \leq m.$$

For small values of β one may replace $\tan \beta$ by $\sin \beta$, and one obtains thus a first-order approximation of $\psi(i, R, \beta)$:

$$\psi(i, R, \beta) \doteq (-id \sin \beta + i^2 d^2/2R)/\lambda; \qquad (74)$$

$$|i| \leq m, \qquad md \ll R, \qquad \tan \beta \doteq \sin \beta \ll 1.$$

The output voltage of the hydrophone i at the location id in Fig. 45 is denoted $v(i, R, \beta, t)$:

$$v(i, R, \beta, t) = V \sin 2\pi[vt/\lambda - \psi(i, R, \beta)]. \qquad (75)$$

Using the approximation of Eq. (74) one obtains the output voltages of Fig. 25:

$$v(i, R, \beta, t) \doteq V \sin [2\pi(vt + id \sin \beta - i^2 d^2/2R)/\lambda].$$

We now define the focusing phase or focusing time $\chi(i, R, \gamma)$ for the hydrophone i, a distance R between the plane of focus and the array, and an observation angle γ:

$$\chi(i, R, \gamma) = R\lambda^{-1}\{[1 + (\tan \gamma - id/R)^2]^{1/2} - (1 - \tan^2 \gamma)^{1/2}\}. \qquad (76)$$

$\chi(i, R, \gamma)$ and $\psi(i, R, \beta)$ are equal except that the continuously variable angle of incidence β is replaced by the observation angle γ, which assumes the values $0, \pm\varepsilon, \pm 2\varepsilon, \ldots$ only. The values of λ, γ, i, and d are fixed, and a certain value of R is chosen for focusing. Hence, the numerical value of $\chi(i, R, \gamma)$ is known. The value of $\psi(i, R, \beta)$, on the other hand, is not known, since β is not known but is determined by the process of image generation.

For small values of β and large values of R one obtains an approximation according to Eq. (74):

$$\chi(i, R, \gamma) \doteq (-id \sin \gamma + i^2 d^2/2R)/\lambda. \qquad (77)$$

Since the numerical value of $\chi(i, R, \gamma)$ is known, one may produce the sine $\sin [\chi(i, R, \gamma)]$ and the cosine $\cos [\chi(i, R, \gamma)]$ of this numerical value. Furthermore, one may phase shift $v(i, R, \beta, t)$ of Eq. (75) by means of an integrator; the shifted voltage is denoted $v^*(i, R, \beta, t)$:

$$v^*(i, R, \beta, t) = V \cos 2\pi[vt/\lambda - \psi(i, R, \beta)]. \qquad (78)$$

Let us now form the following expression:

$$\tfrac{1}{2}\{v(i, R, \beta, t) \cos [\chi(i, R, \gamma)] + v^*(i, R, \beta, t) \sin [\chi(i, R, \gamma)]\}$$
$$= \tfrac{1}{2}V \sin 2\pi[vt/\lambda - \psi(i, R, \beta) + \chi(i, R, \gamma)]. \qquad (79)$$

Summation over all values of i yields the voltage $v(R, \gamma, \beta, t)$:

$$v(R, \gamma, \beta, t) = \tfrac{1}{2}V \sum_{i=-m}^{m} \sin 2\pi[vt/\lambda - \psi(i, R, \beta) + \chi(i, R, \gamma)]. \qquad (80)$$

This voltage represents the image of the point $W\beta$ in Fig. 45 in electric form. A light-emitting diode display will transform it into a visible optical image. Let us show that this claim is true by substituting the approximations of Eqs. (74) and (77) for $\psi(i, R, \beta)$ and $\chi(i, R, \gamma)$:

$$v(R, \gamma, \beta, t) \doteq \tfrac{1}{2} V \sum_{i=-m}^{m} \sin\{2\pi[vt + id(\sin\beta - \sin\gamma)]/\lambda\}$$

$$\doteq \frac{V}{2(2m+1)} \left\{ \sin(2\pi vt/\lambda) \int_{-1/2}^{1/2} \cos[2\pi A\lambda^{-1}(\sin\beta - \sin\gamma)x]\,dx \right.$$

$$\left. + \cos(2\pi vt/\lambda) \int_{-1/2}^{1/2} \sin[2\pi A\lambda^{-1}(\sin\beta - \sin\gamma)x]\,dx \right\}$$

$$= \frac{VA}{2d} \frac{\sin[\pi A\lambda^{-1}(\sin\beta - \sin\gamma)]}{\pi A\lambda^{-1}(\sin\beta - \sin\gamma)} \sin(2\pi vt/\lambda); \qquad (81)$$

$$(2m+1)d = A, \qquad i/(2m+1) = x.$$

The last line of this equation is equal to Eqs. (23) and (24).

One may rewrite Eq. (80) into a form that shows more clearly the diffraction pattern produced as the image of the point $W\beta$:

$$v(R, \gamma, \beta, t) = \tfrac{1}{2} V[A^2(R, \gamma, \beta) + B^2(R, \gamma, \beta)]^{1/2} \sin(2\pi vt/\lambda + \phi);$$

$$A(R, \gamma, \beta) = \sum_{i=-m}^{m} \cos[\chi(i, R, \gamma) - \psi(i, R, \beta)]; \qquad (82)$$

$$B(R, \gamma, \beta) = \sum_{i=-m}^{m} \sin[\chi(i, R, \gamma) - \psi(i, R, \beta)].$$

$[A^2(R, \gamma, \beta) + B^2(R, \gamma, \beta)]^{1/2}$ is the diffraction pattern of the image of the point $W\beta$ in Fig. 45. For small values of β and γ and large values of R one obtains readily the previously used approximation:

$$[A^2(R, \gamma, \beta) + B^2(R, \gamma, \beta)]^{1/2} \doteq A(R, \gamma, \beta) \doteq \frac{A}{d} \frac{\sin[\pi A\lambda^{-1}(\sin\beta - \sin\gamma)]}{\pi A\lambda^{-1}(\sin\beta - \sin\gamma)}.$$

The calculation of the diffraction pattern for any value of γ, β and R/λ is tedious but straightforward with the help of Eqs. (73) and (76).

Let us now derive a circuit diagram. One might think that we have derived a new method of image generation rather than a method of focusing, but this is not so. The image is formed by means of a Fourier transform, as in Section II,B. The difference is only that the multiplications are no longer performed by fixed resistors, as in Figs. 11 and 12, but by multipliers. The input voltages to those multipliers can be varied according to the focusing distance R, while there is no practical way to change the resistors.

Figure 46 shows a block diagram of a processor for one spatial dimension that accepts voltages from the hydrophones at the left and delivers voltages representing the electric image on the right. A two-dimensional processor is obtained by stacking printed circuit cards according to Fig. 46 in the way shown by Fig. 5.

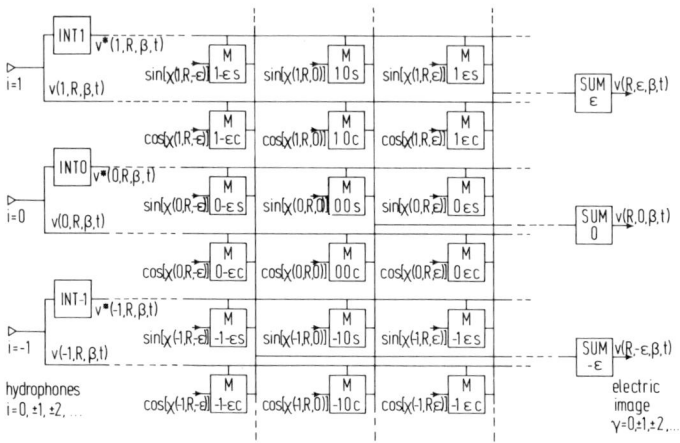

FIG. 46. Generation of focused images by means of multipliers. (INT) integrator; (M) multiplier; (SUM) summing circuit. $i = 0, \pm 1, \pm 2, \ldots; \gamma = 0, \pm \varepsilon, \pm 2\varepsilon, \ldots$

The voltages $v(i, R, \beta, t)$ from the hydrophones are integrated to yield the phase-shifted voltages $v^*(i, R, \beta, t)$. They are multiplied in Fig. 46 with the voltages sin $[\chi(i, R, \gamma)]$ and cos $[\chi(i, R, \gamma)]$ in the multipliers M, and the products are summed in the summers SUM to yield the output voltages $v(R, \gamma, \beta, t)$ that represent the electric image. The circuit is in principle very simple.

However, the practical implementation of the circuit of Fig. 46 is not simple. The multipliers are relatively expensive, and they need many external resistors and capacitors. One may expect that the cost will drop and the external components will vanish in time just as in the case of the operational amplifier. The real difficulty is the many voltages sin $[\chi(i, R, \gamma)]$ and cos $[\chi(i, R, \gamma)]$ that must be fed to each printed circuit card. A card with $2m + 1$ input terminals requires $2(2m + 1)$ contacts to handle the input voltages $v(i, R, \beta, t)$ and the output voltages $v(R, \gamma, \beta, t)$, but it requires $(2m + 1)^2$ contacts to handle the voltages sin $[\chi(i, R, \gamma)]$ and cos $[\chi(i, R, \gamma)]$. How can one get around this problem?

The only known practical way to produce the voltages sin $[\chi(i, R, \gamma)]$ and cos $[\chi(i, R, \gamma)]$ is by means of a microprocessor. The values of i and tan γ can be stored permanently as digital numbers. The values of R/λ must be fed in as digital numbers for the distance one wants to focus on. The microprocessor will then produce sin $[\chi(i, R, \gamma)]$ and cos $[\chi(i, R, \gamma)]$ as digital numbers. These numbers must be transferred to the printed circuit cards. The transfer may be done very slowly since the numbers do not change as long as the focusing distance R/λ remains unchanged. Hence, one may transfer the numbers serially, store them on each printed circuit card, and transform them from digital to analog on the card without destroying the stored numbers.

One may readily conclude from this short discussion that the circuit of Fig. 46 has great potential, but that it would be very hard to implement for a 256 × 256 array with the current technology. We will discuss in the following section how this problem can be overcome by time sharing.

Let us make a few remarks about focusing in two dimensions. One can extend the circuit of Fig. 46 to two dimensions and have all points on a plane in focus. However, such a circuit cannot be broken up in the way shown by Fig. 5 to produce first a transform for the direction x and then one for the direction y. Breaking up the two-dimensional transform into two one-dimensional transforms according to Fig. 5 causes the plane of focus to become the surface of a cylinder with vertical axis. This is perfectly acceptable. If we focus our eyes for a certain distance R and then move our head sideways, we see the points on the surface of a cylinder with vertical axis in focus.

The circuit of Fig. 46 is not only capable of focusing on a plane or cylinder surface, but may also be used to compensate a curvature of the hydrophone array. This is a very important feature, since a plane hydrophone array of several square meters size produces an enormous drag in water.

G. Time Sharing for High Resolution

We need about 25 images per second to create the sensation of a moving image. The analog circuits discussed were capable of generating about 100,000 images per second. Let us try to exploit this high speed by time sharing to simplify the equipment.

The object plane in Fig. 1 with 16 × 16 resolvable points can be extended periodically in the x and y directions. This is shown in Fig. 47 for four periods in the directions x and y, which yields $4^2 = 16$ sections with 16 × 16 points each, for a total of 64 × 64 points. Let the sound projector of Fig. 1 insonificate the section H1V1 in Fig. 47, which contains the columns 16, 17, ..., 31 and the lines 16, 17, ..., 31. An image of this section will be produced on the optical display in Fig. 1.

Let now the sound projector insonificate a different section in Fig. 47, e.g., the section H3V2 with the columns 48, 49, ..., 63 and the lines 32, 33, ..., 47. Now an image of this section will be produced on the same optical

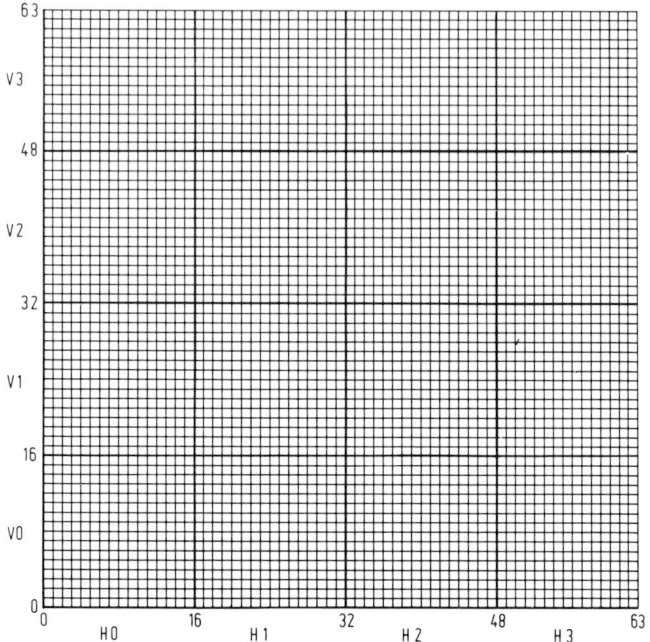

FIG. 47. Generation of an image of 64 × 64 points by scanning 16 sections H0V0 to H3V3 of the object plane with 16 × 16 points each.

display as before in Fig. 1. The ambiguity of the display can be resolved mechanically or electrically. The simplest solution is to use 4 × 4 displays as in Fig. 1 but arranged like the 4 × 4 sections H0V0 to H3V3 in Fig. 47. A switching device is then needed in Fig. 1 between the two-dimensional spatial electric filter and the display that connects the 16 × 16 output terminals of the filter to the 16 × 16 input terminals of one of the 4 × 4 sections of the display.

The insonification of one of the 4 × 4 sections of the object plane can be accomplished by using an array of 4 × 4 projectors. By proper mechanical arrangement one may make each one insonificate one of the sections in Fig. 47. From the standpoint of signal power it is, of course, better to use the 16 projectors as a phased array that electrically scans the 16 sections. All 16 projectors radiate power simultaneously in this case, rather than only one projector.

Let us derive some numerical values. The spatial filter must be able to produce 25 images per second of each of the 16 sections, or a total of 400 images per second. This number is still small compared with the 100,000 images per second that can be produced. Let us replace the 4 × 4 sections in

Fig. 47 by 16 × 16 sections. A total of (16 × 16)(16 × 16) = 256 × 256 points can then be resolved, which corresponds approximately to the quality of a TV image. The number of images that must be produced per second is increased to 25 × 16 × 16 = 6400, which is well within the capability of the spatial filter.*

We have so far ignored the bandwidth required for image generation. According to the sampling theorem one needs a bandwidth of 12.5 Hz to produce 25 images per second. If one operates with an insonification frequency of 100 kHz one can build filters with 1 kHz bandwidth economically. Using frequency conversion one can reduce the bandwidth to about 100 Hz. Smaller bandwidths would require very stable and thus expensive circuits. Hence, the bandwidth is determined by economic considerations and not by the 12.5 Hz frequency required by theory. The 400 images per second required by time sharing according to Fig. 47 call for a bandwidth of 200 Hz, while the 6400 images required for a resolution of 256 × 256 points call for a bandwidth of 3200 Hz. This means that the signal-to-noise ratio and the Doppler resolution are reduced by time sharing.

Of more practical interest than signal-to-noise ratio and Doppler resolution are the contours of the 4 × 4 sections in Fig. 47. One cannot realistically expect to insonificate the points on the columns 0, 1, ..., 15 and the lines 0, 1, ..., 15, but not those on column 16 and on line 16. The (weakly) insonificated points on column 16 will generate *alias* points on column 0 of

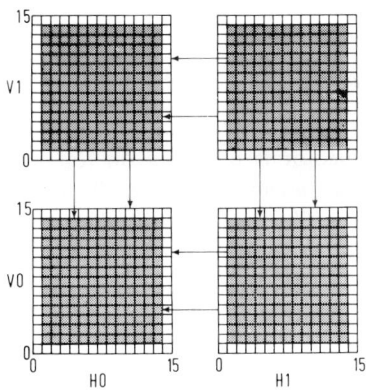

FIG. 48. Use of 14 × 14 inner points of sections with 16 × 16 points to avoid the contour effect. Only the points in the shaded areas are displayed.

---

* The limitation of this time-sharing method is due to the propagation time of the acoustic wave. It may be overcome by amplitude modulating the carrier wave with a set of orthogonal baseband signals, which is the use of a general orthogonal division instead of the special time division (Harmuth, 1972, p. 154).

the display, while line 16 will produce alias points on line 0. There is a simple way to combat this *contour effect*.

Let us produce image sections with 16×16 points but display the 14×14 inner points only as indicated by the shaded areas in Fig. 48 for the sections H0V0 to H1V1. The properly assembled sections of the optical display are shown in Fig. 49. The full display has $(14 \times 14) \times 16 = 56 \times 56$ points rather than 64×64 points as in Fig. 47. Hence, by using only the inner points of each image section one can avoid the contour effect, but one then obtains a smaller total image. These considerations may readily be extended to the use of only 13×13 or 12×12 inner points.

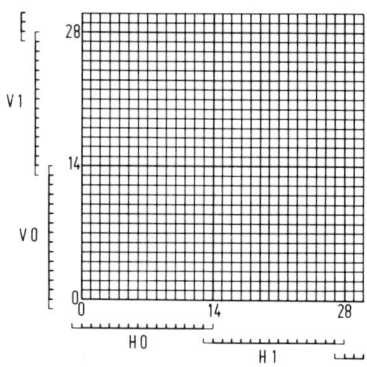

FIG. 49. Properly connected image sections with 14×14 points according to Fig. 48.

VII. EXPERIMENTAL EQUIPMENT AND TEST RESULTS*

Figure 50 shows a hydrophone array with 16×16 hydrophones during assembly. A sound projector is mounted on top. The size of this array is about 140×140 cm. The viewing angle α is $10°$ in the xz as well as in the yz plane as defined in Fig. 5. The resolution angle ε in both planes is about $40'$.

The array of Fig. 50 does not represent the state of the art. Arrays using individually assembled hydrophones as in Fig. 50 have been built with as many as 4608 hydrophones. A good review of techniques for even larger arrays was published by Prokhorov (1972).

An early version of electronic processing equipment according to Fig. 16 is shown in Fig. 51. This illustration is mainly intended to show the size of the equipment, and to make clear that the printed circuit cards do not have to be stacked as in Fig. 16 but are mounted in the usual way and connected by cables.

An enlarged view of the light-emitting diode display in Fig. 51 is shown

---

* This work was supported by the Office of Naval Research, Sensor Systems Group.

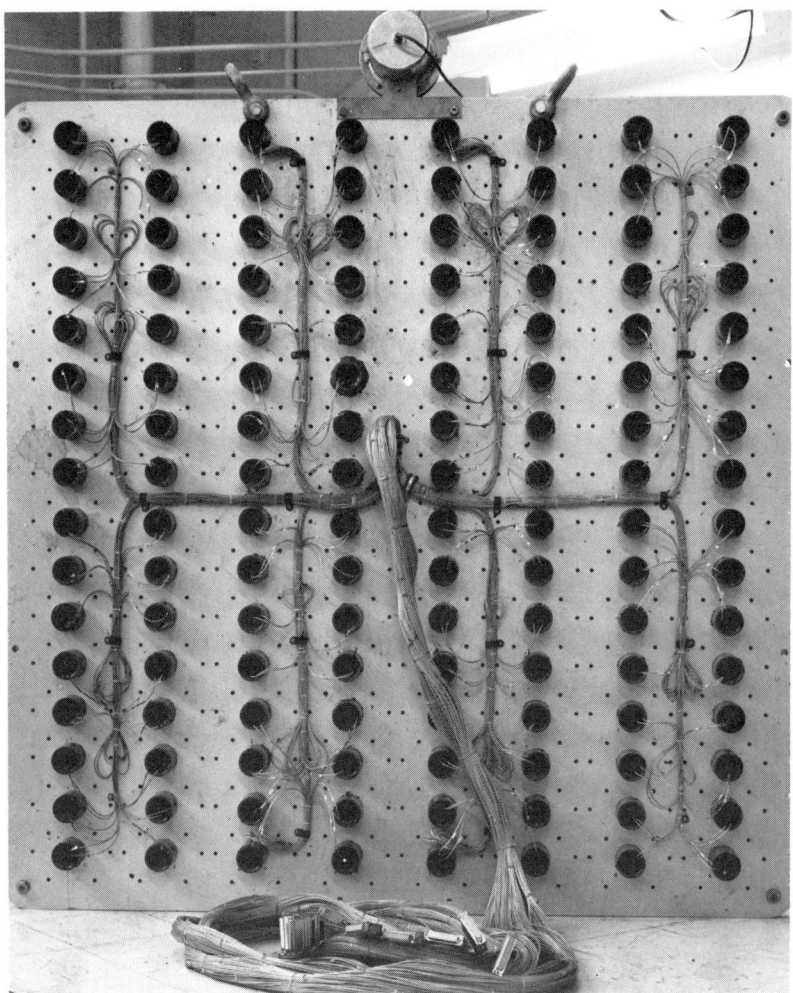

FIG. 50. Rear view of a hydrophone array with 16 × 16 hydrophones during assembly. A projector is mounted on top.

in Fig. 52. It gives a fair impression of the spatial as well as the brightness resolution that can be achieved. Displays with individual light-emitting diodes are not suitable for large displays, due to cost, required currents, and tolerances between diodes. The time-sharing technique discussed in Section VI,G favors very much a TV tube display that uses a scan according to a checkerboard pattern, and this has reduced the interest in light-emitting diode displays.

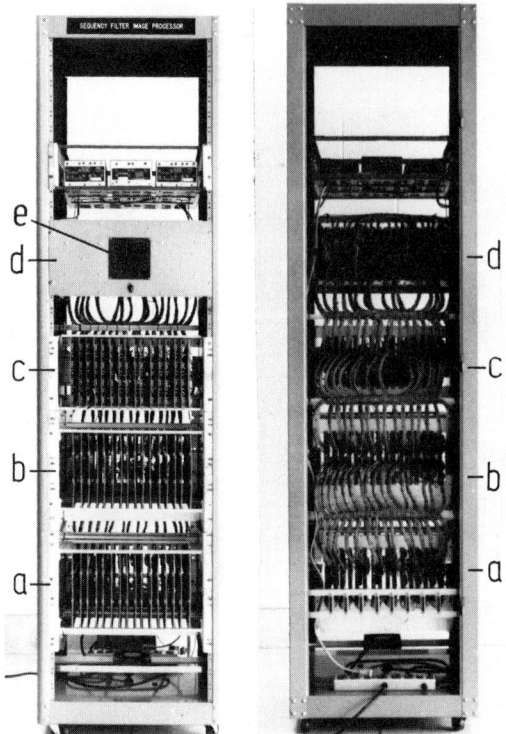

FIG. 51. Front and rear view of electronic processing equipment according to Fig. 16. (a) Fourier transformer for the variable x; (b) Fourier transformer for the variable y; (c) quadrant ambiguity resolver; (d) light-emitting diode driver; (e) display.

Tests of this equipment were run starting in 1974 at the Lake Travis Test Station of the University of Texas at Austin. The achieved resolution was very much in accordance with theory. The side lobes of the diffraction patterns according to Fig. 3 were not 22% of the main lobe as predicted by theory but about 30% due to the tolerances of the hydrophone array and the processor. This is an excellent result for such complicated equipment. Unfortunately, it is not possible to represent in print the impression of a moving image. Those familiar with home movies will readily appreciate the improvement of apparent quality of the moving picture compared with the single still picture. This improvement is just as striking for acoustic imaging.

Test results for equipment using synchronous demodulation and computer processing as discussed in Sections V,B and V,C for still pictures were reported by Thorn et al. (1974).

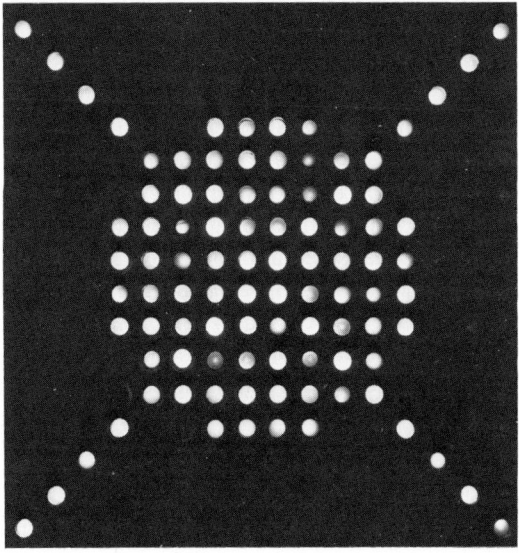

FIG. 52. Light-emitting diode display with test pattern. Two brightness levels are intended to be shown.

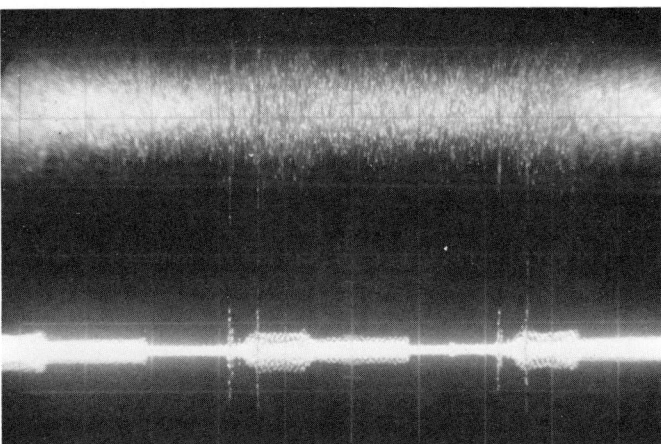

FIG. 53. The suppression of random noise by an array. The output voltage of one hydrophone is shown on top, the sum of the output voltages of 256 hydrophones on bottom. The horizontal scale is 2 msec/division. (Courtesy K. J. Dierks, University of Texas at Austin, APL.)

When testing a new type of equipment one usually runs into effects that had not been anticipated theoretically. Acoustic imaging is no exception. Figure 53 shows such an unexpected effect. The oscillogram on top shows the output voltage of one hydrophone, which clearly is dominated by random noise. The output voltage of the Fourier transformer representing the broadside direction $\beta_x = 0$, $\beta_y = 0$ according to Fig. 5 is shown at the bottom. The random noise has disappeared. The complicated periodic signal shown was traced to imperfect smoothing of the dc supply voltages. This effect occurs only for the broadside direction. According to Fig. 10, the beam for this direction is formed by multiplying both for the variable x and the variable y with the constant of the Fourier series denoted $f(0, x)$ in Fig. 10. Hence, any irregularity of the dc voltage is summed 256 times. The irregularities are suppressed for any other direction, since for any sample of the sine and cosine functions in Fig. 10 there is a sample of equal amplitude but opposite polarity.

Appendix

Computation of the sums in Eqs. (15)–(17) for large values of 2^n can be done by means of integrals:

$$\sum_{k=0}^{2^n-1} 2^{-n} \cos(2\pi k d\lambda^{-1} \sin\beta) \sin(2\pi k d\lambda^{-1} \sin\gamma)$$

$$= \sum_{k=0}^{2^n-1} 2^{-n} \{\sin[2\pi k d\lambda^{-1}(\sin\beta + \sin\gamma)]$$

$$\quad - \sin[2\pi k d\lambda^{-1}(\sin\beta - \sin\gamma)]\}$$

$$\doteq \int_0^1 \sin[2\pi A\lambda^{-1}(\sin\beta + \sin\gamma)2^{-n}k]\,d(2^{-n}k)$$

$$\quad - \int_0^1 \sin[2\pi A\lambda^{-1}(\sin\beta - \sin\gamma)2^{-n}k]\,d(2^{-n}k) = 0;$$

$$A = 2^n d$$

$$\sum_{k=0}^{2^n-1} 2^{-n} \sin(2\pi k d\lambda^{-1} \sin\beta) \sin(2\pi k d\lambda^{-1} \sin\gamma)$$

$$\doteq \int_0^1 \cos[2\pi A\lambda^{-1}(\sin\beta - \sin\gamma)2^{-n}k]\,d(2^{-n}k)$$

$$\quad - \int_0^1 \cos[2\pi A\lambda^{-1}(\sin\beta + \sin\gamma)2^{-n}k\,d(2^{-n}k)]$$

$$= \frac{\sin[\pi A\lambda^{-1}(\sin\beta - \sin\gamma)]}{\pi A\lambda^{-1}(\sin\beta - \sin\gamma)} - \frac{\sin[\pi A\lambda^{-1}(\sin\beta + \sin\gamma)]}{\pi A\lambda^{-1}(\sin\beta + \sin\gamma)};$$

$$\sum_{k=0}^{2^n-1} 2^{-n} \cos(2\pi k d\lambda^{-1} \sin \beta) \cos(2\pi k d\lambda^{-1} \sin \gamma)$$

$$\doteq \frac{\sin[\pi A\lambda^{-1}(\sin \beta - \sin \gamma)]}{\pi A\lambda^{-1}(\sin \beta - \sin \gamma)} + \frac{\sin[\pi A\lambda^{-1}(\sin \beta + \sin \gamma)]}{\pi A\lambda^{-1}(\sin \beta + \sin \gamma)};$$

$$\sum_{k=0}^{2^n-1} 2^{-n} \cos(2\pi k d\lambda^{-1} \sin \beta) \doteq \frac{\sin(\pi A\lambda^{-1} \sin \beta)}{\pi A\lambda^{-1} \sin \beta};$$

$$\sum_{k=0}^{2^n-1} 2^{-n} \sin(2\pi k d\lambda^{-1} \sin \beta) \doteq 0.$$

References

Albers, V. M. (1965). "Underwater Acoustics Handbook." Pennsylvania Univ. Press, University Park, Pennsylvania.
Bergman, D. G., and Yaspen, A. (1968). "Physics of Sound in the Sea." Gordon and Breach, New York.
Booth, N. O., and Sutton, J. L. (1974). Holographic acoustic imaging. Report NUC-TP434, Naval Undersea Center, San Diego, California.
Camp, L. W. (1970). "Underwater Acoustics." Wiley, New York.
Capon, J., and Goodman, N. R. (1971). Probability distribution of estimators of frequency-wavenumber spectrum. *Proc. IEEE* **59**, 112.
Capon, J., Greenfield, R. J., and Kolker, R. J. (1967). Multidimensional maximum likelihood processing of a large aperture seismic array. *Proc. IEEE*, **55**, 192–211.
Cox, H. (1973a). Resolving power and sensitivity to mismatch of optimum array processors, *J. Acoust. Soc. Amer.* **54**, 771–785.
Cox, H. (1973b). Sensitivity considerations in adaptive beam forming. *In* "Signal Processing" (J. W. R. Griffiths *et al.*, eds.), pp. 619–645. Academic Press, New York.
Fritzler, D., Marom, E., and Mueller, R. K. (1969). Ultrasonic holography via the ultrasonic camera. *In* "Acoustical Holography," Vol. 1, pp. 249–255. Plenum, New York.
Funk, C. J., Bryant, S. B., and Heckman, P. J. "Handbook of Underwater Imaging System Design." Technical Publication No. 303 (July 1972), Naval Undersea Center, San Diego, California.
Harger, R. O. (1970). "Synthetic Aperture Radar Systems." Academic Press, New York.
Harmuth, H. F. (1972). "Transmission of Information by Orthogonal Functions." 2nd ed. Springer-Verlag, Berlin and New York.
Harmuth, H. F. (1976). "Sequency Theory, Foundations and Applications." Academic Press, New York (in press).
Harmuth, H. F., Kamal, J., and Murthy, S. S. R. (1974). Two-dimensional spatial hardware filters for acoustic imaging. *In* "Applications of Walsh Functions and Sequency Theory," pp. 94–125. Institute of Electrical and Electronics Engineers, New York.
Havlice, J. F., Kino, G. S., and Quate, C. F. (1973). A new acoustic imaging device. *Proc. IEEE Ultrasonics Symp.*, pp. 13–17.
Hildebrand, B. P., and Brenden, B. B. (1972). "An Introduction to Acoustical Holography." Plenum, New York.

Johns, C. H., and Gilmour, G. A. (1976). Sonic cameras. *J. Acoust. Soc. Amer.* **59**, 74–85.

Lewis, J. B., and Schultheiss, P. M. (1971). Optimum and conventional detection using a linear array. *J. Acoust. Soc. Amer.* **49**, 1083–1091.

Lo, Y. T., Lee, S. W., and Lee, Q. H. (1966). Optimization of directivity and signal-to-noise ratio of an arbitrary antenna array. *Proc. IEEE* **54**, 1033–1045.

Marom, E., Mueller, R. K., Koppelman, R. F., and Zilinskas, G. (1971). Design and preliminary test of an underwater viewing system using sound holography. *In* "Acoustical Holography" (A. F. Metherell, ed.), Vol. 3, pp. 191–209. Plenum, New York.

Mueller, R. K., and Keating, P. N. (1969). The liquid–gas interface as a recording medium for acoustical holography. *In* "Acoustical Holography," Vol. 1, pp. 49–55. Plenum, New York.

Prokhorov, V. G. (1972). Piezoelectric matrices for the reception of acoustic images and holograms. *Akust. Zh.* **8**, 482–484. English translation: (1973). *Sov. Phys.–Acoust.* **18**, 408–410.

Rolle, A. L. (1973). Focused or lens-type ultrasonic imaging for small deep-diving submersibles. *Proc. IEEE Conf. Eng. Ocean Environ.*, pp. 284–300.

Seligson, C. D. (1970). Comments on high resolution frequency-wave number spectrum analysis. *Proc. IEEE* **58**, 947–949.

Skudrzyk, E. (1971). "The Foundations of Acoustics." Springer-Verlag, New York.

Sondhi, M. M. (1969). Reconstruction of objects from their sound-diffraction patterns. *J. Acoust. Soc. Amer.* **46**, 1158–1164.

Sutton, J. (1972). An experimental focused acoustic imaging system. *In* "Acoustical Holography," Vol. 6. Plenum, New York.

Thorn, J. V., Booth, N. O., Sutton, J. L., and Saltzer, B. A. (1974). Test and evaluation of an experimental holographic acoustic imaging system. Report NUC TP398, Naval Undersea Center, San Diego, California.

Vanderkulk, W. (1963). Optimum processing of acoustic arrays, *J. Brit. Inst. Radio Eng.* **26**, 285–292.

von Ramm, O. T., and Thurstone, F. L. (1975). Cardiac imaging using a phased array ultrasonic system. *Circulation* **53**, 258–267.

Woods, J. W., and Lintz, P. R. (1973). Plane waves at small arrays. *Geophysics* **38**, 1023–1041.

Nonvolatile Semiconductor Memories

J. F. VERWEY

*Philips Research Laboratories,
Eindhoven, The Netherlands*

I. Introduction	249
II. Some Nonvolatile Memory Devices	250
A. Magnetic Cores	250
B. Bubbles	253
C. Ovonics	254
III. Semiconductor Memory Devices	255
A. The MOS Transistor	256
B. Static Semiconductor Memory Cell	257
C. Dynamic Semiconductor Memory Cell	259
D. Charge-Transfer Device	260
E. Nonvolatile Semiconductor Memory Devices	261
IV. Reprogrammable Read-Only Memory (RePROM) Devices	262
V. Injection and Conduction Mechanisms	264
A. Tunneling from Si into SiO_2	264
B. Avalanche Injection of Electrons	266
C. Avalanche Injection of Holes	268
D. Nonavalanche Injection	271
VI. MIOS Devices	274
A. Switching Phenomena in MNOS	274
B. Band–Band Tunneling	278
C. Band–Trap Tunneling	281
D. Charge Retention	284
E. Degradation in MNOS	287
F. MAOS Memory Devices	288
G. MIOS Transistors with Interfacial Doping	292
VII. Floating-Gate Devices	294
A. FAMOS Memory Device	294
B. ATMOS Memory Device	297
C. Other Floating-Gate Devices	300
VIII. Discussion and Conclusions	302
IX. Glossary of Symbols Used in Text	304
References	306

I. INTRODUCTION

    A computer, small or large, consists of several subsystems such as input, output, control systems, one or more central processing units (CPU) and a memory (Fig. 1). The latter is for the storage of information of all

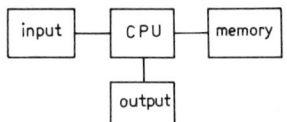

FIG. 1. Schematic drawing of the modular build-up of a computer, showing the memory as one of the modules connected with the central processing unit (CPU).

kinds such as the program, initial and end values of variables, intermediate computational results, and the final answer to the problem to be solved.

Nowadays, the CPU consists of circuitry made of semiconductor material. For reasons of compatibility it would be convenient if the memory were also made with the aid of semiconductors. Indeed, in recent years, from 1965 on, there has been considerable activity in semiconductor memory development. Before that time magnetic materials were (as they still are) widely applied. These materials have the property of nonvolatile storage of information, which means that the information is not lost when the power supply to the memory is switched off. Therefore, in Section II,A we give a brief description of one of the most important memory elements made of magnetic material, namely the magnetic core. In Sections II,B and II,C we describe briefly bubble devices and ovonic devices, respectively. The bubble device is still in the development stage, and illustrates the attempts to tread new paths in making simple memory systems. Bubble devices are still made of magnetic material, namely a single-crystal garnet slab. The ovonic memory device is made of a layer of amorphous material. It has not reached industrial production, mainly because of its limited number of switching operations. Ovonic memory devices are briefly discussed in Section II,C because the amorphous layer is often described as a semiconducting layer and, moreover, possesses the property of nonvolatile information storage. Therefore, these devices must not be omitted in this review of nonvolatile semiconductor memories.

In Section III, semiconductor memories are introduced, namely those memories in the form of integrated circuits in single-crystal silicon. These memories contain a large variety of types, or rather families. One family, that of the reprogrammable nonvolatile semiconductor memory devices, has been chosen from this group, and its physical aspects are treated in more detail in Section IV onwards.

II. SOME NONVOLATILE MEMORY DEVICES

A. Magnetic Cores

A well-known substance for making memory devices is the magnetic material ferrite. It can have one-way magnetization, which it retains when placed in a relatively weak magnetic field of opposite direction (Smit and

Wijn, 1954). An increase in this magnetic field beyond a certain critical value changes the magnetization of the ferrite to the opposite direction. This occurs more or less abruptly as a function of the field, as sketched in Fig. 2. By reversing the field the procedure can be repeated, and the magnetization of the ferrite changes to the opposite direction but remains the same in magnitude. If we assign to the upper saturation value of the magnetization

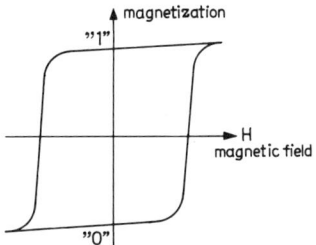

FIG. 2. Ferromagnetic hysteresis loop and indication of the stored digital information.

the memory state "1" and to the lower value the state "0," it is clear that we have a means of storing digital information (Quartly, 1962). The memory element as such can be a core of magnetic material such as ferrite furnished with three crossed wires (Fig. 3). The memory that is made of these elements is 3D-organized (Russell, 1965); two crossed wires are for selection and one is for detection of the information in the elements. A memory needs many elements to store many bits of information. They are arranged in matrices to facilitate selection of a particular bit, a 2 × 2 matrix being shown in Fig. 3 for purposes of illustration. An actual core memory of a big computer may contain millions of bits. If initially the four cores in Fig. 3 are all magnetized in the same direction, then, by sending an electric current through the wires X_1 and Y_1 in the directions indicated by the arrows, the magnetization in one core (where the wires cross) changes sign due to the magnetic field induced by the currents in the wires. This is the "write" operation in a mag-

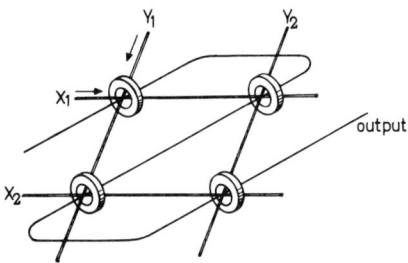

FIG. 3. A 2 × 2 array of magnetic cores.

netic core memory device. The current in one wire is not sufficient to cause a change in magnetization, so the magnetization, i.e., the information in the other three cores in Fig. 3, is not changed. To erase, currents are merely sent through wires X_1 and Y_1 in the direction opposite to that in Fig. 3. The "read" operation is performed by sending current through two wires, say X_2 and Y_2. When this is accompanied by a change in the magnetization of the core at the point where these lines cross, one obtains a current pulse on the third, output sensing wire in Fig. 3.

From this behavior of our very simple memory matrix, two important properties of magnetic core memories can be derived.

(a) The readout of the information is destructive. After the read operation the selected core must be reset in its original magnetization state. This is made possible by some peripheral circuitry.

(b) The storage of the information is nonvolatile; that is, the information remains in the memory element when the power is switched off. This is because the magnetization remains in one of the two states in Fig. 2 when the external field is zero. A consequence of this property is that the information remains stored without the use of energy.

In the early period of development of the digital computer the ferrite material was accepted as a gift of nature because this made the storing of large amounts of information possible and it eased the construction of one of the essential parts in the computer, the memory. But nowadays many people are looking for other materials. The reason for this is the better performance desired with respect to the following.

(a) Speed of the memory. One of the typical parameters is, for instance, the access time. This is the period between the time the information is asked for (i.e., interrogating signals arrive) at the input of the memory and the time the information appears at the output. The development of ever-faster computers requires the development of faster memories.

(b) Cost per bit. In magnetic core memories the peripheral circuitry for driving the signals and interrogating the stored information is relatively expensive.

The requirement of high speed is met by semiconductor memories, but the low cost requirement only in small ones. Hence, at the moment big computers have a multilevel memory structure in which the part (level) nearest to the CPU is made of semiconductors, and the other levels are magnetic systems such as disc and tape. Medium and small computers have all-semiconductor memories. We will briefly discuss them in Section III, inasmuch as they are relevant to our subject of nonvolatile semiconductor memory devices. Before doing so we should like to mention two other interesting developments in the field of nonvolatile memory devices, namely bubbles and ovonics.

B. Bubbles

A single-crystal slab of a magnetic material has arranged its magnetization in a preferred direction. For the device application the slab is made by epitaxial growth in a direction parallel to this preferential direction. If an external magnetic field is now present, the magnetization is divided into many domains which have magnetizations in one of two opposite directions. When an external magnetic field is applied to the slab in a direction perpendicular to the surface of the slab, the domains with magnetization in the same direction as the external field grow at the expense of the other domains. This continues with increasing field until relatively small cylindrical domains remain: the bubbles (Fig. 4). The shape and size of these bubbles are

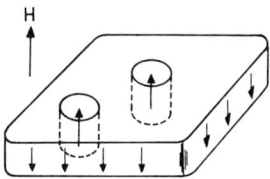

FIG. 4. Schematic representation of two magnetic bubbles in a garnet slab in the presence of a magnetic field H. This situation corresponds to two bits of stored information.

relatively stable in the presence of small variations in the external magnetic field. A weak magnet is able to displace the bubbles in the plane of the slab. This makes it possible to align the bubbles in a matrix of rows and columns by the application of a T and bar structure, as shown in Fig. 5. The Ts and bars are made of a thin layer of a "soft" magnetic material like Permalloy. "Soft" means that the material is easily magnetized in a weak magnetic field, in this case lying in the plane of the slab. The induced north pole attracts the south pole of a bubble. In this way bubbles are arranged in rows. In a bubble memory the presence and absence of bubbles in the row corresponds to stored 1s and 0s. The structure of Ts and bars in Fig. 5 is for the movement of the bubbles in the row. The weak magnetic field lying in the plane of the slab is rotated, causing changing magnetizations of the Permalloy structure, which in turn causes the movement of all bubbles in one direction.

FIG. 5. Permalloy T and bar structure on top of a garnet slab for the alignment of bubbles in rows and for their transport in the slab.

"Bubble detectors" have been developed for the read operation. One of them employs the magnetoresistance effect. The resistance of a Permalloy strip changes when a bubble passes below it. By placing a bubble detector at the end of the row, it is possible to detect the information present in the row. Circular T–bar structures make it possible to bring the bubbles back to their original position. Thus, the readout is not destructive. Bubble generators and bubble annihilators make it possible to construct a complete memory with write and erase properties. A description is outside the scope of this paper. For an excellent review of bubble devices the reader is referred to the paper of Druyvesteyn *et al.* (1975).

It should be noted that the bubble device is a serial memory: the bits are read sequentially, one after the other. This increases the average access time to a randomly chosen bit. The bubble or bit density can be very high and approaches that of the bit density in a dynamic semiconductor memory. It has the advantage of being nonvolatile. When the strong magnetic field perpendicular to the garnet slab is induced by a permanent magnet, the bubbles remain in the slab when the power is switched off.

C. Ovonics

Ovonic devices consist of layers of amorphous material between two suitable electrodes (Fig. 6). The amorphous material changes its conductivity when a voltage pulse is applied. The devices are named after their inventor, Dr. Ovshinsky. A favorable amorphous material is a Ge–Te alloy with composition near the eutectic composition. To this alloy, one or two elements from Group V or VI of the periodic table are added in amounts of a few percent.

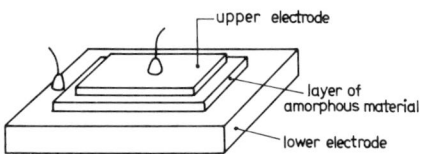

FIG. 6. Structure of an ovonic memory device.

The current–voltage characteristic of a "memory switch" is given in Fig. 7 (Ovshinsky and Fritzsche, 1973). Initially it may be in a high-resistance state. Application of a switching voltage larger than the threshold value, for a sufficiently long time, switches the memory element to the high-conductivity state. Now the high-conductivity state may correspond to a memory 1 state and the low-conductivity state to a memory 0 state.

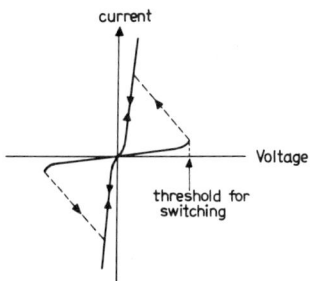

FIG. 7. Current–voltage curve of an ovonic memory device showing the two conduction (memory) states (Courtesy of the Institute of Electrical and Electronics Engineers).

The resetting of the device (i.e., the erase operation) is done simply by the application of a short voltage pulse (lower than the threshold but higher than that in the high-conductivity region in Fig. 7), which reconverts the Ge–Te alloy to its high-resistivity condition.

The physical mechanism is very complicated. There is evidence that crystalline filament growth in the amorphous film causes the high-conductance state. Destruction of this polycrystalline region by vitrification occurs during the erase operation.

Not all the physical processes during switching are fully understood. This is one of the reasons that a degradation process during switching has not been eliminated. The average failure-free life is 2×10^4 set–reset cycles. This result is obtained after testing a large number of 256-bit arrays.

The ovonic memory switch is a nonvolatile memory element. The array needs a relatively large amount of circuitry for write–read–erase operations because of the two-terminal nature of the device, and the device and its circuitry are made with different technologies. The limited number of write–erase cycles has also reduced interest in this device.

Rutz et al. (1973) reported on a switchable resistor made of a sputtered aluminum nitride layer. Switching curves have been obtained analogous to that shown in Fig. 7, with impedance ratios between 10 : 1 and 1000 : 1. Switching times are as low as 500 psec and as many as 10^{11} cycles have been obtained without degradation. However, the physical mechanism is not clear.

III. Semiconductor Memory Devices

Semiconductor memory devices consist of either bipolar transistors or metal–oxide–semiconductor (MOS) transistors. A description of bipolar transistors has been given in this series by Kennedy (1963). These transistors are used in static memories (see Section III,B) with small capacity. A large number of the semiconductor memories, including nonvolatile semiconductor memories, are based on the MOS transistor. Therefore, we start this chapter with a short description of the MOS transistor.

A. The MOS Transistor

For a review of the properties of the MOS transistor we refer to the paper of Kahng and Nicollian (1972). A cross section of the structure is shown in Fig. 8. It is made by the diffusion of source and drain into single-crystal silicon. What is known as a p-channel MOS transistor consisting of p^+-type source and drain regions in n-type silicon is depicted. The metal* electrode (or gate) can be given a negative voltage with respect to the silicon.

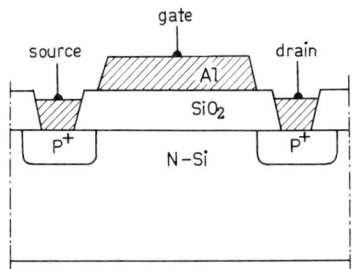

FIG. 8. Schematic cross section of an MOS transistor.

This induces an inversion layer (channel) at the silicon surface between source and drain. The application of a negative voltage to the drain causes an electric current to flow between source and drain. The drain current I_D as a function of drain voltage V_D shows the behavior, as sketched in Fig. 9: the current increases and soon saturates at a value I_{DS} (Pao and Sah, 1966). This I_{DS} can be approximated by a quadratic function of V_g (see Crawford, 1967; Sze, 1969). Thus, the square root of I_{DS} versus V_g (at constant V_D) gives the familiar curve shown in Fig. 10. Extrapolation of this curve to $I_{DS} = 0$ gives the threshold voltage, V_{TH}, also called the turn-on voltage. This is an important parameter, which we will often use to describe processes in nonvolatile semiconductor memories.

Now let us look at a hypothetical situation in which charge is present in the oxide. Assume that much positive charge is introduced into the oxide layer of the MOS transistor. Then, compared to the situation without charge, a higher gate voltage is necessary to bring the transistor into the conducting state (dashed line in Fig. 10). The threshold voltage is increased.

* Polycrystalline silicon is often used as a gate material.

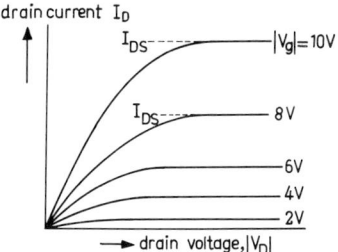

FIG. 9. Drain current I_D as a function of drain voltage V_D in a MOS transistor. I_D saturates to a value I_{DS}.

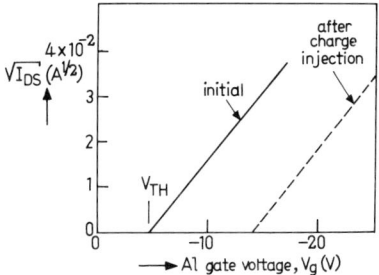

FIG. 10. Saturated drain current I_{DS} as a function of gate voltage V_g and the definition of the threshold voltage V_{TH}. After the injection of positive charge into the oxide, the curve is shifted (dashed line).

If the presence of charge is now a memory 1 state and the absence is a 0, we would have a nonvolatile memory element, since this charge can remain in the oxide for long times. However, the write and erase operations are not quite so simple. Many mechanisms have been tried for these operations, leading to different devices. Before discussing them and their properties, we shall briefly discuss in Sections III,B, III,C, and III,D the semiconductor memories that do not have the property of nonvolatile information storage.

B. Static Semiconductor Memory Cell

The basic element is the bistable flip-flop, an example of which is shown in Fig. 11. It contains two active devices, in this case two MOS transistors T_1 and T_2, and two load devices, in this case two resistors R (Eimbinder, 1971). The points A and B in the circuit are held at a constant voltage difference. If one of the transistors is conducting, say T_1, then the other is switched off. Point D_1 in the circuit attains a potential near B, and D_2 a potential near A. This is a stable condition which can be changed into T_2 conducting only by the application of suitable voltage to D_1 and D_2. This

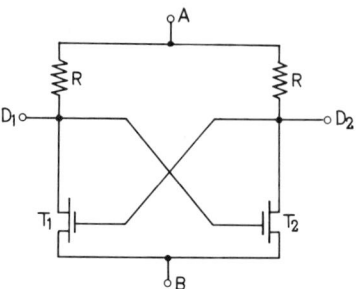

Fig. 11. Example of a static semiconductor memory cell.

T_2-conducting state is another stable condition. These conditions correspond to the memory 0 or 1 state. To complete the memory cell some transistors have to be added for addressing (read, write, erase) purposes. An example is given in Fig. 12 (Luecke et al., 1973). The resistors R in Fig. 11 are replaced by load transistors T_3 and T_4 because in MOS integrated circuits transistors are easier to make than resistors. A suitable voltage to address (select) lines X_1 and Y_2 turns on the transistors T_5 through T_8 and the content of the memory cell can be detected at the data lines D_1 and D_2.

From the description given above the following properties of the static memory cell become apparent: (a) same technology as the peripheral circuitry; (b) nondestructive readout; (c) volatile information storage.

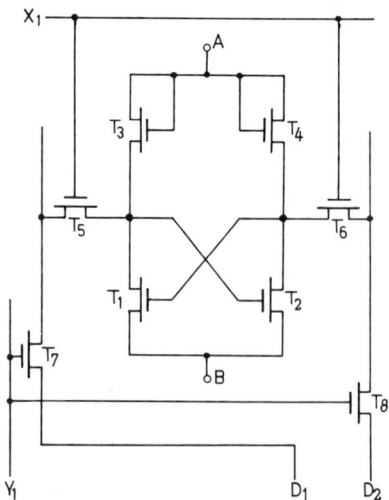

Fig. 12. Complete static semiconductor memory cell including the MOS transistors for addressing purposes in the X_1 and Y_1 address lines.

C. Dynamic Semiconductor Memory Cell

In this memory cell the information is present in the form of charge on a capacitor. The name is due to the fact that the information must be refreshed within a certain time because it is impossible to make a capacitor without any leakage current. As an example, we describe a one-transistor-per-bit MOS* memory cell (see Fig. 13) (Luecke *et al.*, 1973, Riley, 1973). The

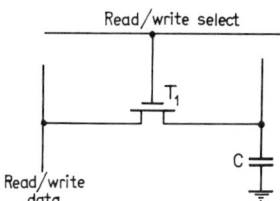

FIG. 13. Example of a dynamic semiconductor memory cell, showing a one-transistor-per-bit MOS cell.

charge containing the information is stored on capacitor C. The refresh, write, and read operations occur via the transistor T_1. From the figure it is clear that the size of the memory cell is considerably reduced compared to the static memory cell. This allows the construction of a large number of memory cells on one silicon crystal (chip). An example is given in Fig. 14. It shows a 4096-bit (4k) one-transistor-per-bit memory (Lambrechtse *et al.*, 1973). The chip dimensions are 3.01 × 4.09 mm. The simplicity of the memory cell is a great advantage; however, the memory chip needs a more or less complicated refresh system. The retention time, that is, the time at which the charge on capacitor C in Fig. 13 is still sufficiently large to be detected, is of the order of 1 msec. Thus the refresh cycle frequency must be at least 1 kHz. Moreover, it must be selective in that only the charged capacitors are recharged. Nevertheless, the larger bit density possibilities outweigh the larger complexity of the circuitry. It should be noted that Fig. 13 is one example of a dynamic memory cell. There are also four-, three-, and two-transistor-per-bit versions. However, there is a tendency in favor of the one-transistor (or 1-MOS)-per-bit cell in order to obtain a large bit density per chip.

The memory chip in Fig. 14 is a random-access memory (RAM). This means that the access times to all bit positions are about equal. The other type of memory with respect to addressing is the sequentially addressed memory. A well-known example is magnetic tape: many bits must be

---

* A bipolar one-transistor-per-bit cell has been proposed by Kasperkovitz (1973).

Fig. 14. 4096-bit, one-transistor-per-bit dynamic memory (Courtesy of the Institute of Electrical and Electronics Engineers, and R. H. W. Salters of Philips Research Laboratories).

scanned before the required information is obtained. The magnetic bubble memory devices (Section II,B) are also of this type. Many bubbles in an array of bubbles in the garnet slab must pass the bubble detector before the required bit is detected. Sequentially addressed memories have been designed in all semiconductor technologies and circuit techniques. They are generally called shift registers. They have a longer access time than RAM circuits but a much simpler surrounding circuitry because x–y addressing is not necessary. We describe a semiconductor memory which can only be sequentially addressed, namely the charge-transfer device, in the next section.

D. *Charge-Transfer Device*

In the charge-transfer device, packets of charge are transported parallel to the Si–SiO$_2$ interface. The device is based on the invention of Sangster and Teer (1969) of the bucket brigade shift register. A very simple form of a charge-transfer device shift register is given in Fig. 15. It is a so-called three-phase surface CTD, because it has three clock pulses, indicated in the figure by ϕ_1, ϕ_2, and ϕ_3, and the charge is transported near the silicon surface. Reviews on CTD devices are given by Collet and Esser (1973) and by Carnes

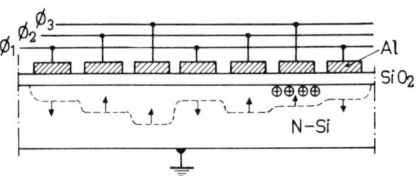

FIG. 15. Charge-transfer device shift register. A negative voltage is sequentially applied to the clock lines ϕ_1, ϕ_2, ϕ_3, and this causes the depletion layer to move in the direction of the arrows and the information bit (packet of holes) to move from the left to the right.

(1974). The application of positive voltage pulses to the metal (Al) electrodes in Fig. 15 develops a depletion layer, indicated by the dashed line. No inversion can be formed by thermal generation of holes, as the duration of the pulses is short compared to the thermal generation time. The three clock times in Fig. 15 are out of phase, causing a movement of the border of the depletion layer in the direction of the arrows. This forces the inversion layer of holes, under the second Al electrode counting from the right-hand side, to move to the right. The inversion layer is generated at the beginning of a line, for instance, by a gated diode, which is fairly similar to one-half of a MOS transistor. The presence of holes corresponds to a memory 1 state and their absence to a 0, or vice versa. We see in Fig. 15 that six electrodes contain two bit positions. The packing density of bits per chip can be very high because of the small electrode dimensions and the simple cell structure.

E. Nonvolatile Semiconductor Memory Devices

The following division of nonvolatile memory devices can be made according to their storage function: (a) read-only memories (ROM) or fixed-program memories; (b) programmable read-only memories (PROM); (c) reprogrammable read-only memories (RePROM).

In the first group of memories the programming is carried out by the manufacturer. A possible array is a matrix of MOS or bipolar transistors in which some transistors are contacted and others are not, corresponding to 1s and 0s. In MOS circuits it is possible to have either a thick or a thin oxide layer below a gate electrode. This corresponds to a difference in threshold voltage and, thus, to a difference in information content. A simple 2 × 2 matrix is shown in Fig. 16 in top view (Luecke et al., 1973). The horizontal bars are aluminum metallization strips, while the vertical bars are diffused source and drain regions. At position A the oxide layer is thinner than in the remainder of the 2 × 2 matrix, giving a MOS transistor with a low threshold voltage corresponding, for instance, to a memory 1 state. Its physical principle is very simple and material requirements do not deviate from conventional MOS technology. These devices are used in code converters, etc.

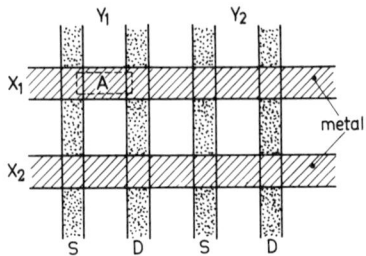

FIG. 16. Simple 2 × 2 matrix in a read-only memory (ROM) formed by four MOS transistors. The MOS transistor at position A has a thin oxide region (dashed line) corresponding to a memory 1 state.

Electronically, they are random-access memory devices and have access times shorter than or equal to those of static and dynamic semiconductor memory devices.

The second group, the programmable read-only memories, have the property that they can be programmed, but only once, by the user of the circuits. An example is an array of fusible links. It is also possible to have an array of conventional bipolar transistors. Programming of the memory is performed by destruction of selected emitter–base junctions by voltages considerably above the emitter–base breakdown voltage (Riley, 1973).

The third group of nonvolatile semiconductor memories, namely the reprogrammable read-only memories, are based on a MOS transistor with two dielectric layers between gate and silicon. Programming is by injection of charge into and subsequent storage in the double-dielectric structure. The storage may be either in discrete centers near the interface of the two dielectrics or on a layer of silicon (floating gate) between the two dielectrics.

The RePROM devices show interesting properties with respect to physical phenomena that are used to obtain the programming and reprogramming features. They are electronically organized as random-access memories with a read cycle much shorter than the write cycles. Therefore they are often referred to as read-mostly memories.

IV. Reprogrammable Read-Only Memory (RePROM) Devices

As mentioned in Section III,E, the reprogrammable read-only memory (RePROM) devices are based on MOS transistors with a two-dielectric layer structure. In this chapter we give an outline of the electronic aspects of these devices.

Two groups of devices may be discerned, depending on the way the charge is injected into and stored in the double-dielectric structure. The first group consists of the MIOS transistors in which the first dielectric layer O on the silicon is thin oxide and the second layer may be any other dielectric I, but is usually either silicon nitride (in the MNOS device) or aluminum oxide (in the MAOS device). Write and erase operations are performed by drawing tunnel currents through the oxide layer. Consequently, the oxide thickness d_{ox} should be small (< 100 Å).

Figure 17 shows the basic structure of the MNOS transistor. This is the most important memory device in this group because it has relatively good properties with respect to switching speed (of the order of 10 μsec), switching voltage (25–30 V), number of write–erase cycles ($> 10^{11}$), and information-retention time (about 1 yr). Nonvolatility of the information is obtained by trapping of the tunneling electrons in relatively deep centers at or near the oxide–nitride interface.

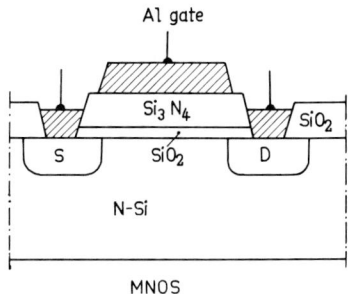

FIG. 17. Basic structure of the MNOS memory transistor.

In the MAOS device the switching and the nonvolatility of the information are obtained in the same way as in the MNOS device. At the time of writing of this review, the performance of the MAOS device is inferior to that of the MNOS device because structures with thin SiO_2 layers have poor interface properties, which necessitates $d_{ox} \geq 80$ Å and, consequently, the application of a relatively high write–erase gate voltage of 40 V. Moreover, the maximum number of write–erase cycles is about 10^4.

The MIOS devices with interfacial doping have metal particles at the I–O interface which are used as the storing medium (Kahng et al., 1974). This increases the speed of the trapping phenomenon, and therefore the writing speed is higher and/or the write gate voltage is lower than in MNOS devices with the same thicknesses of the dielectric layers. The maximum number of write–erase cycles is not yet known, but it determines whether this device will be superior to the MNOS device.

The second group of RePROM devices consists of the floating-gate transistors. These transistors derive their name from a layer of polycrystalline silicon embedded between two relatively thick oxide layers and completely isolated from the remainder of the device. The FAMOS device, the first

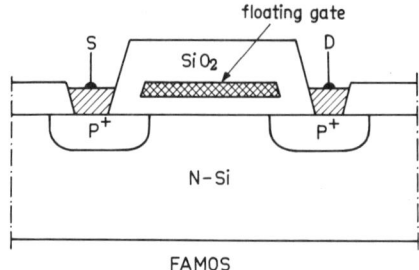

FIG. 18. Cross section (schematic) of the floating-gate avalanche-injection MOS (FAMOS) transistor.

floating-gate device to be realized on an industrial scale, is shown schematically in Fig. 18. It is a p-channel MOS transistor in which the programming is by avalanche injection of electrons from the silicon into the oxide and subsequent trapping on the floating gate. The write pulse voltage of -50 V is applied to one of the junctions for 5 msec. The charge-retention time is about 100 yr. Erasing is not done by electrical means but by UV irradiation or X-rays for a few minutes. The maximum number of write—erase cycles is less important due to this time-consuming erase operation. The FAMOS device finds application mainly in the product development centers, as an alterable ROM.

There are a number of proposals to solve the difficulty of time-consuming and nonelectrical erase. One of them is the ATMOS device. However, the write time is relatively long (10–100 msec) and the erase time is not sufficiently reduced (1 sec). The maximum number of write–erase cycles is expected to be 10^4. The information retention time is 1000 hr at 125°C (Verwey and Kramer, 1974).

An external gate on top of the floating gate can be used to erase the information. The second oxide layer between external and floating gate may become conducting at high fields, thus forming a drain for the stored charge on the floating gate. A drawback is the relatively high gate voltage in the erase operation.

The next section is devoted to the injection and conduction mechanisms of the write–erase operations of RePROM devices.

V. Injection and Conduction Mechanisms

A. Tunneling from Si into SiO_2

The mechanism of tunneling from silicon into silicon dioxide is shown in Fig. 19 in an energy-band scheme near the Si–SiO_2 interface. The quantities E_C and E_V are the bottom level of the conduction band and the top level of

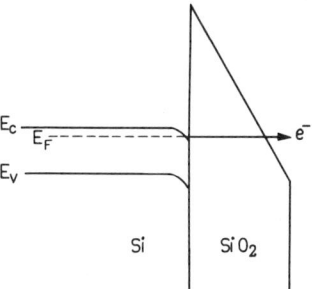

FIG. 19. Fowler–Nordheim tunneling of electrons from Si into SiO_2.

the valence band, respectively; E_F is the Fermi level. The arrow indicates the trajectory of the tunneling electrons. The resulting tunnel current in SiO_2 was studied by Lenzlinger and Snow (1969) in much detail. They used metal–oxide–silicon structures in which a voltage positive with respect to the silicon was applied to the metal. An increase of this voltage decreases the tunnel distance in Fig. 19 and, as the tunnel probability strongly depends on the tunnel distance, the current through the oxide strongly increases. This tunneling from energy band to energy band is called Fowler–Nordheim tunneling. The equation governing the current–voltage (or rather current–field) behavior is

$$J_{ox} = A_{FN} F_{ox}^2 \exp\left[-(32qm_0)^{1/2}\phi_B^{3/2}/3\hbar F_{ox}\right], \qquad (1)$$

where $\hbar$ is Planck's constant divided by 2π, q is the electronic charge, $q\phi_B$ is the barrier height, m_0 is the free electron mass, and F_{ox} is the field in the oxide. We should note that this equation holds for the simplest case of electron emission into vacuum: a triangular barrier and temperature $T = 0°K$. The following corrections can be applied to obtain a less approximative expression for the current: (a) the triangular barrier is lowered by the image force potential; (b) a nonzero temperature is taken into account; (c) the free electron mass m_0 is replaced by the effective mass m_{ox}^*, of the electron in the forbidden gap of the dielectric; (d) the relative dielectric constant is taken into account. These corrections are relatively small and the plot of $\log J_{ox}/F_{ox}^2$ versus $1/F_{ox}$ remains a straight line as predicted by Eq. (1). The current through the oxide is plotted in Fig. 20 in a Fowler–Nordheim plot. From the slope of the curve the effective mass is derived, $m_{ox}^* = 0.42$, provided the value of $q\phi_B = 3.25$ eV from photoemission experiments (Deal et al., 1966) is used.

These experiments on relatively thick oxide layers show how the experimental curves neatly fit the theoretical description in terms of Fowler–Nordheim tunneling. There is only one unknown parameter, namely the effective mass m_{ox}^*. This is not the case when the current–voltage

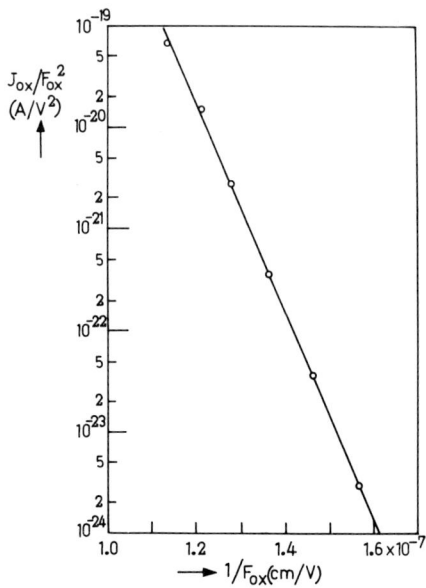

FIG. 20. Fowler–Nordheim plot of the oxide current J_{ox} as a function of oxide field F_{ox} (from Lenzlinger and Snow, 1969, courtesy of the American Institute of Physics).

behavior in MNOS or, generally, in a double-dielectric structure with a thin oxide, has to be described. Then the electron tunnels through the oxide into the nitride conduction band or into centers in the dielectrics, and more parameters enter the expression, such as the effective mass in the nitride or the energy and spatial location of the trapping centers. Different models can be used, and we shall discuss them in connection with the particular devices.

B. *Avalanche Injection of Electrons*

In a *pn* junction which is biased into avalanche breakdown, the charge carriers are able to gain sufficient energy from the field in the junction to cause ionization of silicon atoms. In this way electron–hole pairs are created and these charge carriers are also able to gain energy from the junction field and to form new electron–hole pairs. In this way an avalanche of energetic carriers is formed in the space-charge layer of the junction. When the energy, for instance, of electrons is above 3 eV, measured from the bottom of the conduction band, and when these electrons are near the Si–SiO$_2$ interface, they may then be injected into the SiO$_2$ under the influence of suitable fields. This situation is reached in the structure in Fig. 21, known as a gated diode, designed to get information about the avalanche-injected current in the silicon dioxide (Verwey and de Maagt, 1974). A breakdown voltage V_B is

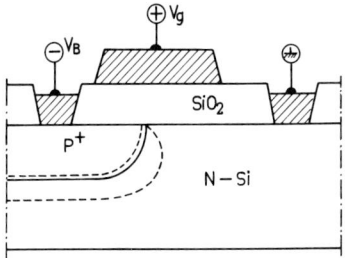

FIG. 21. Gated-diode structure for the study of the avalanche-injected electron current in SiO$_2$ (from Verwey and de Maagt, 1974, courtesy of Pergamon Press).

applied to the junction, with V_B negative with respect to the n side. A voltage V_g is applied to a metal electrode (gate) on top of the SiO$_2$. This V_g is positive with respect to the n side and causes accumulation on this side with subsequent constriction of the space-charge layer (shown dashed in Fig. 21) at the interface. This constriction can be better visualized and understood by drawing the intersections of the equipotential planes with the plane of the figure (de Graaff, 1970). The field in the junction (caused by the application of V_B) is highest at the Si–SiO$_2$ interface, and breakdown occurs there. However, the depletion of the Si surface at the p^+ side pushes the place of the avalanche away from the Si–SiO$_2$ interface. This must be minimized when it is desired to have a high injection density of electrons in the oxide. In the p^+n diodes of Fig. 21 the depletion at the surface of the p^+ side of the function is minimized by a high concentration of dopant.

The avalanche-injected electron current through the oxide decays as a function of time. This is due to the fact that some electrons are trapped during the electron transport through the oxide. This decreases the field at the interface and subsequently the injection density. An example of the current–time curve is shown in Fig. 22. At $t = 0$ a breakdown current of 10^{-2} A is switched on, while a constant voltage V_g is applied to the gate electrode. The oxide current reaches a maximum I_t and then decays due to the trapping effect. One should note that for the charging of a floating gate only a very small part of this curve is needed. For example, the switching time in the FAMOS device (Section VII,A) is of the order of milliseconds. Nevertheless this trapping effect may become important when the memory devices are repeatedly switched.

It has been found (Verwey and de Maagt, 1974) that the current through the oxide is injection-limited. To investigate this, the current–voltage curve (I_{ox}–V_g) has been measured (Verwey and de Maagt, 1974). The examples in Fig. 23 show I_t plotted against V_g, because by measuring I_t the trapping effects are a minimum, and they presumably do not interfere with the current measurement. The solid curve is a semitheoretical curve calculated with the

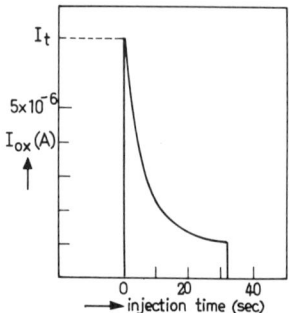

FIG. 22. Example of current-time behavior of the avalanche-injected electron current in SiO_2.

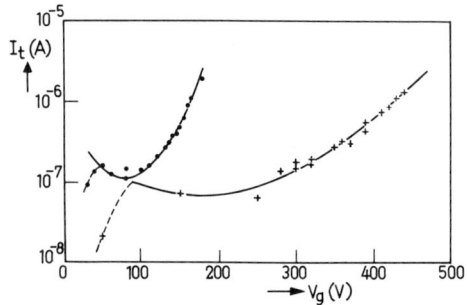

FIG. 23. Maximum value I_t of the avalanche injected electron current as a function of gate voltage V_g in samples with two oxide thicknesses. The solid line is a semitheoretical curve fitted to the experimental points (from Verwey and de Maagt, 1974, courtesy of Pergamon Press).

aid of a one-dimensional model and fitted to the experimental points by choosing a value for two parameters, the surface concentration of the p^+ side and the energy loss of the hot carriers in the silicon. In the model the electrons in the avalanche region are assumed to be exponentially distributed in energy with hot-electron temperature T_e (Bartelink et al., 1963).

With respect to this we mention the work of Bulucea (Bulucea et al., 1974; Bulucea, 1975a,b). He uses a more realistic two-dimensional model, in which the hot electron temperature T_e is a function of the electric field of the junction at the interface.

C. Avalanche Injection of Holes

The avalanche injection of holes from Si into SiO_2 is used in the erase operation of the ATMOS memory device. Its mechanism differs somewhat from that of electrons. The oxide current versus gate voltage curve of the latter is determined by processes in the silicon, as is discussed in Section V,B. This is not the case in the avalanche injection of holes. It has been found (Verwey, 1972a) that the current–voltage curve is governed by a Poole–Frenkel conduction. The current through the oxide has been measured in a gated diode very similar to that in Fig. 21, but with an n^+p junction with a relatively high-doped n^+ side to suppress the depletion of this side by the now negative gate voltage. It is expected that the current becomes injection-limited when the doping of the n^+ side is too low, as is discussed in Section V,B for electrons.

The Poole–Frenkel conduction is the thermal emission of charge carriers from Coulombic centers (in the oxide) in which the barrier is lowered by the presence of the field (Frenkel, 1938). Figure 24 shows the potential energy

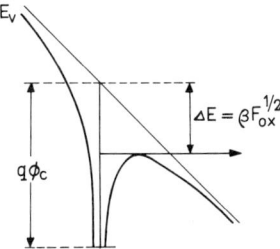

FIG. 24. Energy diagram of a Poole–Frenkel center at a distance $q\phi_c$ from the top of the valence band. The energy of the hole is drawn upward.

versus the distance in the direction of the field. E_V is the top of the valence band, and the energy of the hole increases in the vertical direction. The center with energy depth $q\phi_c$ is caused by a negative charge. The energy barrier lowering by the field is $\Delta E = \beta F_{ox}^{1/2}$ (Sze, 1967), where β is the Poole–Frenkel constant, which can be calculated from simple electrostatics (Frenkel, 1938). The hole in the potential well can be thermally activated into the valence band, and then the Boltzmann factor governing this process changes and the ratio of free-to-captured hole concentration becomes

$$\frac{p}{p_t} = \frac{N_v}{N_t} \exp\left[-\frac{q\phi_c - \beta F_{ox}^{1/2}}{kT}\right], \qquad (2)$$

where p is the concentration of free holes, p_t is the concentration of trapped holes, N_v is the density of states in the valence band, N_t is the density of traps, k is Boltzmann's constant, and T is the temperature. Based on these considerations, the current J_{ox} through the oxide becomes (Sze, 1967)

$$J_{ox} = A_{PF} F_{ox} \exp\left[-\frac{q\phi_c - \beta F_{ox}^{1/2}}{kT}\right], \qquad (3)$$

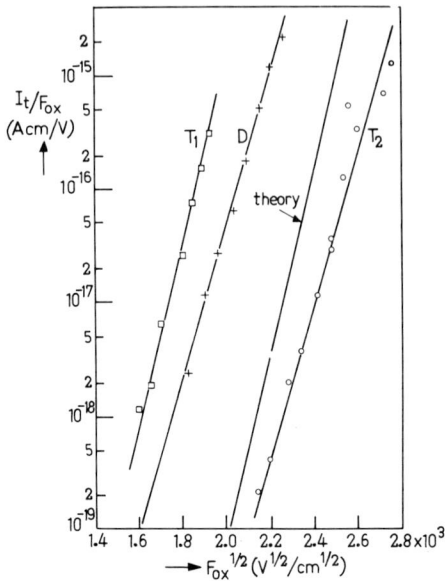

FIG. 25. Poole–Frenkel plots of the avalanche-injected hole current (maximum) I_t in the paper of Verwey (1972), replotted in the manner indicated in the text. The avalanche current of the diodes was $I_R = 10^{-5}$ A.

where A_{PF} is a proportionality constant depending on N_t. A curve of $\log J_{ox}/F_{ox}$ versus $F_{ox}^{1/2}$ should give a straight line. Indeed, this has been found, as can be seen in Fig. 25. This figure is replotted from the literature (Verwey, 1972a) but with

$$F_{ox} = (V_g + V_B)/d_{ox}, \qquad (4)$$

instead of

$$F_{ox} = V_g/d_{ox}.$$

In Eq. (4), V_B is the breakdown voltage of the diode. It is to be noted that the oxide has two field regions, one above the p side with $F_{ox} = V_g/d_{ox}$, and one above the n^+ side with a higher F_{ox} given by Eq. (4). The holes probably move in the high-field region. Comparing the slope of the experimental curve in Fig. 25 with the theoretical slope $\beta/2.3kT$ gives a reasonable agreement. It is possible that the oxide current becomes space-charge-limited, as is reported for avalanche injection of holes in MNOS structures (Verwey, 1972b). This has not yet been investigated.

It is pointed out here that the measurements of the avalanche-injected hole current (and the electron injection currents in Sections V,A and V,B) have not been carried out in double-dielectric structures. They have been carried out in single dielectrics to investigate the injection and conduction mechanisms.

D. *Nonavalanche Injection*

The nonavalanche injection of charge carriers is used in the write operation of the ATMOS memory device. In Section V,B we described the avalanche injection of electrons. In this injection mechanism the charge carriers are generated in the *pn* junction itself and accelerated in the field of the junction. In the nonavalanche injection mechanism the charge carriers are generated outside the accelerating junction in a separate forward-biased supply junction. The electrons are again accelerated in a reverse-biased junction, but the applied voltage is below the avalanche breakdown voltage. So, in the *pn* junction itself, there is hardly any source of charge carriers. Figure 26 shows a structure in which nonavalanche injection of charge carriers can take place. It is the structure used by Bosselaar (1973), who was the first to investigate this phenomenon, and it consists of a bipolar *npn* transistor with a gate on top of the oxide also covering the area where the emitter–base junction intersects the Si–SiO$_2$ interface. A positive gate voltage is applied and the emitter–base junction is reverse-biased. This induces a depletion layer below the gate electrode in the silicon and this forms the accelerating junction. The supply junction in this case is the forward-biased collector–base junction. Electrons are injected into the base region and they diffuse to the emitter–base junction and the field-induced junction below the gate. They are accelerated in the depletion layer of the field-induced junction and they obtain sufficient energy to overcome the Si–SiO$_2$ barrier. Of course, the reverse bias on the emitter–base junction must be sufficiently high (i.e., above 3 V) to give sufficient energy to the carriers (Verwey, 1973).

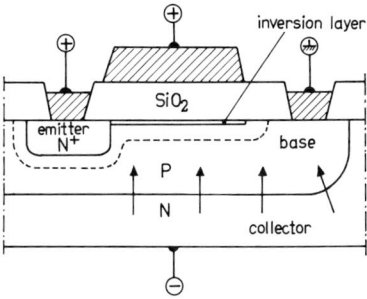

FIG. 26. Bipolar transistor with gated emitter–base junction for the study of nonavalanche injection of electrons into SiO$_2$.

What actually happens in the junction can be shown with the aid of an energy-band diagram in the depletion layer near the Si–SiO$_2$ interface. This diagram is given in Fig. 27. The meaning of the symbols is the same as in Fig. 19. The stepped arrow at the top of the figure symbolizes the electron trajectory. The electrons are thermal at the edge of the depletion layer of thickness W. If the band-bending or surface potential ϕ_s is high enough, the electrons gain sufficient energy to overcome the Si–SiO$_2$ barrier. However, the electrons also lose energy when they traverse the depletion layer. This is indicated in Fig. 27 by the step in the arrow. Hence, the ϕ_s must be sufficiently high to compensate barrier height and energy loss.

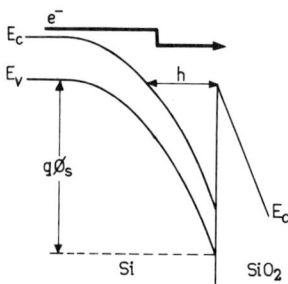

FIG. 27. Energy-band diagram near the Si–SiO$_2$ interface during nonavalanche injection of electrons.

For a discussion of the diode voltage V_D and gate voltage V_g dependences of the oxide current we first give formal descriptions of the injection process. Two approaches are possible. In the first approach the number of electrons arriving at the interface with sufficient energy is described in terms of a certain probability (Pepper, 1973). The basic principle of this model goes back to Shockley (1961), who used it to describe multiplication processes in pn junctions. An electron starting at zero kinetic energy has to travel a distance h (see Fig. 27) in the field direction before it achieves an energy $q\phi_B$. The probability P_h that it does so without collisions is (Shockley, 1961)

$$P_h = \exp(-h/l), \tag{5}$$

where l is the mean free path of the hot electrons. Under the assumption that J_{ox} is proportional to P_h we obtain

$$J_{ox} = P_h J_w, \tag{6}$$

where J_w is the current of electrons injected into the depletion layer. Combining Eqs. (5) and (6) gives

$$J_{ox} = J_w \exp(-h/l), \tag{7}$$

where h is given by (Verwey *et al.*, 1975)

$$h = \frac{2\varepsilon_{\text{Si}}(V_{\text{D}} + 2\phi_{\text{F}})^{1/2}}{qN_{\text{A}}}\left\{1 - \left[1 - \left(\frac{\phi_{\text{B}} - \beta_{\text{s}}(V_{\text{g}} - V_{\text{D}})^{1/2}/d_{\text{ox}}^{1/2}}{V_{\text{D}} + 2\phi_{\text{F}}}\right)\right]^{1/2}\right\}, \quad (8)$$

with ϕ_{F} the Fermi potential and β_{s} the Schottky constant. It should be noted that this equation for h differs somewhat from that in the literature. In Eq. (8) the lowering of the energy barrier by the oxide field F_{ox} is included and F_{ox} is given by $F_{\text{ox}} = (V_{\text{g}} - V_{\text{D}})/d_{\text{ox}}$.

One interesting aspect of the measurement of J_{ox} as a function of V_{D} is that it makes it possible to determine the mean free path l of the hot electrons in the silicon. From Eq. (7) we derive that a plot of $\ln J_{\text{ox}}$ (or $\log J_{\text{ox}}$) versus h gives a straight line. Such a plot is reproduced in Fig. 28 as a plot of $\log I_{\text{ox}}$ (current instead of current density) versus h. From the slope we derive the value of the mean free path $l = 135$ Å. For a discussion of the contributions of different scattering processes, we refer to the original literature (Verwey *et al.*, 1975; Pepper, 1973, 1974).

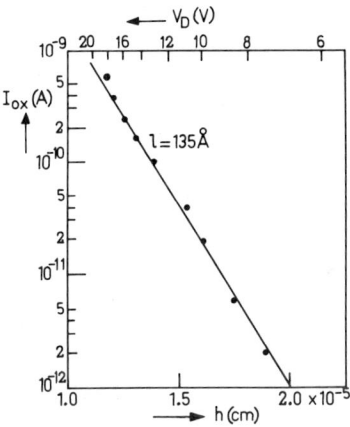

FIG. 28. Nonavalanche-injected electron current plotted as a function of h (and V_{D}), as a means for deriving the mean free path l (from Verwey, 1973, courtesy of the American Institute of Physics).

The other approach to describing the dependence of the oxide current on V_{D} and V_{g} in the nonavalanche injection is as follows. The electrons arriving at the Si–SiO$_2$ interface, after acceleration in the depletion layer, are considered to have a Maxwell–Boltzmann distribution with effective temperature T_{e} (Ning and Yu, 1974). Then the Schottky emission current may be written as

$$J_{\text{ox}} = A_{\text{s}} T_{\text{e}}^2 \exp\left[-q(\phi_{\text{B}} - \beta_{\text{S}} F_{\text{ox}}^{1/2})/kT_{\text{e}}\right], \quad (9)$$

where A_S is a constant and $\beta F_{ox}^{1/2}$ again gives the lowering of the energy barrier by the oxide field. The electron temperature is related to the field in the silicon by (Mönch, 1971)

$$kT_e = \pi(ql_p F_{Si})^2/8E_p, \qquad (10)$$

where l_p is the mean free path of the hot electrons for inelastic phonon collisions and E_p is the energy loss in these collisions. For the field F_{Si}, the field at the interface is taken. However, Ning and Yu (1974) showed that this procedure results in inconsistencies, probably because the electrons do not move in a constant field; they do not reach equilibrium with the field. Therefore the first approach is preferred for a description of the current–voltage behavior.

VI. MIOS Devices

By the name MIOS we indicate a group of devices in which the first dielectric layer O on the silicon is thin oxide and the second layer may be any other dielectric I. As pointed out in Section IV, the MNOS device (N is nitride) is the most important memory device in this group. Therefore the main part of this section is devoted to the MNOS memory cell. The MAOS device (A stands for alumina) is discussed in Section VI,F. At the time of writing of this review its development has not reached industrial level. A development in which the O–N and O–A interfaces are intentionally doped is described in a separate section (VI,G).

We should like to mention here the suggested (Horninger, 1973; Mavor, 1973a,b; Mellor and Dunn, 1973) combination of the MNOS structure with avalanche injection in a short-channel MOS transistor. Goser and Knauer (1974) have suggested the combination of a CTD array with MNOS memory transistors. Shuskus et al. (1973) investigated a transistor in which the second dielectric consisted of rf-sputtered HfO_2 or $SrTiO_3$. Sewell (1974) has described the application of an MNOS transistor as a photosensitive memory element. All these seem interesting developments, and we mention them here before discussing the MNOS device in more detail in the following sections.

A. Switching Phenomena in MNOS

The acronym MNOS stands for metal–nitride–oxide–semiconductor. The basic structure of MNOS transistors is shown in Fig. 17. The figure shows a *p*-channel transistor, but charging and decharging characteristics apply equally well to *n*-channel transistors. The oxide layer is usually of the order of 20–30 Å for proper memory operation by tunneling, but structures with thinner and thicker oxides have been investigated. The nitride thickness is between 100 Å and 1000 Å. This nitride must have a high resistance for two reasons: first, to avoid charge loss or instabilities during switching, and, second, to avoid charge loss through the nitride layer during storage of the device, in order to improve the nonvolatility of the stored information. In order to obtain a nitride layer with high resistance there must be optimal preparation of the layer, usually obtained by the pyrolysis of SiH_4 and NH_3 in hydrogen as a carrier gas at a temperature of 700–800°C. A high NH_3 to SiH_4 ratio (> 100) gives the best result (Brown et al., 1968). In that case the current J_N through the second dielectric can be considered zero, in any case during the initial stage of the switching.

The switching is as follows. A voltage positive with respect to the silicon is applied to the metal gate of the transistor. This causes a high field in the thin oxide layer, whereupon electrons start to tunnel through this layer. They are trapped at or near the nitride–oxide interface. At not too low fields the current density in the SiO_2 can become relatively high, so that switching times may become very short.

When the gate voltage is negative with respect to the silicon, the electrons tunnel from the dielectric into the silicon, or holes tunnel from the silicon into the dielectric. This is further discussed below. In any case the amount of negative charge decreases and the threshold (turn-on) voltage of the MNOS transistor shifts in the opposite direction.

An example of two experimental ΔV_{TH}–log t curves is given in Fig. 29. They are two curves from the work of White and Cricchi (1972) showing the shift in threshold voltage as a function of switching time for two gate voltages ($V_g = -15$ V and $V_g = -20$ V). The shift ΔV_{TH} shows a logarithmic time behavior. The slopes of the curves at the two gate voltages are not

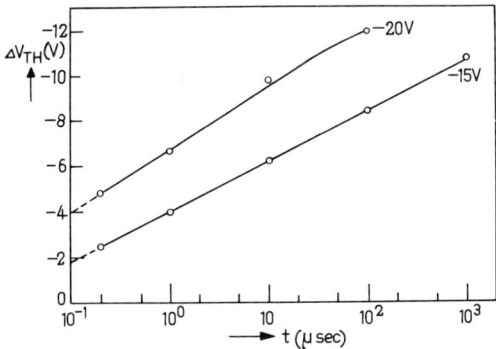

FIG. 29. Threshold voltage shift ΔV_{TH} as a function of write time t for two gate voltages. $d_N = 335$ Å, $d_{ox} \simeq 25$ Å; n-channel. Replotted from the paper of White and Cricchi (1972) (Courtesy of the Institute of Electrical and Electronics Engineers).

completely identical, although the difference is very small. In an earlier publication of Ross and Walmark (1969) the ΔV_{TH} versus $\log t$ curves, measured at several values of the gate voltage, all have the same slopes. A simple equation for the threshold voltage shift $\Delta V_{TH}(t)$ has been derived (Lundström and Svensson, 1972a), based on the assumption that the oxide current J_{ox} depends exponentially on the oxide field F_{ox} and that the nitride current is zero. This equation is

$$\Delta V_{TH}(t) = \left(d_{ox} + \frac{\varepsilon_{ox}}{\varepsilon_N} d_N \right) B^{-1} \ln \left(\frac{t}{t_s} + 1 \right), \qquad (11)$$

where ε_N is the permittivity of the nitride, d_N the thickness of the nitride, t the time variable, and B the slope of the $\ln J_{ox}$ versus F_{ox} curve. It should be noted that a tunnel current can be approximated by an exponential dependence on field, but then the quantity B depends slightly on the oxide field. Therefore the slopes of the curves in Fig. 29 depend on the value of V_g applied in the switching. The constant t_s is given by

$$t_s = \frac{\varepsilon_{ox} + \varepsilon_N \dfrac{d_{ox}}{d_N}}{B J_{ox}(t=0)}. \qquad (12)$$

It is the time at which the time behavior of $\Delta V_{TH}(t)$ changes. For $t < t_s$ the logarithmic term in Eq. (11) is expanded and approximated by t/t_s and one has a linear behavior. For $t > t_s$ the 1 in the logarithmic term may be neglected and the increase in threshold voltage is a logarithmic function of time.

In Fig. 29 the values of t_s may be obtained to a good approximation by extrapolation of the curves to $\Delta V_{TH} = 0$. The values of t_s are a strong function of V_g in qualitative agreement with Eq. (12), remembering that J_{ox} depends exponentially on F_{ox} and $F_{ox}(t=0) \approx \varepsilon_N V_g / (\varepsilon_{ox} d_N + \varepsilon_N d_{ox})$.

For a description of the memory device one often uses the dependence of the threshold voltage shift ΔV_{TH} on gate voltage. From the description of the time dependence given above, it should be clear that this gate-voltage dependence changes with the switching time used in the measurement. An example of ΔV_{TH}-V_g plots is shown in Fig. 30. Similar plots can be found, for instance, in the paper of Walmark and Scott (1969). After each voltage pulse of different amplitude and length (write time) the memory transistor is switched back to the starting value of V_{TH} by a suitable pulse of opposite polarity. A hysteresis loop results when the gate is switched back and forth in equal voltage increments and at constant switching times. An example is shown in Fig. 31. The gate voltage V_g is increased and decreased in steps of 7 V and held at each gate voltage for 1 μsec. The extreme values of V_g were

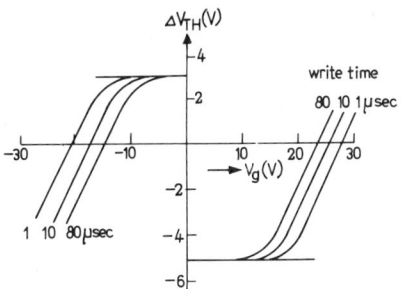

FIG. 30. ΔV_{TH} as a function of gate voltage V_g for several write times (Courtesy of R. H. W. Salters of Philips Research Laboratories).

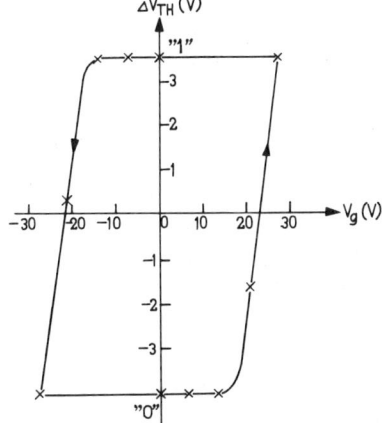

FIG. 31. Closed ΔV_{TH}-V_g curve, with indication of the two resulting memory states. Write time = 1 μsec. (Courtesy of R. H. W. Salters of Philips Research Laboratories.)

−28 V and 28 V. The V_{TH} values as a function of V_g are sequentially obtained, as indicated by the crosses in the loop and the arrows along the loop. The hysteresis loop resembles that of the curve of the magnetization as a function of the magnetic field (Fig. 2). For a magnetic core memory device the loop enables the definition of the digital memory states 0 and 1. In the same way this can be done for the hysteresis loop in Fig. 31. For instance, a memory 1 state is assigned to the high-V_{TH} condition and a memory 0 state to the low-V_{TH} condition. It should be noted that the shape of the hysteresis loop depends on the switching time.

In an MNOS transistor with relatively thick oxide ($d_{ox} \approx 100$ Å) and thin nitride ($d_N \approx 200$ Å) layers the current densities are changed in such a way that a deviating switching behavior is observed (Frohman-Bentchkowsky and Lenzlinger, 1969, Frohman-Bentchkowsky, 1970). The current J_N is increased and J_{ox} is decreased with respect to the thin-oxide/thick-nitride transistors. This makes it possible to observe the accumulation of positive charge at the SiO_2–Si_3N_4 interface when V_g is positive. The switching time is relatively long: 60 sec. This is because the Poole–Frenkel current density J_N in the nitride layer, which is now the larger of the two currents, does not reach as high a value as J_{ox} in the tunneling in thin oxide.

In this section we have described the oxide current in terms of a tunnel current. For a thick-oxide MNOS transistor with $d_{ox} \geq 100$ Å the current can be described by the model in Fig. 19. For the thin-oxide ($d_{ox} = 20$–30 Å) MNOS transistor the electrons from the Si conduction band do not tunnel into the SiO_2 conduction band but into the nitride conduction band (modified Fowler–Nordheim tunneling). This is discussed in Section VI,B. Another model is the tunneling of electrons from the Si valence band into traps at or near the nitride–oxide interface (Section VI,C).

B. Band–Band Tunneling

First, we give a description of the energy-band diagram of the MNOS system (Fig. 32). The figure is taken from the review paper of Balk (1974). It is to be noted that the Al–Si_3N_4 barrier is 2.1 eV. This is considerably lower than the energy difference (3.2 eV) between the bottom edges of the silicon and oxide conduction bands. Moreover, the energy difference between these band edges in the oxide and nitride is 1.1 eV. The latter quantity is important for a proper description of the band–band tunneling in MNOS structures.

Figure 33, in an energy-band diagram, gives the principles of band–band tunneling processes. The upper part of the figure gives the band diagram for

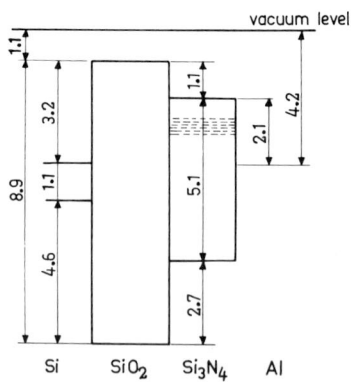

FIG. 32. Energy-band diagram of the MNOS system from the paper of Balk (1974) (Courtesy of Prof. Balk, Aachen, and The Institute of Physics, London).

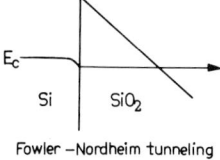

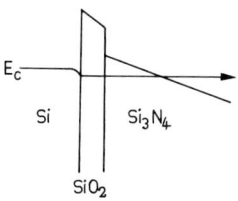

FIG. 33. Band–band tunneling in the Si–SiO$_2$ and in the Si–SiO$_2$–Si$_3$N$_4$ systems.

tunneling into a relatively thick oxide. This is the same as in Fig. 19, where the electron tunnels from the silicon conduction band into the oxide conduction band. In the lower part of the figure the situation in a relatively thin-oxide MNOS structure is sketched. The electron (see arrow in Fig. 33) tunnels from the silicon conduction band into the nitride conduction band. This is the modified Fowler–Nordheim tunneling mechanism (Svensson and Lundström, 1970). It should be noted that in both cases of Fig. 33 the energy diagrams are sketched for the same field in the oxide.

The equation describing the tunnel current in the modified Fowler–Nordheim process is based on an equation very similar to Eq. (1) (Svensson and Lundström, 1970; Lundström and Svensson, 1972a). The tunnel current may be written as

$$J_{ox} = A_{FN} F_{ox} P_{ox} P_N, \qquad (13)$$

where A_{FN} is a constant characteristic of Fowler–Nordheim tunneling (see Section V,A) and is assumed to be independent of the actual form of the barrier. Furthermore, the constant A_{FN} is assumed to be the same for both hole and electron tunneling. P_{ox} and P_N are the transmission probabilities through the oxide and nitride, respectively. The results of the calculation of t_s in Eq. (12) and the measurements on two transistors by Lundström and Svensson (1972a) are given in Fig. 34. It should be noted that t_s is the smallest time constant in the charging in which hardly any charge has yet accumulated. Then the nitride field $F_N(0) \approx V_g/d_N$.

The experimental results showed a disagreement with the calculated curves for the sample with $d_{ox} = 15$ Å and low nitride fields. This has been ascribed to an excess of current arising from a trap-assisted tunneling

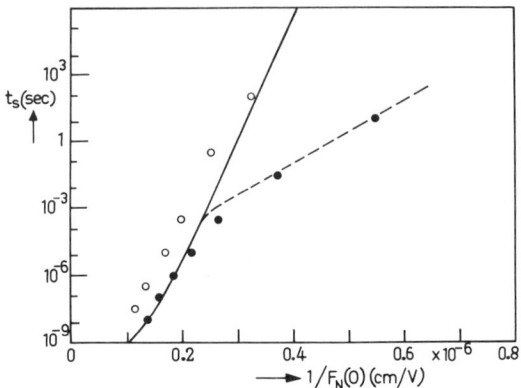

FIG. 34. Measured and calculated values of t_s in the MNOS devices of Lundström and Svensson (1972a). d_{ox} = (●) 15 Å; (○) 30 Å. (Courtesy of the Institute of Electrical and Electronics Engineers.)

process (Svensson and Lundström, 1973). This model of trap-assisted tunneling in MNOS is shown in Fig. 35. Process I is the tunneling from the silicon to traps in the nitride at a single energy level $q\phi_t$ below the nitride conduction band edge. Process II is the tunneling from traps into the nitride conduction band. The latter is shown to be a very fast process for the samples considered. So, in the tunneling equation, two extra parameters appear: the trap depth $q\phi_t$ and the concentration. A good fit of the calculated curves to the experimental points has been obtained by choosing $q\phi_t = 0.7$ eV and a value for the concentration of about 10^{22} cm^{-3} (Svensson and Lundström, 1973).

After the injection into the nitride conduction band the electrons are trapped. It should be noted that electron injection is extensively discussed in terms of the modified Fowler–Nordheim tunneling, but that hole injection has also been described in terms of this theory. After the trapping of the charge carriers, i.e., electrons, the oxide field decreases and the nitride field increases. This leads to a decrease of the oxide current and an increase of the

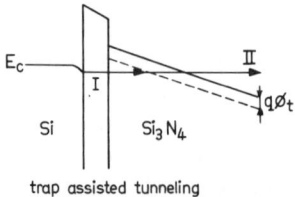

FIG. 35. Trap-assisted tunnel model for calculating the values of t_s at low nitride fields (dashed line in Fig. 34).

nitride current until these two currents are equal. Using the usual equations for the current in the nitride, namely, the Poole–Frenkel conduction current, the Swedish investigators have calculated the maximum charge stored in the MNOS double layer (Svensson, 1971; Lundström and Svensson, 1972a; Carlstedt and Svensson, 1972). A good agreement has been found supporting their model of modified Fowler–Nordheim tunneling for the oxide current J_{ox}.

As usual, we call the electron injection the write operation. In the model of modified Fowler–Nordheim tunneling, the erase operation is performed by the injection of holes in the same way, of course, in this case by tunneling from the silicon valence band to the nitride valence band, where the holes recombine with the electrons from the write operation present in traps. This model is supported by the experiments of Gordon and Johnson (1973), who showed qualitatively that holes can be injected into the nitride valence band. For the hole injection the model has also been extended by a trap-assisted tunnel process (Svensson and Lundström, 1973).

C. Band–Trap Tunneling

Band–trap tunneling is another model developed to describe the charging of the dielectric in the MNOS memory device. In this model the electrons tunnel from the silicon valence band to traps in the nitride. Figure 36 shows schematically the energy-band diagram. The arrow

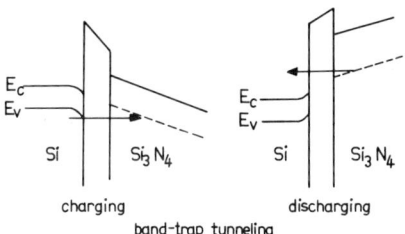

FIG. 36. Energy-band diagram for the band–trap tunneling model in the $Si-SiO_2-Si_3N_4$ system.

indicates the electron trajectory from the silicon into a trapping state in the nitride. This can be considered as the charging in the write operation. Then the erase operation is by back-tunneling from the trapping states to the silicon conduction band. This is indicated by the arrow in the right-hand part of Fig. 36. Therefore, in this model of band–trap tunneling, the tunneling of holes does not occur.

The trap states are drawn spatially distributed in the nitride. They are also drawn at one energy level. This is the model used by Ross and Walmark

(1969), who were the first to give a quantitative treatment of the charging processes in the MNOS transistor. Ferris-Prabhu (1973) generalized the model to include the case in which the trapping centers are distributed in energy. In this treatment the change in charge density $dQ(x, E, t)$ is given by

$$dQ(x, E, t) = q\{N(x, E, t)P(x, E, t)\}\, dx\, dE\, dt, \qquad (14)$$

where x is the spatial variable counted from the oxide–nitride interface, E is the energy, t is the time variable, $N(x, E, t)$ is the concentration of electrons in trapping states, and $P(x, E, t)$ is the tunneling probability of the electrons. The rate equation (Ross and Walmark, 1969) is:

$$dN(x, E, t)/dt = -N(x, E, t)P(x, E, t). \qquad (15)$$

The tunneling probability is approximated by

$$P(x, E, t) = P_B \exp(-x/\lambda) \equiv P(x), \qquad (16)$$

where P_B is a constant with the dimensions of reciprocal time and which depends only on the shape of the potential barrier. An expression for λ is

$$\lambda = \hbar/(8qm_{ox}^* \phi_{eff})^{1/2}, \qquad (17)$$

where $q\phi_{eff}$ is the effective magnitude of the energy barrier. The approximation for $P(x, E, t)$ allows one to find a solution of the rate equation. This solution is substituted into Eq. (14) and the time integral evaluated, giving

$$Q_1 = q\lambda \int_E dE \int_x dx\, N(x, E)\{1 - \exp[-tP(x)]\}. \qquad (18)$$

A further simplification is that the distribution of centers is considered homogeneous in space, giving

$$N(x, E) = N_0 f(E), \qquad (19)$$

where $f(E)$ indicates the energy distribution function of the trapping centers.

For not too short charging times and monoenergetic traps one again obtains (Ross and Walmark, 1969), as in the modified F–N model (Section VI,B),

$$\Delta V_{TH} \sim \log(t/t_s). \qquad (20)$$

This can be understood qualitatively because the electrons must tunnel through a continuously increasing distance x which gives a dispersion of time constants (Lundström et al., 1970) and consequently a log t dependence in the charging. Ross and Walmark's expression for t_s, of course, contains P_B from Eq. (16), and no explicit function for P_B has been derived, but only the way in which t_s (or better, t', for which $\Delta V_{TH} = 0$ in their $V_{TH}-t$ expression) varies with V_g. Ross and Walmark (1969) derived $t' \sim (V_g)^{-1}$ and in their Fig. 7 a good agreement is found between experimental results on t' and the $(V_g)^{-1}$ dependence.

The saturation of the ΔV_{TH}–log t curve in the model of Ferris-Prabhu occurs because the electrons have a maximum tunnel distance; hence, beyond this distance no charge can be trapped. This distance has been estimated to be about $x = 34$ Å. However, this is not in agreement with the measurements of Yun (1973, 1974), who found that the spatial distribution of the trapped electrons extends deep into the nitride layer. Actually, he found (Yun, 1974) that the centroid of the spatial distribution of the trapped charge in a Si_3N_4 film about 500 Å thick can be as much as 150 Å from the SiO_2–Si_3N_4 interface.

Dorda and Pulver (1970) use a model with all traps at $x = 0$ (at the oxide–nitride interface) but distributed homogeneously in energy. This gives log t dependence over one time decade, which is not in agreement with the experimental results of others. The model of White and Cricchi (1972) also considers the trap as being located at the oxide–nitride interface but having one energy level. In this way it does not seem either a very realistic model in the light of the experimental results mentioned above. However, White and Cricchi (1972) assumed that the nitride current $J_N \neq 0$, which is not the case in the other band–trap tunneling models.

In these models it is calculated how the transition probability and, hence, the oxide current vary with the oxide field. Thus the constant B in the oxide current versus field relation in Eq. (11) is calculated. The preexponential factor is not known. In the case of band–band (modified Fowler–Nordheim) tunneling, this factor was simply assumed to be equal to the preexponential factor A_{FN} in the standard Fowler–Nordheim tunneling. Lundström and Svensson (1972b) calculated the preexponential factor for the case of band–trap tunneling. Actually they calculated the transition probability for tunneling. They start by calculating the matrix element for the transition with the aid of the appropriate wave function for the electron in the semiconductor and in the trap. The trap is characterized by a δ-function potential in three dimensions. A relation similar to Eq. (16) has been derived, namely

$$P = P_0 \exp\left(-d_{ox}/\lambda\right). \tag{21}$$

An expression for P_0 has been derived and the numerical value calculated: $1/P_0 = 6.6 \times 10^{-14}$ sec. This is in good agreement with the experimental results (Lundström and Svensson, 1972a).

Now the crucial question arises as to which model is to be used in a description of the tunnel processes in the MNOS memory devices: the band–band or the band–trap tunneling model. It is possible that the actual tunnel process depends heavily on the properties of the oxide and nitride layers, namely the concentration and distribution in space of the trapping centers. The transition probability depends on the density of the states to which the electrons tunnel (Dorda and Pulver, 1970) and on the tunnel distance (see above). Thus, if the density of traps is low or when they are distributed into the nitride, the direct tunnel transition probability will be low and the band–band tunneling will have a higher probability. In this way it can be understood that the memory behavior depends not only on the oxide thickness (Ross et al., 1970), but also on the nitride properties (Goodman et al., 1970; Kobayashi and Ohta, 1973; Naber and Lockwood, 1973; Tanabashi and Kobayashi, 1973).

D. Charge Retention

The MNOS memory devices have a linear and a logarithmic part in the threshold voltage decay curve (White and Cricchi, 1972, Lundkvist et al., 1973). In the case of the MNOS device the charges in the trapping states at or near the oxide–nitride interface are able to tunnel back to the silicon (Walmark and Scott, 1969). This gives a change of the threshold voltage in the direction of the original threshold. Consequently, information may be lost from the memory device when the shift in threshold voltage becomes too low to be detected as a memory state different from the original state.

Two quantities are important in the charge retention: the time t_d at which the decay curve changes from linear to logarithmic decay and the slope $\partial(\Delta V_{TH})/\partial(\log t)$ of the latter. Two quantitative models have been developed to account for the discharge. In the model of Lundkvist et al. (1973) the charge carriers tunnel from traps in the nitride to the silicon conduction band. For $t > t_d$ (see below) Lundkvist et al. derived

$$\Delta V_{TH}(t) = \Delta V_{TH}(\text{initial})[1 - (\alpha_N d_T)^{-1} \ln (t/t_d)], \qquad (22)$$

where $\Delta V_{TH}(t)$ is now the shift in threshold after a time t (storage time). The parameter α_N is from the relation for the transition probability [see Eq. (16)]

$$[P(x)]^{-1} = P_0^{-1} \exp (\alpha_{ox} d_{ox}) \exp (\alpha_N x). \qquad (23)$$

The constants α_{ox} and α_N contain the effective barrier heights and trap depths. The parameter d_T in Eq. (22) takes into account that the traps in the nitride are not all filled in this model. The electrons, supposedly injected by the modified Fowler–Nordheim tunneling, have a certain mean free path d_T in the nitride before they are trapped. In this way, the concentration of trapped electrons decays from a high concentration at the oxide–nitride interface ($x = 0$) to a zero concentration deep in the nitride. The decay length is d_T. It should be clear that this phenomenon strongly depends on the concentration and capture cross section of the trapping centers.

Lundkvist et al. (1973) derived for t_d in their model

$$t_d = 0.56 P_0^{-1} \exp(\alpha_{ox} d_{ox}). \tag{24}$$

In Fig. 37 we have replotted the results of Lundkvist et al. (1973) on t_d as a function of oxide thickness d_{ox}. We see that Eq. (24) is in good agreement with the experimental results. From the extrapolation to oxide thickness $d_{ox} = 0$ the value of P_0^{-1} is determined: $P_0^{-1} \approx 10^{-12}$ sec. The value calculated by Lundström and Svensson (1972b) and mentioned in Section VI,C is $P_0^{-1} = 6.6 \times 10^{-14}$ sec.

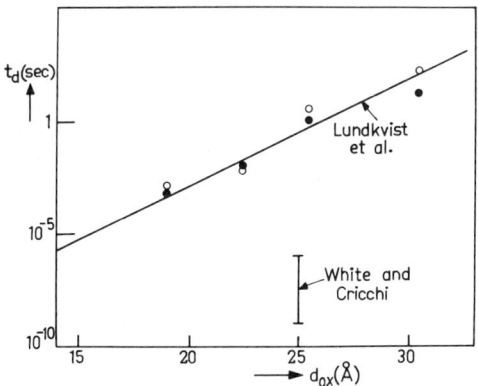

FIG. 37. Experimental results of Lundkvist et al. (1973) and White and Cricchi (1972) on t_d as a function of oxide thickness d_{ox} in MNOS devices. (●) Electrons; (○) holes. (Courtesy of Pergamon Press.)

In Fig. 37 we have also given the experimental result of White and Cricchi (1972) on t_d. Their point does not lie on the experimental curve of Lundkvist et al. (1973). This is probably due to a difference in the constant α_{ox} in Eq. (24) originating from a difference in trap level, although this point is not clear from the experimental results. One should note that White and Cricchi have measured transistors with only one oxide thickness of 25 Å.

The slope $\partial(\Delta T_{TH})/\partial(\log t)$ in the decay curves depends on the initial value of the threshold voltage V_{TH}. In Fig. 38 we have again plotted some experimental results of the Swedish investigators (Lundkvist et al., 1973) combined with those of White and Cricchi (1972). The dashed line drawn through the experimental points of the latter is purely tentative because more results are lacking. The dashed lines are drawn parallel to the curves of Lundkvist et al. (1973) because there is possibly a difference in the initial charge present in the dielectric layers in the form of fixed oxide charge. This

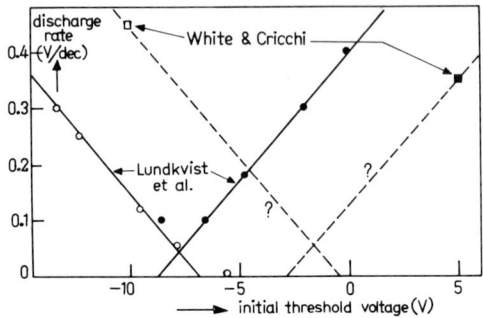

FIG. 38. Experimental results of Lundkvist et al. (1973) and White and Cricchi (1972) on the discharge rate as a function of the initial threshold voltage. (●, ■) Electrons; (○, □) holes. (Courtesy of Pergamon Press.)

causes the points of minimum decay rate, determined by the crossing of the two solid lines on the one hand and the two dashed lines on the other, to lie at different values of the initial threshold voltage. Lundkvist et al. (1973) calculate a value of $d_T \approx 40$ Å from the slopes of their curves for the charge decay length in the nitride.

In the model of White and Cricchi (1972) the charge is also determined by back-tunneling of electrons from traps to silicon, but the traps are thought to be at the oxide–nitride interface and tunneling is not to the silicon conduction band but to states at the oxide–silicon interface. The interface state density ρ_{st} is assumed to be a function of the surface potential ϕ_s as

$$\rho_{st} = \rho_0 \exp (\phi_s/\phi_0), \qquad (25)$$

where ρ_0 and ϕ_0 are characteristic constants of the interface state distribution. In the case of electron back-tunneling, this is the distribution in the upper half of the Si band gap.

It is interesting to estimate the time $t_{1/2}$ for which in the decay $\Delta V_{TH}(t)/\Delta V_{TH}(\text{initial}) = \frac{1}{2}$. From Fig. 14 in the paper of White and Cricchi (1972) we estimate by extrapolation $t_{1/2} = 6 \times 10^8$ sec for $\Delta V_{TH}(\text{initial}) = 8$ V. One should note that the value of $t_{1/2}$ is not equal to the information-retention time before which the information must be refreshed, but is only an indication of the retention capability. The information-retention time is determined by the lowest level (lowest ΔV_{TH}) that can be detected, the decay from the other memory state, the spread in switching properties [spread in $V_{TH}(\text{initial})$ in the memory array] and the influence of gate voltage on the back-tunneling of charge carriers during the read operation.

The retention properties can be varied by varying the oxide thickness and by varying the nitride properties. A thicker oxide decreases the back-tunneling of electrons from traps, thus improving the charge retention, but also increasing the write time because the tunnel probability in the charging process also decreases. Hence one has to find a compromise, which seems to be $20 < d_{ox} < 30$ Å for most applications of the MNOS device.

The nitride can be enriched in silicon by a deviation in stoichiometry. Then many trapping centers may be present at the oxide–nitride interface, increasing the trapping at that interface and thus increasing the writing speed. But this increases the decay, so that in this case also the best choice depends on the type of application and specification desired. Lockwood *et al.* (1972) and Naber and Lockwood (1973) studied the nitride growing systematically, and they found that (write time) × (decay rate) = constant. For their application they achieved a write time of about 10 μsec and an information-retention time of at least 1 yr.

E. Degradation in MNOS

In MNOS devices a peculiar degradation occurs after many write–erase operations: the threshold voltages in the memory 0 and 1 states drift slowly under the influence of (identical) write–erase operations. In the MNOS devices this may also give a decrease in the difference between the 0 and 1 states. The degradation phenomena are sketched in Fig. 39. Creation of

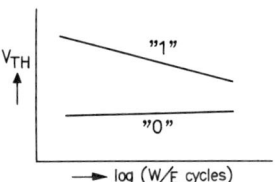

FIG. 39. Switching degradation (schematic) in the MNOS device resulting in a decrease of the logic window after many switching operations.

Si–SiO$_2$ interface states occurs (Woods and Tuska, 1972) which results in a reduced charge-retention capability and a reduced channel conductance. It has been shown that both the number of interface states near the valence band and near the conduction band in the silicon are increased (Woods and Tuska, 1972).

The actual reaction scheme in the degradation processes is not clear. Relatively high fields strongly increase the degradation. High fields are necessary for fast switching in the usual random-access memories. However, in this application the number of required write–erase operations is many orders larger than the 10^6 cycles obtainable in conventional MNOS memory devices.

A considerable improvement in device performance with respect to degradation can be obtained with the use of a double gate-insulator thickness structure, which is shown in Fig. 40 (Naber and Lockwood, 1973; Brewer, 1974; Koo, 1974). The thickness of the SiO_2 layer near the source and drain regions is at least 400 Å, so that the tunnel processes take place in the middle of the channel region where the oxide thickness is about 25 Å. In this structure and using moderate write and erase voltages (write time about 10 μsec) it is possible to obtain at least 10^{11} write–erase cycles.

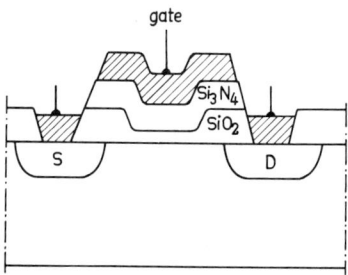

FIG. 40. Double gate–insulator thickness MNOS with partial thin oxide in order to diminish the effect of the switching degradation.

F. MAOS Memory Devices

The metal–alumina–oxide–silicon (MAOS) transistor also has two dielectric layers; the oxide layer is relatively thick (50–100 Å) compared to this layer in the MNOS transistor (20–30 Å), but the alumina layers are in the same range of thicknesses as those of the nitride layers. The oxide layer is still sufficiently thin for electron injection by tunneling from the silicon under the influence of a positive gate voltage. The electrons are trapped at or near the oxide–alumina interface and this gives a shift ΔV_{TH} in the threshold or turn-on voltage of the transistor. We call this again the write operation. ΔV_{TH} is again a logarithmic function of the write time and the curves of ΔV_{TH} versus log t at different write gate voltages are nearly parallel (Sato and Yamaguchi, 1974) in the same way as in the MNOS devices (Ross and Walmark, 1969). Lundström and Svensson (1972a) have shown that the model of modified Fowler–Nordheim tunneling (Section VI,B) is also applicable to the write operation in the MAOS device.

The question arises: what is the advantage of the MAOS transistor? The answer is that the dielectric constant of the aluminum oxide is higher than that of the nitride layer. At the same gate voltage and the same thicknesses of the dielectric layers this gives a higher field in the oxide layer of the MAOS structure. For the oxide field F_{ox} (at $Q_I = 0$) in the MNOS device the following relation exists:

$$F_{ox} = (\varepsilon_N/\varepsilon_{ox})F_N . \qquad (26)$$

Therefore, the oxide field is proportional to the ratio of the two permittivities, $\varepsilon_N/\varepsilon_{ox}$. For the nitride–oxide double layer this ratio is 1.8 and for alumina–oxide it is 2.3. This illustrates the importance of the aluminum oxide in the MIOS devices, because high SiO_2 fields facilitate the tunnel process.

However, three phenomena at the moment cause inferior performance of the MAOS compared to the MNOS device: (1) the poor Si–SiO_2 interface properties in thin SiO_2 ($d_{ox} < 80$ Å) structures, which necessitate structures with $d_{ox} \geq 80$ Å and the application of relatively high write and erase gate voltages; (2) injection of electrons from the negative Al electrode and subsequent conduction through the Al_2O_3 layer results in an asymmetric hysteresis loop (Fig. 41); (3) a switching degradation gives a maximum of about 10^4 write–erase cycles. These phenomena are discussed briefly.

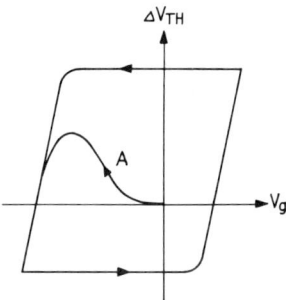

FIG. 41. Hysteresis loop (schematic) in the MAOS memory transistor, showing a strong asymmetry. Curve A is observed in the as-grown device.

The interface properties in the MAOS device are poor compared to those in the MNOS device, but they strongly improve with increasing SiO_2 thickness. For instance, the effective mobility of the electrons in a transistor channel increases from 40 cm^2 V^{-1} sec^{-1} at $d_{ox} \approx 0$ to 460 cm^2 V^{-1} sec^{-1} at $d_{ox} = 110$ Å (Nakagiri and Wada, 1972). The latter value of the effective mobility can be increased to the value in a MOS-transistor channel (without alumina as a second dielectric) by a heat treatment of the MAOS ($d_{ox} = 110$ Å) structure in a hydrogen ambient.

It is possible to start from a structure with one single layer of Al_2O_3 in order to obtain an MAOS structure. The Al_2O_3 layer is annealed in oxygen at 800°C to 900°C and the oxygen atoms diffuse through the Al_2O_3 layer to form an SiO_2 layer at the silicon surface (Duffy et al., 1971; Iida and Tsujide, 1972). The growth of an SiO_2 layer below the Al_2O_3 layer in the oxygen heat treatment is proved by the He^+ back-scattering technique (Kamoshida et al., 1972).

At this point a few remarks are made concerning the peculiar shape of the hysteresis loop in Fig. 41. As stated above, the switching at positive gate voltages is ascribed to the tunneling of electrons from the silicon to trapping centers at or near the SiO_2–Al_2O_3 interface. At negative gate voltages two injection processes occur. In an as-grown sample a bump is seen in the ΔV_{TH} versus V_g curve. This means that for negative gate voltages, the threshold voltage moves in the same direction as under the application of a positive gate voltage. It has been found that electrons are injected from the Al electrode to centers in the Al_2O_3 and transported through this layer by Poole–Frenkel conduction. That this is correct is proved by capacity measurements with a mercury drop (Tsujide, 1972), platinum dots (Tsujide and Iida, 1972a), or gold dots (Balk and Stephany, 1971) instead of aluminum. In this case the bump (now in the V_{FB} versus V_g curve) is absent. Moreover, for $d_{ox} \geq 100$ Å the bump is nearly independent of the oxide thickness (Tsujide and Iida, 1972b), which makes injection at the SiO_2 side less likely as a mechanism for the negative charge accumulation. In Fig. 42 we have

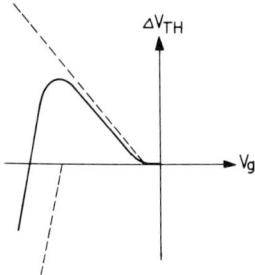

FIG. 42. Possible explanation of curve A in Fig. 41 in terms of two injection mechanisms (dashed lines) with different V_g dependences.

sketched, following Tsujide (1972), the two shifts which are thought to cause the bump in the ΔV_{TH}–V_g curve in Fig. 41. The upper (dashed) curve is due to the electron injection through the Al electrode followed by transport through the Al_2O_3 layer. The electrons are possibly injected via traps. There are many centers present in the Al_2O_3 in the energy range of 2.3 to 4.5 eV below the conduction band (Mehta et al., 1972; Harari and Royce, 1973). The electron injection gives a shift in flat band voltage (or threshold voltage) corresponding with negative charge in the double layer. The lower dashed curve represents the hole injection by tunneling from the silicon. At a certain voltage the current in this process becomes dominant and the flat band voltage shifts in the opposite direction, indicated by the solid line in Fig. 42. In the second and the following write–erase cycles this bump is not observed because the field in the alumina is too low due to the negative charge from the write operation.

It should be noted that the electron injection at the Al side can be reduced by annealing in oxygen (Balk and Stephany, 1971). Moreover, MAOS structures with relatively thick SiO_2 layers can be made that are stable against bias stress (Gnadinger and Rosenzweig, 1974). Therefore it seems possible to make MAOS memory devices with sufficiently low electron injection at the Al side. This would significantly improve the symmetry of the hysteresis loop in Fig. 41.

The following performance has been obtained in switching (Sato et al., 1972; Sato and Yamaguchi, 1974). Write within 10 μsec, a shift $\Delta V_{TH} = 10$ V at $V_g = +40$ V and erase within 10 msec at -40 V. This is done in a structure with an oxide thickness of 50 Å and an alumina thickness of 700 Å.

The charge retention in the MAOS memory devices is reasonably good. This is possibly due to the relatively thick SiO_2 layer which makes back-tunneling to the Si less probable (Salama, 1971). Tsujide (1970) found an initial fast decay within 10 min, followed by a very slow decay. The charge retention seems to be determined by the Poole–Frenkel conduction through the Al_2O_3 layer (Sato and Yamaguchi, 1974). It should be noted that in the as-grown MAOS device, negative charge is present at the A–O interface, as is shown by etch-off experiments (Aboaf et al., 1973; Kalter et al., 1971). This may be useful in other than memory devices (Kalter et al., 1971).

In the MAOS devices, a degradation occurs under the action of many write–erase cycles. The logic window decreases, i.e., the threshold voltages in the memory 0 and 1 states approach each other. Moreover, Si–SiO_2 interface states are formed (Salama, 1971). The degradation is very pronounced under negative gate bias (Sato and Yamaguchi, 1974). In this way it has some resemblance to the switching degradation in MNOS (White and Cricchi, 1972). It should be noted that in both devices the injection of holes is often assumed as an erase mechanism. It is possible that the injection of holes and the subsequent trapping in the silicon oxide is the origin of the degradation (see also Section VII). The maximum number of write–erase cycles is 10^4 to 10^5 (Sato et al., 1972; Sato and Yamaguchi, 1974). This is much lower than in the MNOS devices (10^{11} times or more).

G. MIOS Transistors with Interfacial Doping

In this memory device a few atomic layers of metal are present between the insulator I and the oxide O (Kahng et al., 1974). These metal layers give rise to many trapping states at the I–O interface. In this way the interfacial doping by metal atoms considerably increases the writing speed at a given tunnelable oxide thickness. Laibowitz and Stiles (1971), who report on interfacial doping by a layer of Pt or Ag (≈ 30 Å), have stated that the charge trapping takes place on small metal particles with a diameter of 30 to 50 Å. The number of trapped electrons is about 1.5 per particle.

Kahng et al. (1974) studied the properties of the MIOS devices with interfacial doping in much detail. They used several metals for doping, and silicon nitride and aluminum oxide as the outer insulator layer. However, all their quantitative results were obtained on tungsten (W) doping in MAOS and MNOS structures. The doping levels suitable for the proper memory function are between 10^{14} and 5×10^{15} atoms per cm^2. The tungsten atoms have the desired properties of not migrating during deposition of dopant and outer layer and of having a low vapor pressure. The last named avoids evaporation of tungsten during the first phase of the outer layer growth. The thicknesses of the oxide layer are the same as those used in other MAOS devices, 50–150 Å (see Section VI,F). The write operation is again performed by tunneling through the SiO_2 under the influence of a gate voltage bias. When a positive V_g is applied, the electrons tunnel from the silicon to the trapping states either directly or, more likely, via the SiO_2 conduction band by the Fowler–Nordheim mechanism.

The increase in write speed by the interfacial doping is spectacular, as can be seen in Fig. 43, which is taken from the paper of Kahng et al. (1974). It should be noted that the slope of the curves is about 2 V/decade. A shift in the threshold voltage of about 8 V is obtained in 10 µsec under $V_g = 30$ V. Without interfacial doping the shift is negligible. We mention this write time and voltage because this is comparable to MNOS practical devices (Lockwood et al., 1972). In the latter devices the shift ΔV_{TH} is about the same after this write operation. Hence in the MIOS device of this section we have the same write time, but in a structure with a thicker SiO_2 layer.

The erase operation again shows a spectacular improvement compared to that in MAOS devices. Starting with an as-grown device, the trapping of negative charge under negative gate bias is absent. The negative charge present from a write operation is removed in times comparable to that in the write operation.

In MAOS devices with W interfacial doping the charge retention is very good. The quantitative aspects are dealt with in the paper of Thornber et al. (1974). The decay of the charge in the trapping centers is found to be due to a

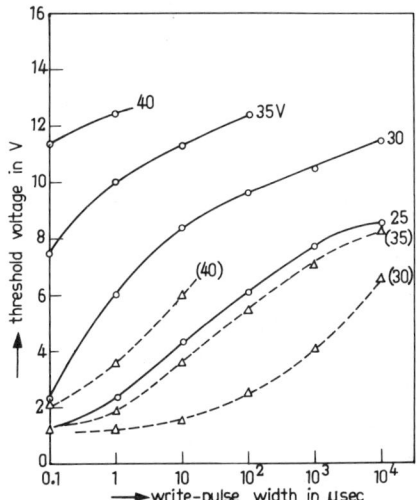

FIG. 43. Threshold voltage as a function of injection time in MIOS devices with (solid lines) and without (dashed lines) interfacial doping, with gate voltage as a parameter. Al_2O_3, 520 Å; SiO_2, 70 Å; W, 1.5×10^{15} cm^{-2}; n channel. (From Kahng et al., 1974, courtesy of the American Telephone and Telegraph Company.)

current through the aluminum oxide layer. The origin of the decay current is thought to be due to Poole–Frenkel conduction. However, the measured temperature dependence of the slope $\partial(\Delta V_{TH})/\partial(\log t)$ of the decay curve does not fit in this model. Thornber et al. (1974) state that this may be due to high field effects which make the carrier velocity a rapidly varying function of field. This has not been verified, so that the physical aspects of the decay current remain unclear. It is to be noted that knowledge of the actual conduction mechanism is not essential in the prediction of the charge retention at operating temperatures.

To predict the charge retention, and hence to estimate the nonvolatility, the charged and uncharged devices have been tested at temperatures up to 300°C and at several gate voltage biases. Emphasis was placed on the measurement of time t_d (see Section VI,D) because the slope of the logarithmic part of the decay curve is relatively steep. Therefore the charge retention is mainly determined by t_d. Thornber et al. (1974) derive from their tests a charge-retention time of hundreds of years at 80°C. This may be considered as a very good figure in terms of nonvolatility of the information storage.

In MNOS and MAOS devices we encountered the phenomenon of switching degradation. Nothing has been reported about this degradation in devices with interfacial doping.

In conclusion, it can be stated that the interfacial doping gives a considerable improvement in writing speed. However, investigation of switching degradation and determination of the possible number of write–erase cycles is necessary. It will be interesting to test the effect of interfacial doping in thin oxide (25 Å) MNOS devices.

VII. Floating-Gate Devices

In floating-gate devices, a conducting layer of metal or, usually, silicon is embedded between two oxide layers and is completely insulated from the remainder of the device. Kahng and Sze (1967) suggested a structure similar to the MIOS devices in Section VI, but with this floating gate as the electron trapping medium. Hence the first layer of oxide should be relatively thin in order to have tunneling of electrons to the floating gate. However, this device has never been realized, possibly because such thin oxide layers are difficult to make without any weak spot with low resistance. Such weak spots, of course, decrease the nonvolatility of the memory device. In the MIOS devices with interfacial doping (Section VI,G), there is no conducting path between the trapping centers parallel to the silicon surface. Then, a weak spot in the thin-oxide layer is not prejudicial to the nonvolatility because charge is lost only locally.

For the above reasons, all floating-gate devices have a relatively thick first oxide layer of about 1000 Å. Therefore they must have other means for the write and erase operations. The FAMOS device, the first floating-gate device realized on an industrial scale, has avalanche injection of the electrons into SiO_2 as the write operation. A detailed description and discussion of this device is given in the next section. However, it cannot be erased electrically. Sections VII, B and C describe structures in which this difficulty has been solved more or less successfully.

A. *FAMOS Memory Device*

The FAMOS memory device is shown schematically in Fig. 18. FAMOS stands for *F*loating-gate *A*valanche-injection *MOS* transistor. It is a *p*-channel MOS transistor without an external gate, but with a floating silicon gate completely insulated from the remainder of the device. On the floating gate a chemical vapor-deposited layer of SiO_2 1 μm thick is present.

In the write operation, electrons are injected from the *pn* junction, either source or drain, into the SiO_2 layer in which they drift to the floating gate. In Section V,B we discussed the avalanche injection of electrons in more detail. In that section we described the measurement of the avalanche-injected electron current through the oxide when a voltage was applied externally to the gate. In the FAMOS device this is not possible, so that the question arises as to how a sufficient field is built up to obtain the avalanche injection. The answer is that the floating gate is capacitively coupled to the silicon, i.e., to the source, channel, and drain regions. A voltage is applied to the drain region, bringing the drain–substrate junction into breakdown. The floating gate initially keeps about the same potential as the substrate because the substrate–gate capacitance is much larger than the drain–gate capacitance. Therefore, above the drain region there is a field in the oxide of about V_D/d_{ox}, where V_D is the drain voltage. The equipotential lines are drawn in Fig. 44. A constriction of the space-charge layer of the avalanching junction is obtained (de Graaf, 1970) and, consequently, a high concentration of energetic carriers at the Si–SiO$_2$ interface (see Section V,B). The electrons with sufficient energy may be injected into the SiO$_2$, and then drift to the floating gate because this is positive with respect to the drain region.

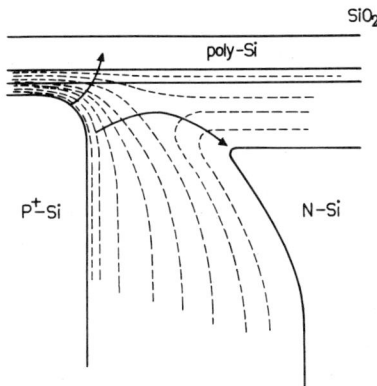

FIG. 44. Equipotential lines (dashed) in the cross section of the avalanche breakdown region in the FAMOS transistor. The arrows indicate two possible electron trajectories; the left-hand one contributes to the switching operation.

In Fig. 45 we reproduce Fig. 3 from the paper of Frohman-Bentchkowsky (1971a). It shows the amount of charge transferred to the floating gate as a function of write time (pulse time). The charge $\Delta Q_I/q = 2.0 \times 10^{12}$ cm^{-2} corresponds to $\Delta V_{TH} = 10$ V. It is observed that the characteristic time t_s (Section VI,A) is a strong function of the pulse amplitude V_D. For $V_D = 35$ V the value of t_s is about 1 μsec. The slope of the curves is about 5 V/decade. Thus a shift of 10 V in threshold voltage is obtained after an injection (write) time of 100 μsec. In the actual 2048-bit memory array a voltage pulse of -50 V for 5 msec is used for programming one bit. The difference from the data in Fig. 45 is ascribed to the additional voltage drop across the decode circuits (Frohman-Bentchkowsky, 1971b).

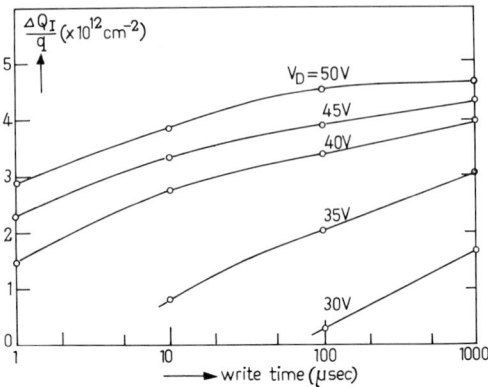

FIG. 45. Amount of charge transferred to the floating gate of a FAMOS device as a function of write time. The junction voltage V_D is the parameter (from Frohman-Bentchkowsky, 1971a, courtesy of the American Institute of Physics).

The erase is by UV irradiation or X-rays, and thus not by electrical means. The UV-erase operation has been thought to occur by photoemission of electrons at that side of the polycrystalline floating gate facing the single-crystal silicon. However, the (polycrystalline) silicon strongly absorbs the UV irradiation, so that emission may also take place at the other side of the floating gate, emitting electrons into the layer of 1-μm thick SiO_2. The electrons are ultimately trapped in this layer because there is no means of draining them from it. Also in the thermal oxide layer below the gate there may be some trapping in the erase operation as well as in the write operation. Presumably, these trapped electrons are not removed in the subsequent switching operation and they may degrade the switching operations by gradually decreasing the oxide field. It should be noted that the erase operation requires a couple of minutes of UV irradiation. This limits the use of the FAMOS memory device to applications for which not many write–erase cycles are required, and then a small decrease in the efficiency per switching operation is not observable.

Charge retention in the FAMOS device is relatively good. Plots of the familiar V_{TH} versus $\log t$ curves show straight lines with a slope of 0.2 V/decade at 125°C. The decrease of the charge on the floating gate is thermally activated. Therefore it is not a tunnel mechanism like Fowler–Nordheim emission of electrons into the SiO_2 conduction band, but it may be another conduction mechanism like Poole–Frenkel conduction. In any case, the value of t_d is expected to be very high and the curves in Fig. 4 of the paper of Frohman-Bentchkowsky (1971a), showing V_{TH} as a function of $\log t$, are still in the linear region, in other words, $t < t_d$. Frohman-Bentchkowsky (1974) has ascribed the information loss to hole trapping at the single-crystal side of the thermal oxide layer (Hofstein, 1967; Deal et al., 1967), giving as the sole reason equality of the activation energies. The measurement of t_d, possibly at higher temperatures, and the slope of the V_{TH}-log t curves for $t > t_d$ should give more information about the actual charge-loss mechanism. It should be noted that Breed and Kramer (1972) and Breed (1975) recently reinvestigated the hole-trapping phenomenon. They came to the conclusion that it was due to a mechanism of electron emission from the SiO_2 into the Si. In any case, apart from the actual charge-loss mechanism, the measurements of the charge retention in the FAMOS device yield an information storage time of about 100 yr.

At the time of writing this review, the realization of a 2048-bit (2k-bit) fully decoded memory array has been published (Frohman-Bentchkowsky, 1971b) while an 8k-bit array has been commercially announced. These arrays find application mainly in the product development centers as alterable ROMs and for small-quantity use. For use in larger quantities one has a mask-programmable ROM array (Section III,E) with the same number of bits and pinning, so that the array of FAMOS transistors is compatible with the mask-programmable ROM-array.

B. ATMOS Memory Device

The name of the ATMOS memory device is derived from *A*djustable-*T*hreshold *MOS* transistor (Verwey and Kramer, 1974). Figure 46 shows schematically a cross section of the device. An n-channel MOS transistor is furnished with an underlying supply junction A and a floating polycrystalline silicon gate C. The latter is embedded between two oxide layers, each about 0.1 μm thick. It should be noted that an external metal gate is neces-

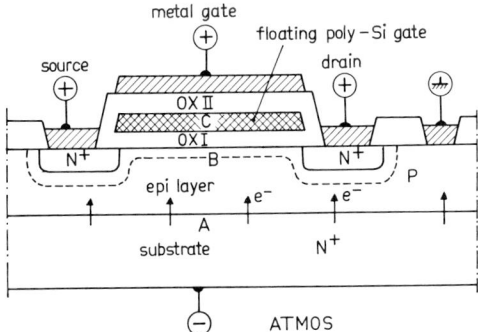

FIG. 46. Cross section (schematic) of the ATMOS memory transistor. The arrows indicate the electron injection during the write operation (from Verwey and Kramer, 1974, courtesy of the Institute of Electrical and Electronics Engineers).

sary in the write and erase operations, which is not the case in the FAMOS memory cell.

The memory action arises from a change in the charge content of the floating gate. In the ATMOS device the required negative charge is injected into the oxide by the nonavalanche injection described in Section V,D. The supply junction A (epilayer–substrate junction) is forward-biased, and the electrons injected into the epilayer diffuse to the depletion layer B in Fig. 46. In this layer they obtain sufficient energy to cross the Si–SiO$_2$ barrier (Verwey, 1973), and are injected into the oxide. Under the influence of the electric field in the oxide (gate voltage positive) they drift to, and are subsequently trapped on, the floating gate. This we call the write operation, and the threshold voltage is shifted in this case to higher positive values.

Figure 47 shows the shift ΔV_{TH} in threshold voltage as a function of injection time in two transistors, one with an epilayer resistivity $\rho_{epi} = 0.10$ Ω cm and the other with $\rho_{epi} = 1.1$ Ω cm. The source and drain voltage was 15 V. It can be clearly seen in Fig. 47 that the injection rate of

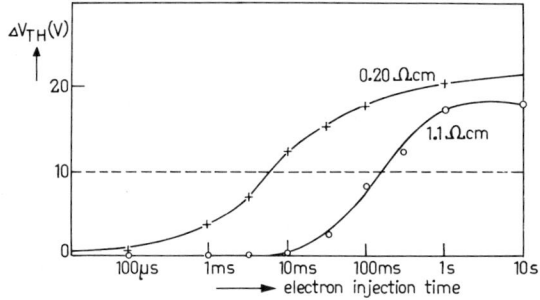

FIG. 47. Threshold voltage shift ΔV_{TH} as a function of injection time in the ATMOS transistor for two epilayer resistivities. The dashed line is for the derivation of the write time (from Verwey and Kramer, 1974, courtesy of the Institute of Electrical and Electronics Engineers).

electrons is highest for the lowest ρ_{epi}. This corresponds to the higher probability for injection in thinner depletion layers, as is discussed in Section V,D [see Eqs. (7) and (8)]. It is possible to derive the write time for an ATMOS device. If the write state of a transistor is defined by a shift $\Delta V_{TH} = 10$ V (dashed line), we derive from the intersection of the dashed and the measured curves a write time of about 10 msec for $\rho_{epi} = 0.20$ Ω cm and about 100 msec for $\rho_{epi} = 1.1$ Ω cm.

The dependence of write time on drain (and source) voltage V_D and on gate voltage V_g has not been measured. Saturation in the curves in Fig. 47 occurs because the voltage drop in the silicon has gone below the minimum value necessary for electron injection.

The erase operation in the ATMOS device is performed by means of the avalanche breakdown of the source or drain junction. Hot holes are injected into the oxide (see Section V,C), they drift to the negatively charged floating gate, and then neutralize the electrons present from the write operation. The threshold voltage shifts, but now in the direction opposite to that in the write operation. It has been found that the erase time in the ATMOS memory device is about 1 sec (Verwey and Kramer, 1974).

The information-retention time in the ATMOS cell is comparable to that in the FAMOS device. Some information about this property can be found in Fig. 48. This figure shows the relative change in threshold voltage as a function of storage time at three temperatures, and it is taken from the paper of Verwey and Kramer (1974). An analysis in terms of t_d (see Section IV,C) or the actual charge-loss mechanism has not been given.

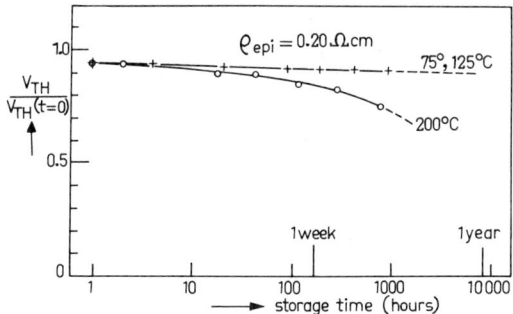

FIG. 48. Charge retention in ATMOS. The initial threshold voltage $\Delta V_{TH}(t=0)$ was about 20 V (from Verwey and Kramer, 1974, courtesy of the Institute of Electrical and Electronics Engineers).

The ATMOS memory device shows a switching degradation, namely a decrease of the logic window (Verwey and Kramer, 1974). The sweep in threshold between the written and erased condition decreases with increasing number of write–erase steps. This is to be expected because some of the charge carriers become trapped during their transport through the oxide layer. These trapped charge carriers decrease the field for the charge injection in the following switching operation, thus gradually decreasing the current density as a function of the number of write–erase steps. The trapped electrons are not neutralized in the subsequent hole injection and the trapped holes are not neutralized in the electron injection. This is because electron and hole injection occur at different places. If one out of 10^4 injected charge carriers is trapped in the oxide, the maximum number of write–erase steps is about 10^4. This is completely different from the situation in the MNOS devices, where the electrons trapped in the oxide or the nitride during the write operation may be neutralized in the next erase operation (see Section IV,B). This explains why the maximum number of write–erase operations in the MNOS devices is higher than in the ATMOS devices.

C. *Other Floating-Gate Devices*

The first device we discuss in this section is a modification of the FAMOS transistor. It is also a *p*-channel transistor but with an external gate (Card and Worrall, 1973a,b). The dielectric layer between the floating gate and the external gate is made of silicon nitride with a thickness of 1000 Å. The write operation is in the standard manner of avalanche injection of electrons, but the erase is by conduction through the nitride layer at relatively high fields, namely the application of +80 V for 1 sec restores the original threshold value before write. In this way, an electrically erasable memory device is made. It should be noted that a relatively high conduction is required through the nitride layer for the erase operation, but that, at low fields occurring during storage, this layer should have relatively low conduction in order to have a good information-retention time. Thus good information-retention time may not be compatible with low gate voltages during erase and with short erase times.

Iizuka *et al.* (1972) reported on a similar device, but without mentioning the actual material of the layer between the floating gate and the external gate. An array of 2048 bits has been made. It has been noted that a voltage on the external gate helps the write operation. The number of write–erase cycles is more than 100 (Masuoka *et al.*, 1973a). The erase time is 100 msec at $V_g = 70$ V and the information-retention time is longer than 10 yr at 125°C (Masuoka *et al.*, 1973a). It seems that the actual device has a first layer of 2000 Å thermal oxide below the floating gate and a second layer, also of 2000 Å thermal oxide, on top of it (Masuoka *et al.*, 1973b).

The second device to be discussed in this section is actually a group of devices developed by Tarui and his co-workers (Tarui *et al.*, 1972a,b). The first one in this group is of the double-junction type (Fig. 49a). A *p*-channel transistor is furnished with an extra n^+p junction which, in breakdown, supplies holes in the erase step.

The second device in this group is of the type known as channel injection (Tarui *et al.*, 1972a,b). In an *n*-channel MOS transistor the hole injection is from an avalanching source or drain junction (Fig. 49b), and in this way it is better organized than in the device shown in Fig. 49a. The electron injection is from avalanche processes at the high-field region around the channel, which can be considered as a step junction, where, as a consequence of the curvature, the highest field strength is found at the edges when a voltage is applied to source or drain (Conti and Conti, 1972). If this voltage is high

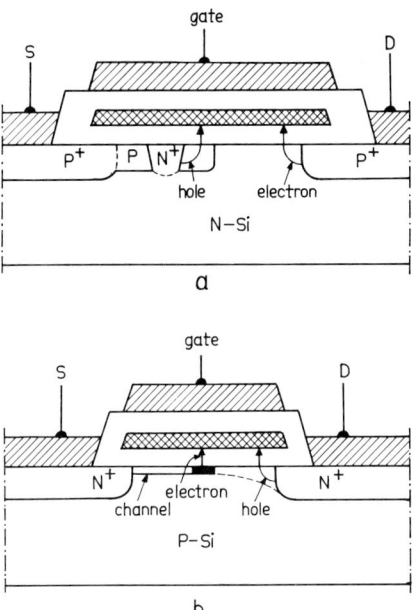

FIG. 49. (a) Double-junction type memory transistor. (b) Channel-injection type memory transistor (courtesy of the Institute of Electrical and Electronics Engineers).

enough, avalanche breakdown occurs and many electrons generated by impact ionization are injected into the oxide. This process seems to be poorly reproducible (Verwey and Kramer, 1974). It is represented schematically in Fig. 49b.

A substantial improvement in the electron injection is achieved by the plane injection mode (Tarui et al., 1973). An n-channel transistor has a local p diffusion in its channel region (Fig. 50). At this diffusion the depletion layer, induced by a positive gate voltage, is much thinner than in the remain-

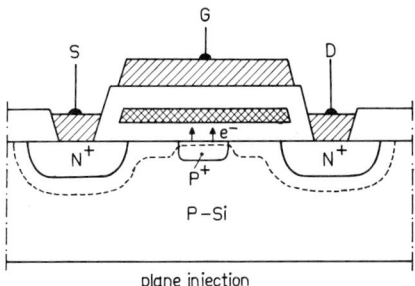

FIG. 50. Plane-injection type memory transistor.

der of the channel region. If now an increasing voltage is applied to the source and drain junction, the avalanche breakdown starts at this p diffusion, with the subsequent injection of electrons into the oxide.

It has been stated (Tarui et al., 1973) that this plane injection mode has some disadvantages with respect to the necessary gate voltages. It has also been stated that the bipolar injection mode should give an improvement in this respect. Electrons are injected from a forward-biased n^+p junction and accelerated in the depletion layer below the floating gate, thus making the electron injection similar to the nonavalanche injection in the ATMOS memory cell, but this time laterally. However, the write time in the ATMOS device is expected to be shorter, based on the geometrical considerations for the electron diffusion from the n^+p supply junction to the accelerating depletion layer.

VIII. Discussion and Conclusions

In the preceding sections we have discussed the physics of nonvolatile semiconductor memory devices. For the application of these devices the following properties are important.

1. Cost per bit.
2. Performance: speed, required voltages, maximum number of write–erase cycles, and reliability of operation.
3. Power consumption.
4. Compatibility with other technologies in the computer.

Nonvolatile semiconductor memory devices have to be compared with other nonvolatile memory devices described in Section II: magnetic cores have lost the competition with the semiconductor memories because of the speed and cost per bit, as is discussed in Section II,A. The bubble memories are described in Section II,B. It is possible that they score over semiconductor memories in those areas of application where:

a. Dimensions are important.
b. Nonvolatility is required.
c. Compatibility with CPU technology is not important.
d. Requirements as to access time are relatively low.

Ovonics have not yet found application because of their limited number of write–erase operations and the noncompatibility of the technology of devices and surrounding circuitry.

The requirement for nonvolatile information storage in general is imposed for the following two reasons:

a. to permit the storage of information without energy consumption;
b. to avoid loss of information when the power supply temporarily fails.

There are two other alternatives to nonvolatile semiconductor memories. First, the use of a semiconductor memory in combination with a (rechargeable) battery, and second, the use of semiconductor memories together with a back-up memory made of magnetic material (disc, tape, or drum). The latter, of course, underlines the necessity for a nonvolatile memory (at least part of it). In large computers the usual manner of construction is semiconductor memories in combination with large (and relatively very cheap) magnetic back-up memories. In small computers this is not possible, and there the first alternative enters the picture. It should be noted that this alternative has some disadvantages. These are the presence of the acid in the battery, which may be harmful to the electronic circuits in an apparatus, and the current that is needed by the memory together with leakage current of the battery, which limits the information-retention time. In this respect the minimum power from the battery that keeps the information in the memory is important.

The static semiconductor memory cell (Section III,B) always has one active device conducting. If the active device is a MOS transistor, then its current is lower than in the case of a bipolar transistor, but in both cases it is much higher than in the dynamic memory cell (Section III,C). In the dynamic cell a relatively short current pulse is required to charge a capacitor in order to refresh a memory state. The frequency of the refresh cycle can be diminished, for instance, from about 1–10 MHz in normal operation to 10 kHz in stand-by operation, so that the power consumption is considerably reduced.

Promising with respect to power consumption in normal and stand-by operation are CTD devices (Section III,D), complementary MOS, and integrated injection logic. In the CTD shift register the charge packets must be continuously shifted, detected, amplified, and fed back into the register. The frequency v of the voltage pulses on the electrodes can be diminished, but this depends strongly on the length of the register because each charge packet must be refreshed within a few milliseconds. For a not-too-long three-phase surface CTD with an electrode area of about 100 μm^2 and $v = 10$ kHz, a power consumption of 1–5 μW per bit in stand-by operation has been calculated (Boonstra and Sangster, 1972). In a complementary MOS (which is a combination of p-channel and n-channel MOS transistors) 1024-bit static memory, the power consumption in stand-by operation can be as low as 10 nW for the whole circuit (Cole, 1974). This is five orders of magnitude lower than that of a CTD device. This makes stand-by operation from a battery supply feasible. However, the complementary MOS circuits are more expensive due to a complicated technology and a six-transistor-per-bit cell. When low-power operation is also required, it is expected from Fig. 13 in Berger's paper (1974) that static semiconductor memories based on bipolar integrated injection logic (Hart and Slob, 1972; Berger and Wiedman, 1972) are still superior to complementary MOS circuits.

For many cases, true nonvolatile information storage is required, which cannot be achieved with the magnetic mass (drum, tape, or disc) memories because the apparatus containing the memory is relatively small. In these cases the RePROMs (Section IV) of semiconductor memories can be used. In the memory elements the information is stored in the form of charge in the gate dielectric of MOS transistors, while the charge injection is either by tunneling or by injection from a reverse-biased junction. The injection and conduction mechanisms in the dielectric are now fairly well known.

For good charge retention in the dielectric the dielectric layers between the storage side in the dielectric (traps, metal, or polycrystalline Si) and the outside world (gate or Si) must have a high resistance. A relatively thick layer seems favorable for obtaining good charge retention (i.e., nonvolatile information storage). However, switching degradation is observed in these devices: the efficiency of the switching operation decreases with increasing number of switching cycles. The origin of this effect is not clear at present. Presumably, the effect is due to the phenomenon of charge trapping when charge carriers are transported through (insulating) dielectric layers. The effect seems to be lowest when the layer is thin. This is and will remain in conflict with the other requirement of thick layers for high nonvolatility. The best compromise between high nonvolatility and low switching degradation is at the moment to be found in the thin-oxide ($d_{ox} = 30$ Å) MNOS transistor with a special gate structure (Fig. 40).

ACKNOWLEDGMENTS

The author is indebted to D. Breed, R. P. Kramer, J. H. Lohstroh, and J. G. van Santen of the Philips Research Laboratories for critically reading the manuscript and for useful comments. He especially wishes to thank R. H. W. Salters for his numerous contributions.

GLOSSARY OF SYMBOLS USED IN TEXT

A_{FN}	Constant in the equation for Fowler–Nordheim tunneling, Eq. (1)
A_{PF}	Preexponential factor in the Poole–Frenkel current–voltage relationship, Eq. (3)
A_S	Preexponential factor in the Schottky emission current–voltage relationship, Eq. (9)
B	Slope in the semilogarithmic current–voltage (field) relationships
d_N	Thickness of the second dielectric layer (nitride) in an MNOS memory transistor
d_{ox}	Thickness of silicon oxide layer on silicon
d_T	Decay length of the trapped charge in the nitride in the MNOS device

E	Energy variable
E_C	Bottom level of conduction band
$E_F = q\phi_F$	Fermi level
E_p	Energy loss of hot electron in a phonon generation process
E_V	Top level of valence band
F_N	Electric field in nitride layer in an MNOS memory transistor
F_{ox}	Electric field in silicon oxide layer on silicon
h	Shortest distance the electron travels in the nonavalanche injection to gain an energy $q\phi_B$
$\hbar$	Planck's constant divided by 2π
H	Magnetic field
I_D	Drain current in MOS transistor
I_{DS}	Saturated drain current
I_{ox}	Current in SiO_2
I_R	Diode avalanche current
I_t	Maximum value of avalanche-injected oxide current
J_N	Current density in the nitride layer of the MNOS memory transistor
J_{ox}	Current density in SiO_2 layer
J_w	Current density of electrons entering the depletion layer in the nonavalanche injection
k	Boltzmann's constant
l	Mean free path of hot electrons in silicon
l_p	Mean free path for phonon generation
m_0	Free electron mass
m_{ox}^*	Effective mass of electron in SiO_2
$N(x, E, t)$	Concentration of electrons in trapping centers in the Si_3N_4 layer of the MNOS transistor
N_0	Concentration of trapping centers at $x = 0$, Eq. (19)
N_t	Concentration of Poole–Frenkel centers, Eq. (2)
N_V	Effective density of states in the valence band
p	Hole concentration in SiO_2
p_t	Concentration of holes in Poole–Frenkel traps
$P(x, E, t) = P(x)$	Electron tunneling probability
P_B, P_0	Preexponential factors in the equations for $P(x)$, Eqs. (16) and (21), respectively
P_h	Injection probability in the nonavalanche injection
P_N	Transmission probability for tunneling through Si_3N_4
P_{ox}	Transmission probability for tunneling through SiO_2
q	Elementary charge
Q_I	Charge in trapping centers in the insulator or on floating gate
t	Time variable
$t_{1/2}$	Half-value time of charge decay in MNOS memory transistor
t_d	Smallest time constant in charge decay
t_s	Smallest time constant in charging of memory transistor
T	Temperature
T_e	Electron temperature
V_B	Breakdown voltage of silicon diode
V_D	Drain voltage in MOS transistor
V_g	Gate voltage in MOS transistor or gated diode
V_{TH}	Threshold (turn-on) voltage of MOS transistor

W	Depletion layer width in silicon
x	Space variable
α_N, α_{ox}	Constants in the tunneling model for charge decay in MNOS transistors, Eq. (23)
β	Poole–Frenkel constant
β_S	Schottky constant $\equiv (q/4\pi\varepsilon_{ox})^{1/2}$
ε_{ox}	Permittivity of SiO_2
ε_N	Permittivity of Si_3N_4
λ	Parameter in the direct tunneling model, Eq. (16)
v	Clock frequency in CTD
ρ_{epi}	Epitaxial-layer resistivity
ρ_0	Preexponential factor in the expression for ρ_{st}
ρ_{st}	Si–SiO_2 interface state density
ϕ_0	Constant in the expression for ρ_{st}, Eq. (25)
ϕ_c	Depth of Poole–Frenkel center
ϕ_1, ϕ_2, ϕ_3	Potentials of the three clock lines in the CTD
ϕ_B	Height of the potential barrier at the Si–SiO_2 interface
ϕ_{eff}	Effective potential barrier in tunneling
ϕ_F	Fermi potential
ϕ_t	Depth of trapping centers in Si_3N_4
ϕ_s	Surface potential

REFERENCES

Aboaf, J. A., Kerr, D. R., and Bassous, E. (1973). *J. Electrochem. Soc.* **120**, 1103.
Balk, P. (1974). *In* "Solid State Devices, 1973" (H. Weiss, ed.), p. 51. The Institute of Physics, London.
Balk, P., and Stephany, F. (1971). *J. Electrochem. Soc.* **118**, 1634.
Bartelink, D. J., Moll, J. L., and Meyer, N. J. (1963). *Phys. Rev.* **130**, 972.
Berger, H. (1974). *In* "Solid State Devices, 1973" (H. Weiss, ed.), p. 109. The Institute of Physics, London.
Berger, H., and Wiedman, S. K. (1972). *IEEE J. Solid-State Circuits* **7**, 340.
Boonstra, L., and Sangster, F. L. J. (1972). *Electronics* **45**, Feb. 28, 64.
Bosselaar, C. A. (1973). *Solid-State Electron.* **16**, 648.
Breed, D. J. (1975). *Appl. Phys. Lett.* **26**, 116.
Breed, D. J., and Kramer, R. P. (1972). *Thin Solid Films* **13**, 1.
Brewer, R. E. (1974). *In* "Proceedings of the National Aerospace and Electronics Conference, NAECON '74," p. 32. Institute of Electrical and Electronics Engineers, New York.
Brown, G. A., Robinette, W. C., and Carlson, H. G. (1968). *J. Electrochem. Soc.* **115**, 948.
Bulucea, C. (1975a). *Solid-State Electron.* **18**, 363.
Bulucea, C. (1975b). *Solid-State Electron.* **18**, 381.
Bulucea, C., Rusu, A., and Postolache, C. (1974). *Solid-State Electron.* **17**, 881.
Card, H. C., and Worrall, A. G. (1973a). *Electron. Lett.* **9**, 14.
Card, H. C., and Worrall, A. G. (1973b). *J. Appl. Phys.* **44**, 2326.
Carlstedt, G., and Svensson, C. M. (1972). *IEEE J. Solid-State Circuits* **7**, 382.
Carnes, J. E. (1974). *In* "Solid-State Devices, 1973" (H. Weiss, ed.), p. 83. The Institute of Physics, London.
Cole, B. (1974). *Electronics* **47**, Dec. 26, 111.
Collet, M. G., and Esser, L. J. M. (1973). *Festkörperprobleme* **13**, 337.

Conti, F., and Conti, M. (1972). *Solid-State Electron.* **15**, 93.
Crawford, R. H. (1967). "MOSFET in Circuit Design," p. 51. McGraw-Hill, New York.
Deal, B. E., Snow, E. H., and Mead, C. A. (1966). *J. Phys. Chem. Solids* **27**, 1873.
Deal, B. E., Sklar, M., Grove, A. S., and Snow, E. H. (1967). *J. Electrochem. Soc.* **114**, 266.
de Graaff, H. C. (1970). *Philips Res. Rep.* **25**, 21.
Dorda, G., and Pulver, M. (1970). *Phys. Status Solidi (a)* **1**, 71.
Duffy, M. T., Carnes, J. E., and Richman, D. (1971). *Trans. Metall. AIME* **2**, 667.
Druyvesteyn, W. F., van den Enden, W. A. M., Kuijpers, F. A., de Niet, E., and Verhulst, A. G. H. (1975). *In* "Solid State Devices, 1974" (L. A. Thomas, ed.), p. 37. The Institute of Physics, London.
Eimbinder, J. (1971). "Semiconductor Memories," p. 12. Wiley, New York.
Ferris-Prabhu, A. V. (1973). *IBM J. Res. Develop.* **17**, 125.
Frenkel, J. (1938). *Phys. Rev.* **54**, 647.
Frohman-Bentchkowsky, D. (1970). *Proc. IEEE* **58**, 1207.
Frohman-Bentchkowsky, D. (1971a). *Appl. Phys. Lett.* **18**, 332.
Frohman-Bentchkowsky, D. (1971b). *IEEE J. Solid-State Circuits* **6**, 301.
Frohman-Bentchkowsky, D. (1974). *Solid-State Electron.* **17**, 517.
Frohman-Bentchkowsky, D., and Lenzlinger, M. (1969). *J. Appl. Phys.* **40**, 3307.
Gnadinger, A. P., and Rosenzweig, W. (1974). *J. Electrochem. Soc.* **121**, 700.
Goodman, A. M., Ross, E. C., and Duffy, M. T. (1970). *RCA Rev.* **31**, 342.
Gordon, N., and Johnson, W. C. (1973). *IEEE Trans. Electron Devices* **20**, 253.
Goser, K., and Knauer, K. (1974). *IEEE J. Solid-State Circuits* **9**, 148.
Harari, E., and Royce, B. S. H. (1973). *Appl. Phys. Lett.* **22**, 106.
Hart, K., and Slob, A. (1972). *IEEE J. Solid-State Circuits* **7**, 346.
Hofstein, S. R. (1967). *Solid-State Electron.* **10**, 657.
Horninger, K. H. (1973). *Bull. Schweiz. Elektrotech. Ver.* **64**, 1258.
Iida, K., and Tsujide, T. (1972). *Jap. J. Appl. Phys.* **11**, 840.
Iizuka, H., Sato, T., Masuoka, F., Ohuchi, K., Hara, H., and Takeishi, Y. (1972). *Int. Electron. Dev. Meeting, Washington, late news paper* 7.6.
Kahng, D., Nicollian, E. H. (1972). *In* "Applied Solid State Science" (R. Wolfe, ed.), Vol. 3, p. 1. Academic Press, New York.
Kahng, D., and Sze, S. M. (1967). *Bell Syst. Tech. J.* **46**, 1288.
Kahng, D., Sundburg, W. J., Boulin, D. M., and Ligenza, J. R. (1974). *Bell. Syst. Tech. J.* **53**, 1723.
Kalter, H., Schatorjé, J. J. H., and Kooi, E. (1971). *Philips Res. Rep.* **26**, 181.
Kamoshida, M., Mitchell, I. V., and Mayer, J. W. (1972). *J. Appl. Phys.* **43**, 1717.
Kasperkovitz, D. (1973). *IEEE J. Solid-State Circuits* **8**, 251.
Kennedy, D. P. (1963). *Advan. Electron. Electron Phys.* **18**, 167.
Kobayashi, K., and Ohta, K. (1973). *Jap. J. Appl. Phys.* **12**, 881.
Koo, T. K. (1974). *In* "Proceedings of the National Aerospace and Electronics Conference NAECON '74," p. 37. Institute of Electrical and Electronics Engineers, New York.
Laibowitz, R. B., and Stiles, P. J. (1971). *Appl. Phys. Lett.* **18**, 267.
Lambrechtse, C. W., Salters, R. H. W., and Boonstra, L. (1973). *In* "IEEE International Solid-State Circuits Conference, Digest of Technical Papers," p. 26. Institute of Electrical and Electronics Engineers, New York.
Lenzlinger, M., and Snow, E. H. (1969). *J. Appl. Phys.* **40**, 278.
Lockwood, G. C., Naber, C. T., and Koo, T. K. (1972). *Wescon Tech. Papers* **16**, 1.
Luecke, G., Mize, J. P., and Carr, W. N. (1973). "Semiconductor Memory Design and Application," pp. 28, 29, 155. McGraw-Hill, New York.
Lundkvist, L., Lundström, I., and Svensson, C. (1973). *Solid-State Electron.* **16**, 811.

Lundström, I. J., and Svensson, C. M. (1972a). *IEEE Trans. Electron Devices* **19**, 826.
Lundström, I., and Svensson, C. (1972b). *J. Appl. Phys.* **43**, 5045.
Lundström, I., Christensson, S., and Svensson, C. (1970). *Phys. Status Solidi (a)* **1**, 395.
Masuoka, F., Ishikawa, M., Sato, T., and Iizuka, H. (1973a). *Int. Electron Dev. Meeting, Washington, late news paper* 7.8.
Masuoka, F., Iizuka, H., and Sato, T. (1973b). *In* "Extended Abstracts of the 143rd Electrochemical Society Meeting, Chicago, May 13–18, p. 221. The Electrochem. Soc., Princeton.
Mavor, J. (1973a). *Electron Lett.* **9**, 111.
Mavor, J. (1973b). *Electron. Lett.* **9**, 349.
Mehta, D. A., Butler, S. R., and Feigl, F. J. (1972). *J. Appl. Phys.* **43**, 4631.
Mellor, P. J. T., and Dunn, P. J. (1973). *Electron. Lett.* **9**, 252.
Mönch, W. (1971). *Festkörperprobleme* **9**, 172.
Naber, C. T., and Lockwood, G. C. (1973). *In* "Semiconductor Silicon, 1973" (H. R. Huff and R. R. Burgess, eds.), p. 401. The Electrochem. Soc., Princeton.
Nakagiri, M., and Wada, T. (1972). *Jap. J. Appl. Phys.* **11**, 1484.
Nakanuma, S., Tsujide, T., Igarashi, R., Onoda, K., Wada, T., and Nakagiri, M. (1970). *IEEE J. Solid-State Circuits* **5**, 203.
Ning, T. H., and Yu, H. N. (1974). *J. Appl. Phys.* **45**, 5373.
Ovshinsky, S. R., and Fritzsche, H. (1973). *IEEE Trans. Electron Devices* **20**, 91.
Pao, H. C., and Sah, C. T. (1966). *Solid-State Electron.* **9**, 927.
Pepper, M. (1973). *J. Phys. D., Appl. Phys.* **6**, 2124.
Pepper, M. (1974). *IEEE Trans. Electron Devices* **21**, 174.
Quartly, C. J. (1962). "Square-Loop Ferrite Circuitry," p. 2. Iliffe, London.
Riley, W. B. (1973). *Electronics* **46**, Aug. 2, 75.
Ross, E. C., and Walmark, J. T. (1969). *RCA Rev.* **30**, 366.
Ross, E. C., Goodman, A. M., and Duffy, M. T. (1970). *RCA Rev.* **31**, 467.
Russell, A. (1965). *Advan. Electron. Electron Phys.* **21**, 249.
Rutz, R. F., Harris, E. P., and Cuomo, J. J. (1973). *IBM J. Res. Develop.* **17**, 61.
Salama, C. A. T. (1971). *J. Electrochem. Soc.* **118**, 1993.
Sangster, F. L. J., and Teer, K. (1969). *IEEE J. Solid-State Circuits* **4**, 131.
Sato, S., and Yamaguchi, T. (1974). *Solid-State Electron.* **17**, 367.
Sato, S., Yamaguchi, T., and Aoki, T. (1972). *In* "IEEE International Solid-State Circuits Conference, Digest of Technical Papers," p. 188. Institute of Electrical and Electronics Engineers, New York.
Sewell, F. A. (1974). *IEEE Trans. Electron Devices* **20**, 563.
Shockley, W. (1961). *Solid-State Electron.* **2**, 35.
Shuskus, A. J., Quinn, D. J., and Cullen, D. E. (1973). *Appl. Phys. Lett.* **23**, 184.
Smit, J., and Wijn, H. P. J. (1954). *Advan. Electron. Electron Phys.* **6**, 69.
Svensson, C. (1971). *Proc. IEEE* **59**, 1134.
Svensson, C., and Lundström, I. (1970). *Electron. Lett.* **6**, 645.
Svensson, C., and Lundström, I. (1973). *J. Appl. Phys.* **44**, 4657.
Sze, S. M. (1967). *J. Appl. Phys.* **38**, 2951.
Sze, S. M. (1969). "Physics of Semiconductor Devices," p. 539. Wiley, New York.
Tanabashi, K., and Kobayashi, K. (1973). *Jap. J. Appl. Phys.* **12**, 641.
Tarui, Y., Hayashi, Y., and Nagai, K. (1972a). IEEE International Solid-State Circuits Conference, Digest of Technical Papers, p. 52. Institute of Electrical and Electronics Engineers, New York.
Tarui, Y., Hayashi, Y., and Nagai, K. (1972b). *IEEE J. Solid-State Circuits* **7**, 369.
Tarui, Y. Hayashi, Y., and Nagai, K. (1973). *In* "International Conference on Solid-State Devices, Digest of Technical Papers," p. 203. Chamber of Commerce, Tokyo.

Thornber, K. K., Kahng, D., and Neppell, C. T. (1974). *Bell. Syst. Tech. J.* **53**, 1741.
Tsujide, T. (1970). *J. Electrochem. Soc.* **117**, 703.
Tsujide, T. (1972). *Jap. J. Appl. Phys.* **11**, 62.
Tsujide, T., and Iida, K. (1972a), *Jap. J. Appl. Phys.* **11**, 600.
Tsujide, T., and Iida, K. (1972b). *Jap. J. Appl. Phys.* **11**, 1599.
Verwey, J. F. (1972a). *J. Appl. Phys.* **43**, 2273.
Verwey, J. F. (1972b). *Appl. Phys. Lett.* **21**, 417.
Verwey, J. F. (1973). *J. Appl. Phys.* **44**, 2681.
Verwey, J. F., and Kramer, R. P. (1974). *IEEE Trans. Electron Devices* **21**, 631.
Verwey, J. F., and de Maagt, B. J. (1974). *Solid-State Electron,* **17**, 963.
Verwey, J. F., Kramer, R. P., and de Maagt, B. J. (1975). *J. Appl. Phys.* **46**, 2612.
Walmark, J. T., and Scott, J. H. (1969). *RCA Rev.* **30**, 335.
White, M. H., and Cricchi, J. R. (1972). *IEEE Trans. Electron Devices* **19**, 1280.
Woods, M. H., and Tuska, J. W. (1972). *In* "Proceedings of The Annual Reliability Physics Symposium, 10th, p. 97. Institute of Electrical and Electronics Engineers, New York.
Yun, B. H. (1973). *Appl. Phys. Lett.* **23**, 152.
Yun, B. H. (1974). *Appl. Phys. Lett.* **25**, 340.

High-Power Electronic Devices

GEORGE KARADY

*Hydro-Québec Institute of Research,
Varennes, Québec, Canada*

   I. Introduction 311
  II. Semiconductor Devices 313
     A. The *pn* Junction 313
     B. Diodes 314
     C. Transistors 314
     D. Thyristors 316
     E. New Power Electronic Devices 328
     F. Accessories 332
 III. Thyristor Systems 336
     A. Thyristor Strings 337
     B. Thyristor Arrays 338
     C. Firing of Thyristor Systems 338
 IV. Development of Calculation Methods 341
     A. Thyristor Models 342
     B. Thyristor System Models 347
  V. Fields of Application 354
     A. Rectification (ac to dc Converters) 354
     B. Power Control 358
     C. Frequency Conversion 359
     D. High-Voltage dc Transmission 362
     E. Special Thyristor Applications 366
 VI. Conclusion 367
     References 368

I. Introduction

The introduction of the thyristor or silicon controlled rectifer (SCR) in 1957 was to revolutionize the field of electric power control. Since then, solid-state power electronics has established a foothold throughout the electric power industry, from high-voltage direct-current transmission systems, exciters for large generators, and ac and dc motor drives, to lighting control systems and household appliances.

Although the thyristor is the most widely used of these power electronic devices, power diodes and transistors are finding increasing application.

This paper will concentrate on the thyristor and related special devices, but the diode and transistor will also be discussed briefly.

The extraordinary speed with which semiconductor devices have developed, thanks to a constantly improving technology, has resulted in a marked decrease in the price of thyristor components despite worldwide inflation. Another interesting indication of the growing role played by SCR devices in power electronics is the upsurge of literature related to thyristors and their application, as shown in Fig. 1. Its quasilinear increase in the last

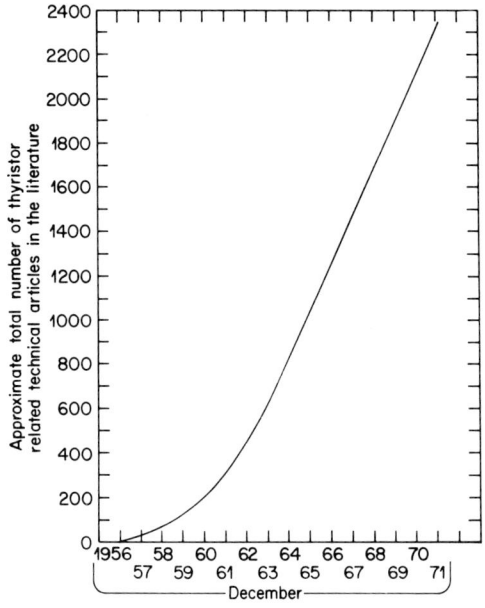

FIG. 1. Thyristor literature growth (not including patents, application notes, or internal company reports) (*1*).

few years suggests that thyristor-based power electronics has attained that particular stage of development in which new devices and new applications are constantly coming to light. Indeed, from the viewpoint of its literature, the thyristor seems far from attaining saturation point, which normally indicates that utilization of the device or technology concerned has reached its full potential. Indeed the volume of sales of thyristors as well as of thyristor-based equipment has maintained a steady rise over the years (*1*).

The only aim a review paper can modestly hope to achieve at this stage in the rapidly evolving history of the thyristor is to determine present trends in both the development and the application of the device and to describe improvements in the design and calculation techniques. Since a complete review of the literature is, from a practical point of view, impossible (there are over 2300 papers on the subject to date), the present work is based on a subjective selection of papers.

Various papers in these *Advances in Electronics and Electron Physics* (2) deal with the physics of semiconductors and related subjects, but do not cover the development of the power electronics field. For the purpose of this paper, it is felt that a short summary of the manufacturing process, operating principles, major parameters, and failure modes might be expedient before discussing development trends and new devices.

II. SEMICONDUCTOR DEVICES

A. The pn Junction

The operating principle of semiconductor devices is based on the properties of the *pn* junction. The *n*-type semiconductor contains impurity atoms (donor type) in its crystal structure which have one more valence electron than the atoms of the original semiconductor material. Since this extra electron cannot form an electron-pair bond, the *n*-type material consequently contains excess electrons bonded very loosely by the impurity atoms (2). On the other hand, the *p*-type material contains impurity atoms (acceptor type) having one less valence electron than the original semiconductor atoms, which results in a vacancy or hole in the structure (3). When a *pn* junction is formed, the electrons are driven by diffusion from the *n* region to the *p* region, where the excess electrons combine with the holes; this produces a free space charge on each side of the junction, as shown in Fig. 2. The free

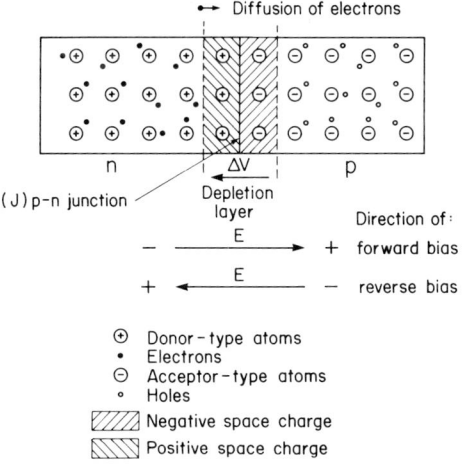

FIG. 2. A *pn* junction with depletion layer (3).

space charges build up a potential difference (ΔV) across the junction which balances the effect of diffusion and brings the system to equilibrium. The result of this process is the formation of a transition region or depletion layer at the junction. The junction is said to have a "reverse bias" when the voltage source is connected to the semiconductor in such a way that the n region is positive with respect to the p region. In this case, the electric field removes the free charges from the semiconductor, the depletion layer widens, and very little leakage current flows through the pn junction, which offers a high resistance in this direction. A "forward bias" of the pn junction is obtained when the voltage source is so connected that the p region is positive with regard to the n region. Here, the electric field drives the free charges through the junction, the depletion layer narrows, and new electrons enter from the electrodes on the n side while a high current flows through the pn junction, which in this direction offers a low resistance.

B. Diodes

The power diode consists of two alternating layers of n- and p-type semiconductor materials to form a pn junction. When the junction is reverse-biased (the n layer, cathode, is positive to the p layer, or anode), the diode blocks the voltage and a current of only a few mA flows through. When the junction is forward-biased, the diode is conducting and the current depends on the external circuit.

The power diode consists basically of an n-type silicon disc attached to a molybdenum substrate plate. A thin layer of p-type material is diffused on or alloyed to the disc to form the junction. The upper layer is metallized to form the anode. Finally the wafer is encapsulated and sealed in a porcelain cylinder.

Power diodes with a blocking voltage of up to 3 kV are available commercially, and devices rated up to 10 kV have been built. The current rating of the diode is in the range of 600–800 A; however, devices of up to 4–5 kA already exist. Owing to their similarity with regard to cooling, mounting, etc., diodes will be discussed together with thyristors.

C. Transistors

The power application of transistors is a recent development, but already they have begun to replace thyristors in certain types of inverters, converters, and switching regulators for ac and dc machine control. However, the transistor is not a typical device in power engineering, although its application is expected to become more widespread in the future.

The transistor consists of a *pnp* or *npn* silicon wafer with three electrodes: an emitter, a base, and a collector. Several types of transistor exist, offering various applications. The single-diffused construction, for example (Fig. 3) is used for high current and low voltage. It comprises a *p*- or *n*-type silicon wafer, both sides of which are diffused simultaneously by *n*- or *p*-type material, respectively. A mesa, diffused on one side of the wafer, forms the emitter, a ring-shaped metal deposit forms the base, and a copper substrate on the other side of the wafer forms the collector.

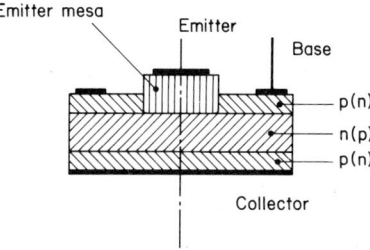

FIG. 3. Cross section of a single-diffused power transistor.

In most power applications, the small base current controls the much larger collector current, which increases from a small leakage current value almost linearly with the base current but reaches saturation above a certain base-current value. This type of transistor is used as a power amplifier within the linear range and as a switch when it is driven into saturation. In the latter case, removal of the gate current turns off the device immediately. Unlike the thyristor, it is able to interrupt the current, but requires constant drive for conduction. Its turn-off time is very short and the voltage can be reapplied almost immediately after it has ceased to conduct. This is particularly advantageous in medium-power inverter circuits.

The transistor requires a much higher base drive current than the thyristor. Most high-power thyristors can be turned on by a current peak in the range of 1–5 A with a duration of about 20–100 μsec, whereas the high-power transistor with a collector current of, say, 250 A typically requires a constant base current of $\approx$ 20–30 A during the entire conduction period. The base-current requirement may be reduced by applying a second transistor in a Darlington arrangement; a typical circuit arrangement is shown in Fig. 4. Recently high-current modulator transistors have been built with the driver transistor and the high-current transistor together on the same wafer and forming a single unit.

The present state of the art (4) of power transistors may be summarized as follows:

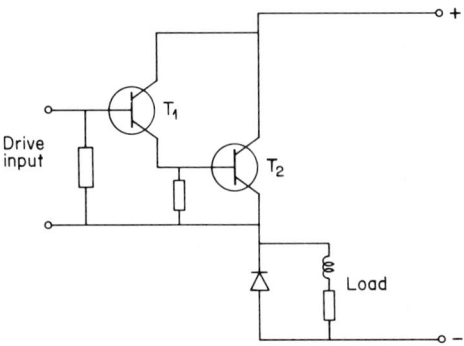

FIG. 4. Methods for decreasing the drive current of a power transistor with the Darlington arrangement. (The two transistors are built on one chip.)

(a) Single-diffused transistors have a voltage of up to 100 V and a collector current of up to 250 A. The voltage may be raised to 140 V, but if this is done the current drops to 150 A. The saturation-voltage drop is about 1.8 V with a 250-A collector current; the typical turn-on time is 2.5 μsec.

(b) Triple-diffused transistors are built for higher voltages—up to 400 V—but for current of only 10 A, while still lower current devices are available for up to 2200 V. The typical turn-on time here is 0.5 μsec. This type of transistor is suitable for up to 20 kHz if used as a switch.

(c) Multiple-epitaxial devices using the latest double-epitaxial double-diffused technology are suitable for a collector current of 50 A at 300 V.

One of the most frequent failures in transistors is the "second breakdown," in which localized energy concentrations produce hot spots and subsequent burnout of the silicon material. Present-day devices now contain a built-in resistance in the emitter structure which prevents excessive local power concentrations, while expected increases in the operating voltage and current should make power transistors competitive with thyristors in many applications.

D. Thyristors

1. Structure and Principle of Operation

The thyristor consists of four alternating layers of n- and p-type semiconductor (silicon) materials (5) to form a *pnpn* structure or wafer. The cross section and plan view of a typical power thyristor are shown in Fig. 5.

The basic stages of the manufacturing process are shown schematically

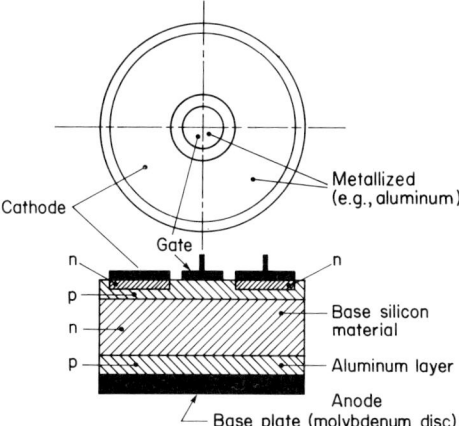

FIG. 5. Cross section of a power thyristor.

in Fig. 6. In the final stage, not illustrated in this figure, the wafer is encapsulated and hermetically sealed against environmental effects by placing it in a porcelain cylinder closed by a metal plate at each end. The plates serve as the anode and cathode. For low-resistance electric and thermal connections, the thyristor heat sink is pressed to the device with a pressure of 500–1000 kg. Since the operation of the device depends critically on the temperature, the low-resistance thermal connection and a suitable cooling technique are essential to ensure the effective cooling of the wafer.

The foregoing outline of the manufacturing process, although brief, is nevertheless sufficient to permit following of the development trend described in the pages that follow. Meanwhile, more detailed information may be obtained from the literature (6).

The thyristor comprises a three-junction arrangement (Fig. 7). When the anode is negative with respect to the cathode and J_1 and J_3 are reverse-biased and J_2 forward-biased, the two reverse-biased junctions block the voltage and the thyristor is in the off state (Fig. 7a). When the anode is positive with respect to the cathode (Fig. 7b), and J_1 and J_3 are forward-biased and J_2 reverse-biased, the reverse-biased J_2 blocks the voltage and the thyristor again is in the off state (current very low, practically zero). For the device to turn on, the voltage must be connected between the gate (positive) and the cathode (Fig. 7c). The gate voltage then drives the current through the forward-biased function, J_1, and electrons enter layer n_1 and are driven through junction J_1 into layer p_1. Some of these electrons flow to the gate, while others are affected by the positive anode and are driven through junctions J_2 and J_3. This process destroys the blocking capability of junc-

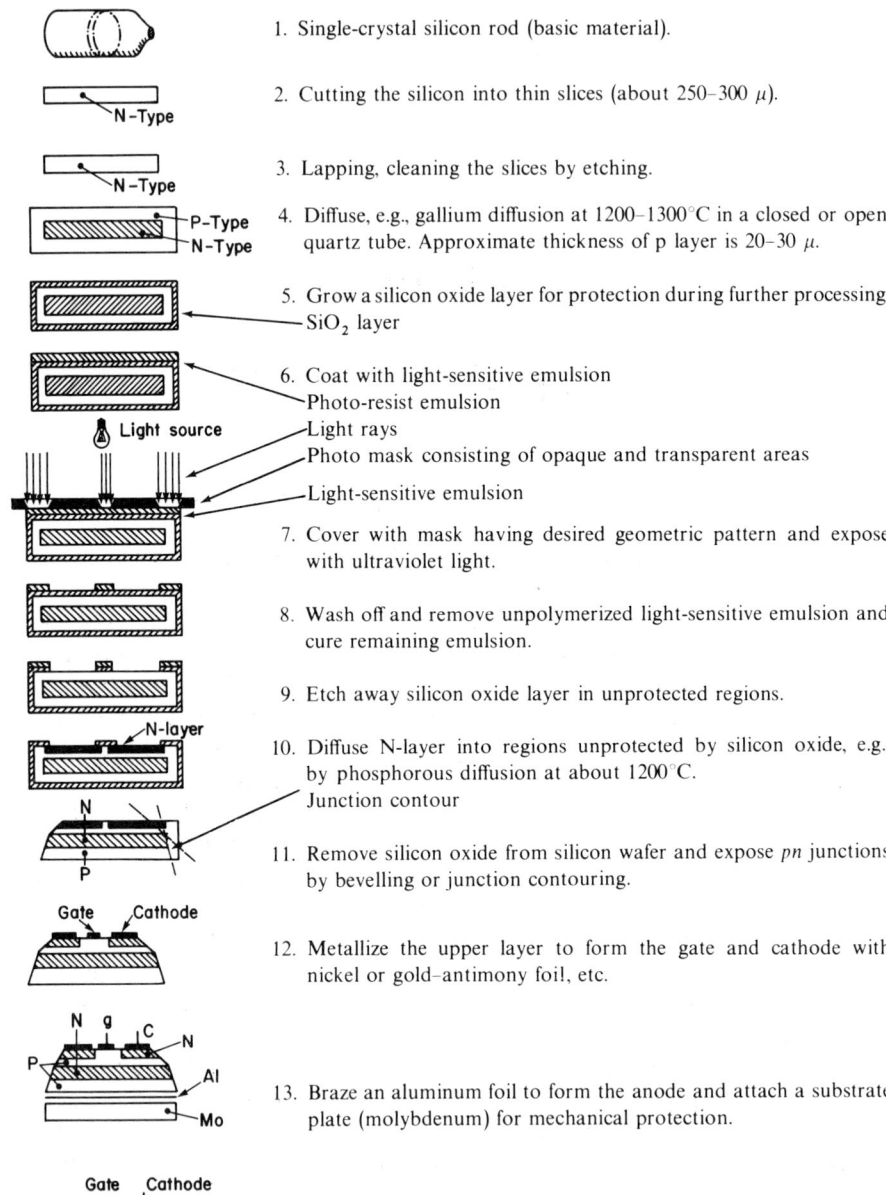

1. Single-crystal silicon rod (basic material).

2. Cutting the silicon into thin slices (about 250–300 μ).

3. Lapping, cleaning the slices by etching.

4. Diffuse, e.g., gallium diffusion at 1200–1300°C in a closed or open quartz tube. Approximate thickness of p layer is 20–30 μ.

5. Grow a silicon oxide layer for protection during further processing.

6. Coat with light-sensitive emulsion

7. Cover with mask having desired geometric pattern and expose with ultraviolet light.

8. Wash off and remove unpolymerized light-sensitive emulsion and cure remaining emulsion.

9. Etch away silicon oxide layer in unprotected regions.

10. Diffuse N-layer into regions unprotected by silicon oxide, e.g., by phosphorous diffusion at about 1200°C.

11. Remove silicon oxide from silicon wafer and expose pn junctions by bevelling or junction contouring.

12. Metallize the upper layer to form the gate and cathode with nickel or gold–antimony foil, etc.

13. Braze an aluminum foil to form the anode and attach a substrate plate (molybdenum) for mechanical protection.

14. Cover the edges with organic protective material; attach or weld the electrodes to the metallized surfaces.

FIG. 6. Basic steps in the manufacture of a thyristor.

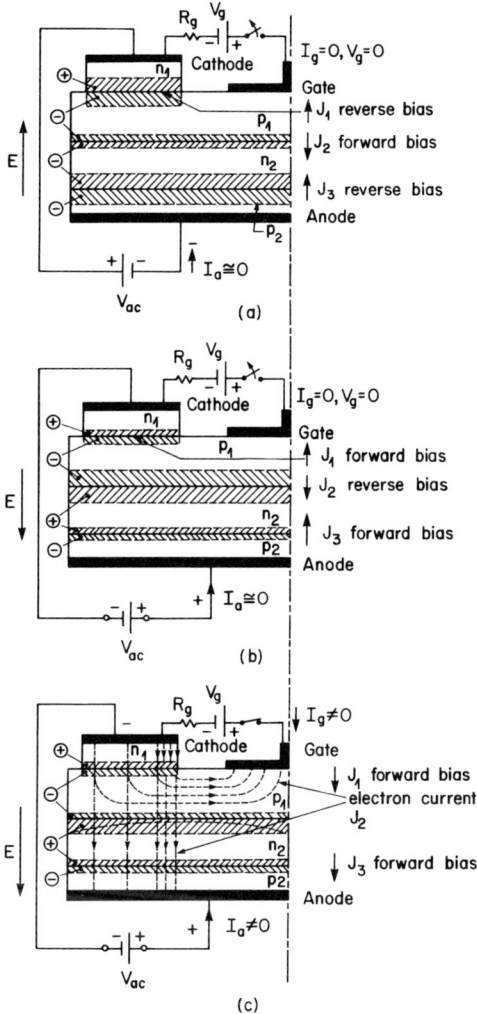

FIG. 7. *pnpn* junctions in a thyristor. (a) Reverse bias without gate drive; (b) forward bias without gate drive; (c) forward bias with gate drive.

tion J_2 and the current increases. All junctions then become forward-biased and the device remains in the on state until the anode voltage becomes negative. The gate current does not affect conduction after turn-on. Due to the uneven distribution of the gate current, first the edge of the cathode turns on, and then the whole device is gradually brought into conduction. A typical I–V characteristic of a power thyristor is shown in Fig. 8.

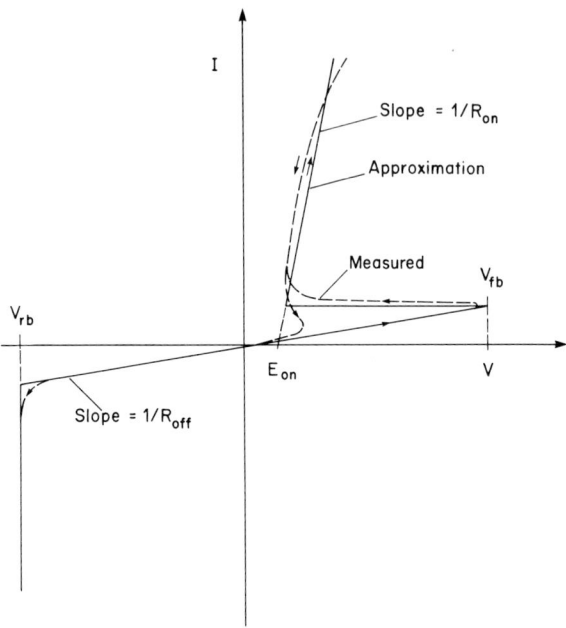

FIG. 8. I-V characteristic of a power thyristor. V_{fb} = forward blocking voltage; V_{rb} = reverse blocking voltage.

2. Development Trends in Thyristor Technology

Paralleling the expanding industrial applications of power electronics, the ratings of presently available devices are rapidly increasing. Meanwhile, the greater reliability required for such high power levels calls for the development of new technologies and devices. We might mention here too that the application of computers is producing more accurate design methods which cannot fail to contribute to the reliability of future SCR devices.

The *voltage rating* of a thyristor is determined mainly by the thickness, resistivity, and surface treatment of the silicon wafer: the precision of the technology is what in fact determines whether the device will be suitable for 12 V or 2 kV, for instance. While higher voltages require higher resistivity and thicker base material (silicon plate), these together lead to an increase in the losses occurring during the conduction period, thus reducing the permitted load current. Furthermore, the high resistivity of the wafer has the effect of prolonging the carrier lifetime, which increases the turn-off time of the high-voltage thyristor, consequently limiting its operating frequency. The high-voltage property of the thyristor can be improved by properly bevelling or contouring the thyristor to reduce the surface gradient at the edges of the device. A further possible improvement is to coat the contour with an insulating material having a high dielectric constant (Fig. 6).

Power thyristors with a blocking voltage of 2.4 kV to 2.5 kV are available commercially, while some companies offer experimental devices rated 4 kV. Meanwhile thyristors have been produced in Japan, using the "double-diffusion" technique, which have a blocking voltage of as high as 10 kV (7). Considering the present limitations of silicon materials and SCR technology, the author believes that the blocking voltage adopted in the near future will be between 3 kV and 4 kV and that this would appear to be the limit, practically speaking, for several years to come.

The *current rating* of a thyristor is determined essentially by the diameter and resistivity of the wafer (8). An increasing wafer diameter causes the rated current to rise almost linearly. Lower wafer resistivity reduces the power loss in the wafer and produces an increase in the current rating of the device. The largest wafer diameter offered by commercially available thyristors today is in the range of 33–40 mm, which corresponds to a rated current of 600–800 A. However, thyristors have been produced with wafer diameters of 52 mm and 102 mm, permitting a rated current of about 1.5–1.8 kA and 4 kA, respectively, although the larger diameter is suitable only for a peak blocking voltage of 600 V.

The fabrication of thyristors with large diameters requires an extremely accurate technology, as the slightest imperfection results in either failure or a drastic decrease in the rated voltage. The output of high-voltage and high-current thyristor fabrication is fairly low; for example, an output of 100 thyristors rated 2500 V, 600 A requires the production of about 200 units; the other hundred can be used for lower voltages. Thyristors rated 2400 V have components identical to those of 300-V thyristors, the differences in their performance being due to imperfections introduced during the manufacturing process. Following fabrication, the devices are tested and classified according to their voltage.

The current rating of the device inversely influences the voltage rating. This may lead to saturation of the kVA rating despite both the increasing voltage and current ratings.

High-frequency application of the thyristor demands lower turn-on and turn-off losses calculated as the product of the current and voltage. The voltage, current, and losses in the device during turn-on are shown in Fig. 9, where it is seen that the thyristor voltage collapses in response to the gate signal. However, the finite speed of the carriers creates a delay after the initiation of the gate pulse, and this delay prevents the voltage collapse and the current rise until a finite time, t_d, is reached. Furthermore a finite time (t_f) is also required for the voltage to collapse to the conduction value (ΔV) and for the current to increase to the load current value (I_L). The carriers

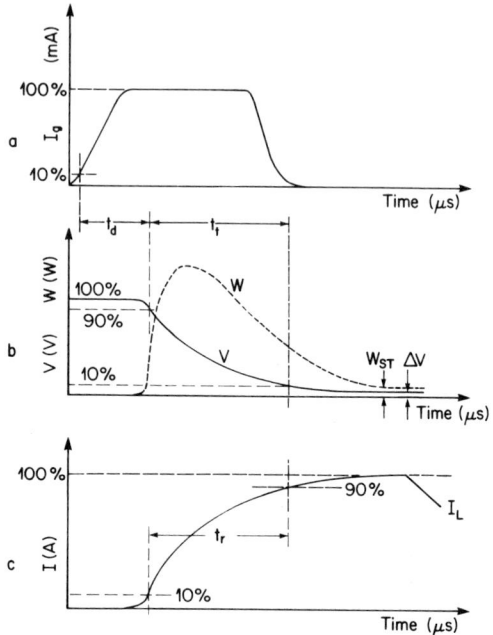

FIG. 9. Turn-on of a thyristor. (a) Gate current; (b) voltage and losses; (c) current.

emitted by the gate current from the cathode initially turn on only the area at the edge of the cathode nearest the gate, but during the t_f period conduction spreads to the entire device. One disadvantage of this process, however, is that it produces an unequal current distribution owing to the initially high current density occurring near the cathode, which decreases as the device turns on.

An excessively high dI/dt may produce high current density and overheating in the cathode area, possibly resulting in the destruction of the silicon wafer. The variation of losses during turn-on is shown in Fig. 9b. Greater dI/dt capability can be achieved by decreasing the losses, which is possible once t_f has been reduced. This may be attained by applying the field-assisted turn-on principle to a thyristor of cross-field construction. In this way the area initially turned on (Fig. 10) is increased, while the current density at the edge of the cathode is reduced (9).

It can be seen in Fig. 10 that the cathode covers only part of the n_1 layer. The gate current flowing through the forward-biased J_1 junction emits electrons from the n_1 layer at point A. The electrons produce an anode current, but the current is limited by the resistance offered by the AB part of the semiconductor. The voltage drop across AB produces a cross electric field

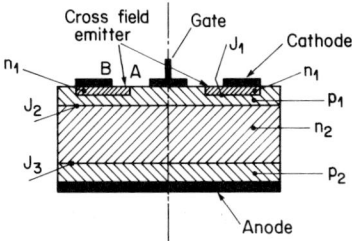

FIG. 10. Cross section of a thyristor with cross-field emitter.

which initiates a further emission of electrons from the cathode and turns on a larger cathode area. The effect of the cross field is to reduce the turn-on time and increase the dI/dt capability of the thyristor. The turn-on area can be increased and the speed of the turn-on consequently accelerated by interdigitating the cathode and the gate. In this manner, the distance between cathode and gate is minimized and the turn-on area on all sides of the cathode digits expands at a uniform rate. However, the increase in turn-on area calls for higher gate-current requirements. Typical interdigitated cathode-gate geometry is shown in Fig. 11.

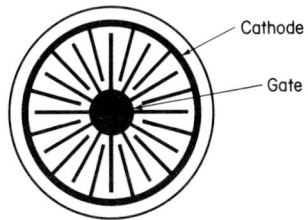

FIG. 11. Interdigitated cathode–gate geometry.

The amplifying gate structure (Fig. 12) employs an auxiliary n layer as a small auxiliary thyristor within the main device to amplify the turn-on current by initiating a multipoint turn-on (10, 11). The gate current emits electrons from n_3; these electrons are driven by the anode field and initiate a current through n_3, n_1, and the cathode. This current in turn acts as an increased gate current which emits a large number of electrons from the gate region. Thus an extended area is turned on, thereby increasing the dI/dt capability and reducing the turn-on time and losses. The auxiliary gate amplifies the gate current by drawing current from the anode to speed up the turn-on of the device.

Reimers (12) describes a 2-kV, 470-A device with a regenerative gate structure. The cross section of such a device is shown in Fig. 13. The dip in the main emitter lip structure causes turn-on to always start at this point,

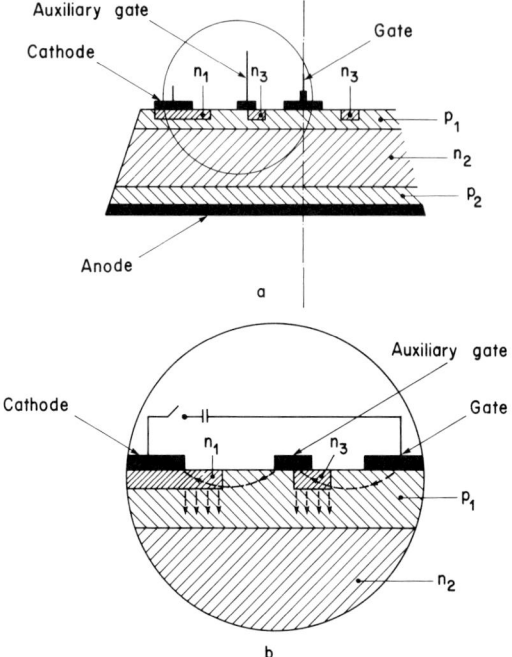

FIG. 12. (a) Amplifying-gate structure; (b) enlargement of circled area in (a).

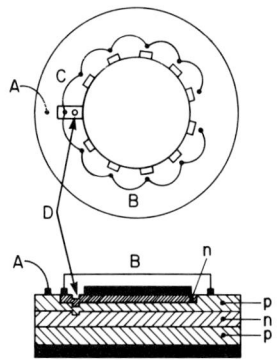

FIG. 13. Thyristor with regenerative gate structure (12). (A) Main gate; (B) regenerative gates connected in parallel by a wire; (C) extended emitter lips; (D) dip on extended emitter lip.

independently of whether the device is triggered by a gate pulse, excessive voltage, or dI/dt. The current produced by the turn-on of the lip area is distributed by the regenerative gate signal wire and produces turn-on in many additional areas, simultaneously drawing the follow current away from the initial conduction areas. This method increases the dI/dt capability of the device and reduces the turn-on losses; however, the dip in the emitter may occasion a slightly lower holding current and critical gate current, although it does not affect other parameters. The major advantage of this method is that the dI/dt rating is independent of the gate drive.

Further losses occur when the thyristor turns off, owing to the storage charge effect. The conducting device contains excess minority and majority carrier charges. When the device turns off, therefore, the stored charge (Q_{st}) in the wafer enables the thyristor to continue to conduct in the reverse direction, as shown in Fig. 14a, until the charge is neutralized. The thyristor

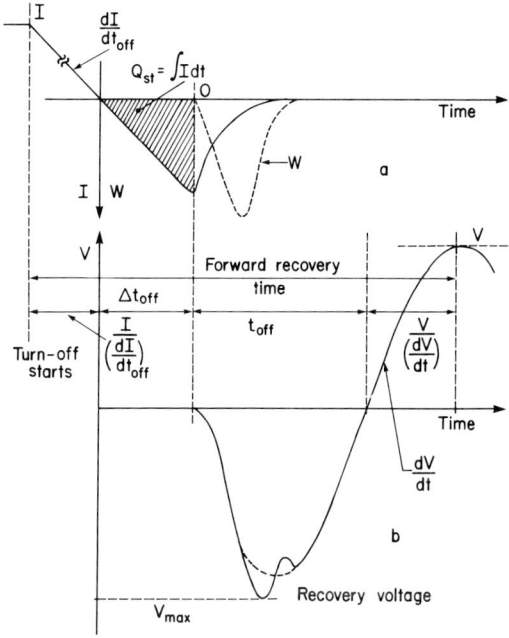

FIG. 14. Turn-off of a thyristor. (a) Current and losses; (b) recovery voltage.

current then gradually falls to zero and the voltage across the device increases. The turn-off losses, calculated as the product of the current and voltage (Fig. 14a), do not produce serious overheating, although they could be significant in high-frequency applications.

The turn-off of the thyristor requires a finite time, t_{off}, to limit the operating frequency, since the forward voltage cannot be applied before the device has turned off. Even then, if the dV/dt of the reapplied voltage is high, unexpected turn-on may occur.

The turn-off time can be reduced by speeding up the removal of carriers from the wafer. This can be achieved by various means:

(a) by reducing the resistivity of the wafer, e.g., by doping it with gold atoms (*13*);

(b) by reducing the thickness of the silicon base. The disadvantage here is that both methods lower the blocking voltage of the device;

(c) by applying three layers of silicon with different impurity content levels in the base material. This enables both the thickness of the base and the turn-off time to be reduced without lowering the blocking voltage (*14*);

(d) by varying the carrier lifetime. Somos and Piccone (*15*) applied different silicons arranged in such a way that carriers in adjacent areas have different lifetimes. Areas with a long carrier lifetime assist forward conduction, while areas with short-lifetime carriers help to reduce the turn-off time;

(e) by applying a reverse-conducting thyristor, which acts as a thyristor in one direction and a diode in the other, to a common wafer (*7*);

(f) by applying a negative-bias voltage on the gate just after the current zero crossing.

A new device with an interdigitated gate structure has been developed, with a 2-μsec recovery time rated for 1 kV, 100 A (*16*).

The high dV/dt of the reapplied voltage after the turn-off of the thyristor may produce an unwanted turn-on. The dV/dt sensitivity of the device can be reduced by the shorted emitter construction shown in Fig. 15. The cathode of the thyristor is enlarged so that it partially covers the p region where the gate is connected. The gate signal drives the current through the p layer directly to the cathode. The voltage drop along the p layer produces a voltage difference (ΔV) between the right-hand edge of the n and p layers of the cathode. At this point, if the gate drive is high enough, electrons are emitted and the device turns on. However, when the high dV/dt produces a capacity current or when a thermally generated leakage current occurs, both smaller than the gate drive current, they will flow directly through the low-impedance path in the p layer and not turn on the device. This construction has brought about a significant improvement in the high-temperature characteristics and dV/dt capability of thyristors.

The high-frequency application of the thyristor called for a reduction in turn-on and turn-off losses, increased dV/dt and dI/dt capabilities, and a shorter turn-off time. Roberts and Wilkinson (*14*) give the following equation for calculating the maximum operating frequency of a thyristor:

$$f_{\max} = \tfrac{1}{2}\{[I_m/(dI/dt_{\text{off}})] + t_{\text{off}} + [V/(dV/dt)]\}, \tag{1}$$

where: I_m is the peak forward-anode current, V is the forward blocking voltage, t_{off} is the turn-off time, dI/dt_{off} is the rate of decrease of anode current, and dV/dt is the rate of rise of reapplied voltage.

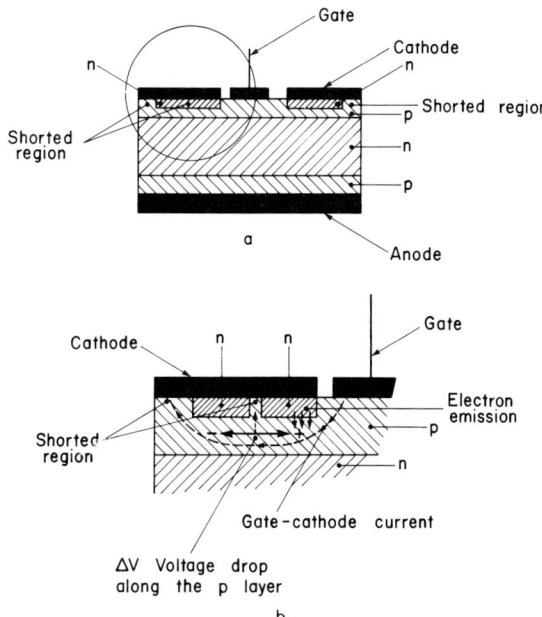

FIG. 15. (a) Shorted emitter construction for thyristors; (b) enlargement of circled area in (a).

The derivation of Eq. (1) can be followed from Fig. 14.

The methods of increasing the operating frequency of the thyristors outlined earlier are not without certain drawbacks. As pointed out, the requirements of the high-frequency thyristor (amplifying gate arrangement, shorted-emitter geometry, etc.) make this a fairly complex and expensive device to produce. According to the present state of the art, new techniques have brought the dI/dt capability of the power device from the 50–100-A/μsec range up to 1 kA/μsec. The turn-on time is now in the range of 1–2 μsec, while the turn-off time is generally in the range of 15–20 μsec, although more recent devices with their more sophisticated gate geometry (interdigitation) and structure operate with a turn-off time of 1–2 μsec (11). The shorted emitter construction has improved the dV/dt capabilities, the former rating of 100–200 V/μsec having risen to 1 kV/μsec.

The frequency range of high-frequency thyristors is 20–25 kHz, but special devices are now being built for up to 100 kHz (11).

Improved mounting and cooling techniques have increased the overload rating. The highest rating recorded to date is 20 kA, using a single-cycle, 10-msec duration half-sine wave for a thyristor with an average current rating of 720 A (7), while a typical value lies in the range of 4–5 kA. The extensive application of thyristors to switching functions and the introduction of the amplifying gate device have changed the short-circuit requirement, however. Thyristors are still tested using a standard 10-msec half-sine wave (5), whereas new devices are in fact capable of conducting several thousands of amperes for a time range of 2–200 μsec. This impulse current represents a danger to the thyristors more because of the dI/dt than because of the heating effect.

A review of the literature shows two major trends in thyristor development:

(1) increasing rated voltage and current values, which results in more powerful thyristors for low-frequency applications (below 400 Hz);

(2) increasing frequency range and switching capability. This offers a wide application of thyristors in high-frequency circuits but reduces their rated current and voltage. Furthermore, the increasing frequency requires more complex and more expensive thyristor construction.

E. New Power Electronic Devices

Today's wider use of power electronics, together with the development of semiconductor technology, has introduced several new devices which in turn are opening up new fields of application. Some of these devices are described below.

1. Gate Turn-Off Thyristors

The gate turn-off thyristor (17) consists of a conventional *pnpn* wafer with an interdigitated cathode–gate arrangement which ensures a higher operating frequency despite the fact that the larger gate area calls for a much higher current than standard thyristors for both turn-on and turn-off. The device is turned on by a positive-polarity gate pulse and turned off by a negative-polarity gate pulse.

Turn-off is achieved by squeezing the conduction region to the dimension of a diffusion length of the minority carriers, using the negative gate current. However, the final area has to remain large enough to withstand the increased current density, while the current distribution must be uniform to avoid hot spots. The effect of the gate current can be increased and the current distribution improved if an interdigitated structure with narrow emitter fingers is used and by intensive gold doping which reduces the carrier lifetime. A 200-A, 1.4-kV device has been built having this interdigitated structure and narrow spiral elements together with an amplifying gate to reduce gate current requirements (17).

Both the reverse blocking voltage and current-carrying capacity are smaller than those of the conventional thyristor. Meanwhile the current density during conduction is much lower, so that the negative gate current reduces the anode current and turns off the device.

The primary use of the gate turn-off thyristor is to replace the transistor in higher power inverter circuits. Presently available devices have a rated-current range of 1000 A, a blocking voltage of 1 kV, and a gate current of 20–40 A to turn off the device.

2. *Radiation-Sensitive Thyristors*

The sensitivity of thyristors to radiant energy is due to the incidence of radiant energy creating holes and electrons in the silicon which function as a gate current and turn on the device. Light-activated thyristors have been used for several years (*5*), while recently the development of a magnetic-field-activated device was reported (*18*).

The structure of the light-activated thyristor is similar to that of the conventional thyristor: it consists basically of a standard *pnpn* silicon wafer with anode, gate, and cathode connections and a glass window, lens, or hole installed in the casing to allow light to penetrate into the cathode–gate region. It can be fired in the normal way using a gate signal or by a light signal brought to the thyristor either directly or via light pipes. In the latest experimental device, the light-conducting fiber is directly attached to the thyristor above the most sensitive areas. The light sensitivity depends on the silicon material impurity content, the light intensity and wavelength, etc. Commerically available devices work with a higher than 70% relative efficiency with a light wavelength of from 0.7 to about 1 μm, the most efficient wavelength being around 0.9 μm.

The following light sources are the most popular: the tungsten lamp (3400°K), which emits a wide spectrum of light; the light-emitting diode with an emission wavelength matching that of the thyristor; and the laser diode which is suitable for emitting high-intensity light pulses (e.g., wavelengths of 0.9 μm). Only low-voltage (200 V) and low-current (1.6–2 A) thyristors of this type are commercially available at the present time. On the other hand, experimental devices rated 0.6 kV to 1 kV, 25–200 A are already being produced, but, owing to the fact that they are turned on by the relatively low energy density, their dI/dt capability is very low. Furthermore, the open gate makes the device sensitive to the forward dV/dt, although this can be improved by connecting the gate to the cathode with a suitable resistance. The latter limitations confine the uses of light-fired devices to the signaling field, where major applications include light-controlled relays and the triggering of higher-power thyristors in high-voltage circuits. Recently the search for rapid solid-state switching devices opened up a new sphere for light-activated thyristors when it was discovered that firing by neodymium laser drastically improves the thyristor's switching capability (19). Because of its high energy and short duration (40 nsec), the laser impulse creates a much higher energy density in the silicon than the gate pulse, and consequently generates more electrons and results in faster turn-on. A further improvement is the fact that the laser impulse turns on a larger region of the device than the gate.

Laser firing has resulted in significant reductions in the turn-on delay, risetime, and turn-on loss as well as an increase in dI/dt capability. Beauséjour and Karady (19) reported a reduction in the turn-on delay time from 5–10 μsec to less than 5 nsec and a risetime from 1–2 μsec to 63 nsec when a commercially available light-activated thyristor was fired by a gate signal or a neodymium laser beam.

Experimental devices rated about 1 kV and 500 V have been built with a delay time of less than 10 nsec, a risetime of 10–15 nsec and a dI/dt capability of over 1 kA/μsec. It is believed that with the improvement of laser reliability these devices will be used as modulators and fast switches, and the author foresees their application in hvdc thyristor valves (19).

Arai (18) reported the development of a magnetosensitive thyristor. The device is built in a planar arrangement. Boron and phosphorus are selectively diffused in a high-resistivity n-type silicon wafer to form the p- and n-type regions, the latter being exceptionally long (100 μm). The device can be triggered either by a gate signal or a magnetic field, the latter requiring a gate bias current of just a few nanoamperes. The gate voltage produces the emission of carriers from the anode and gate electrodes which are driven by the electric field and deflected by the magnetic field. Depending on the magnetic field direction, deflection increases the number of carriers at the reverse-biased junction, which amounts to an increase in the gate current. Thus a sufficiently strong magnetic field turns on the device.

3. Bidirectional Thyristors (Triacs)

The triac type of thyristor (20) conducts current in both directions and blocks the voltage of both polarities. It can be triggered by either the positive or the negative gate current, irrespective of the polarity of the voltage between the terminals (1, 2). As seen in Fig. 16, the triac may be divided into three parts: part T_1 is equivalent to a *pnpn* device, part T_2 to an *npnp* device (these two parts may be considered as two thyristors connected in inverse parallel), and part T_3 is equivalent to an *npnp* device (which could be considered as an auxiliary thyristor).

The turn-on mechanism is as follows.

(1) Terminal 2 is positive with regard to terminal 1, in which case only

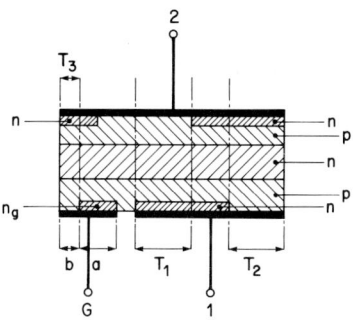

Fig. 16. Construction of a triac.

the T_1 region can conduct and terminal 2 acts as an anode, terminal 1 as a cathode. The positive gate current turns on the device as in a conventional thyristor. The negative gate current also turns on the device, in which case it operates as a junction gate thyristor. The negative gate current flows through the a region of the gate and returns through the p region of electrode 1. This current emits electrons from the cathode region and turns on the device.

(2) When terminal 2 is negative with respect to terminal 1 only the T_2 region can conduct, while terminal 2 acts as an anode and terminal 1 as a cathode.

The positive gate current first turns on the auxiliary thyristor T_3, which initiates the turn-on of region T_2 and brings the device into conduction. The negative gate current emits electrons from the b region of the cathode and turns on region T_2.

Turn-off is not the same operation as in the conventional thyristor: when the voltage across the conducting triac is reversed, it turns on in the other direction. In order for the device to turn off, the current must be lower than the holding current—which means reducing the applied voltage—and the voltage must be near zero in sufficient time to permit recombination of the charges. This restricts the operating frequency to 60 Hz because the dV/dt near current zero is sufficiently slow to allow turn-off. Owing to its bidirectional characteristic, overvoltages do not destroy the device but instead turn it on. A further characteristic is that a far shorter gate pulse is required to trigger the turn-on.

According to the present state of the art, the general rating of these devices is up to 25 A and several hundred volts (600 V), while a number of special models are available in the 200-A rated current range (21).

4. Gateless Thyristors

The development of a new thyristor without a gate was recently reported by Steimel (22). A cross section of this new *pnpn* device in Fig. 17 shows the cathode cut across the middle by a shallow groove (g); thus the cathode short-circuits several concentric *p* layers (S). This type of thyristor can be

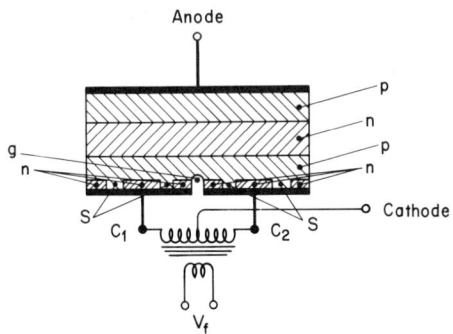

FIG. 17. Gateless thyristor (22).

fired by a voltage connected between the two halves of the cathode. This is achieved in Fig. 17 by a pulse transformer. The V_f voltage drives a current through the two parts of the cathode and causes electron emission and fires the thyristors. First the C_1 part of the cathode turns on; this creates a cross field and turns on the C_2 part of the cathode. The turn-on process is more intensive in the second case because the transformer between the two parts of the cathode provides a feedback and accelerates the complete turn-on of C_1. As a result, in this thyristor, a much larger area of the cathode turns on than in a conventional thyristor, and the turn-on process is more intensive. The major advantages of this new thyristor are that compared with a conventional model its dI/dt capability is much higher and its turn-on time much shorter, a typical value being in the range of 0.5 μsec. It requires a firing pulse in the range of 10-μsec duration. The firing transformer can be made out of a small ring-type ferrite core with a cross section of 1 cm^2, for instance.

This device is believed to represent a promising new family of thyristors which, at a further development stage, could be used for rapid-acting switches.

F. Accessories

The current rating and the voltage-blocking capability of a semiconductor device depend largely on the temperature; effective cooling results in fact in a lower wafer temperature and a higher rating. The need for new high-power devices has directed research toward new mounting and cooling techniques, while sensitivity to overvoltages and currents has led to the development of more effective protection circuits.

1. Mounting Attachments

For high-power applications, the disc-type construction with pressure mounting is used. This allows cooling of the thyristor on both sides and avoids the use of soldered contacts which have often proved unreliable. The heat sinks are clamped to each side of the thyristor, and pressed together with a force of 500–1000 kg. This high pressure ensures a low thermal resistance between the wafer and the heat sinks together with low electric resistance and fatigue-free electric connections. A further advantage of the disc construction is that it facilitates the series connection of thyristors. A typical high-power high-voltage thyristor stack is shown in Fig. 18.

FIG. 18. High-voltage and high-power thyristor stack.

For low- and medium-power applications, the latest development is the integrated mounting method (23), in which wafers are mounted on an electrically insulated but thermally conducting substrate of beryllium oxide fixed to a metal plate which can be connected to a heat sink. The wafers are interconnected by a wire frame or by thick-film techniques and encapsulated. This arrangement reduces the thermal resistance and leads to improved cooling. Units of three or four diodes or thyristors are built with a rating of 250 V, 50 A, which is suitable for medium-power applications.

2. Cooling Attachments

The most popular cooling method to date has been by air, using heat sinks with fins pressure-clamped to the thyristors. In more recent developments, however, the heat sink is an integral part of the semiconductor device and forms a stack with the busbars, terminals, and, in many cases, the fuses, reactors, and RC networks. In addition to the space-saving factor, the advantage of this construction is that it ensures more efficient cooling. This efficiency is further increased by forcing air through the fins, the expected thermal resistance being 0.1°C/W (23).

Liquid cooling has led to important improvements (24). The heat sink here consists of spiral or labyrinthine passageways through which liquid (water or oil) is pumped. This arrangement reduces the thermal resistance to around 0.04 °C/W but has certain disadvantages: the possibility of leaks and the relatively low reliability of the pumps. For extra-high-power applications with space limitations, liquid cooling is nevertheless the most feasible method. Recently this technique was adopted for cooling the 2000-A, 150-kV thyristor valves used in the Cabora-Bassa high-voltage dc transmission system (25).

Since the current rating of low-voltage thyristors goes hand in hand with the progress made in cooling methods, research in this field is active. The very latest development is the heat-pipe type of cooling system (23) (Fig. 19). Here the volatile liquid evaporates near the surface of the thyristor and absorbs heat. The vapor travels through the pipe and condenses at the heat exchanger, and the liquid returns to the thyristor end in response to the capillary effect of the wick. The phase change accounts for the high rate of heat transfer with a negligible temperature difference.

Further development of the application of heat pipes is reported by McKechnie et al. (26) in their work on a transcalent rectifier and thyristor. Two heat pipes are attached directly to the silicon wafer. The gate of the

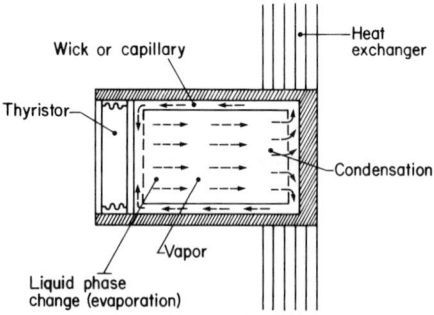

FIG. 19. Heat pipe for thyristor cooling.

thyristor is arranged on the periphery of the wafer; the heat pipe is joined to the gate area. To avoid the short circuit produced by the electrically conducting internal structure of the heat pipe, a thin film (5000–6000 Å) of silicon dioxide insulation is applied between the wafer and the water and metal parts of the heat pipe. This new method provides the most efficient cooling of the thyristor to date. Transcalent thyristors rated 400 A, 1.2 kV, diodes rated up to 5000 A, 500 V, and transistors rated 400 A, 450 V have already been developed. However, the method is still in the experimental stages and limited to relatively low-power applications, although it does seem to hold promise as a future method for cooling semiconductors.

3. Protection Devices

It was shown in a previous section that dI/dt, dV/dt, overvoltages, etc., may destroy the SCR device. The protection devices most frequently used are linear or nonlinear inductances connected in series with the thyristor for protection against high dI/dt and an RC circuit connected in parallel (5) for reducing dV/dt and equalizing the voltage distribution.

An important development in overvoltage protection is the avalanche-type self-protecting thyristor (27) in which the overvoltage turns on the thyristor in both directions without damage. However, this effective form of protection is restricted to short-duration transients, and only relatively low-power devices have been developed to date. Some high-voltage thyristor valves use a Zener-diode type of device consisting of two high-power, high-voltage (1.5–2 kV) Zener diodes connected back-to-back, which clamps overvoltages above a certain limit (25).

A varistor-type overvoltage suppression was developed by Golden (28). The disc-shaped device, with metallized electrodes on both sides, is built out of zinc oxide grains with a diameter in the range of 12 μm. The grains are covered by a very thin (500 atoms) boundary layer of bismuth oxide and additional material. The grains are pressed into a disc and baked at a high temperature for one carefully controlled cycle, during which the electrical characteristics are obtained. The small grains on the thin boundaries form several millions of polycrystalline junctions which make the voltage–current characteristics of the device similar to those of a pair of avalanche diodes. The characteristics of the device can be described by Eq. 2:

$$I = KV^\alpha \qquad (2)$$

where K is the device constant and α is the reciprocal of the slope of log–log V–I characteristics.

The efficiency of the device increases with α, which is the measure of its nonlinearity and clamping or overvoltage suppression capability.

Although the most effective suppressor in use is the Zener type, which has a nearly horizontal characteristic ($\alpha = 35$), the varistor with its $\alpha = 25$ value is also an efficient device. Its principal advantage over the Zener-type suppressor is its higher energy-absorbing capability, which is derived directly from the large number of multiple junctions. The voltage capability is determined by the length of the path between the terminals and by the size and composition of both the zinc oxide and the grain boundaries. The current-handling capability, on the other hand, depends on the area, while the power absorption is proportional to the device's active volume. The suppressor is connected directly in parallel with the thyristor.

Available devices are rated up to 510 V rms and about 721 V recurrent peak voltage. The energy rating is 80 J and the peak current is 1.2 kA, for a pulse of less than 7-μsec duration.

One method of overcurrent protection is the current-limiting fuse (29). Connected in series with the thyristor, the fuse limits the energy flowing through the device in a fault condition to the withstand energy level. The active part of the modern current-limiting fuse consists of a thin-necked silver ribbon attached to each end to a slug of copper and placed in a tube filled with quartz sand. The current starts to melt the silver at the necks first and produces several arcs which are extinguished by the interaction of the thermal and electrical phenomena. The fuse interrupts the high short-circuit current within milliseconds and prevents the thyristors from overheating. The divergence in the characteristics of the fuse and the thyristor creates the need for coordination of the fuse–thyristor relation. Schonholzer (29) developed a mathematical model and practical method for simulating the thermal behavior of the thyristor and the fuse in order to achieve a more accurate coordination. However, further work is required to increase the accuracy of the models and to achieve a better understanding of the physical phenomena involved, particularly the mechanism of fatiguing during cyclic loading.

III. Thyristor Systems

Owing to the limited capability of the individual device, most industrial applications demand the use of thyristor systems (valves) comprising a large number of thyristors connected in series and in parallel.

The general arrangement of a typical thyristor system is shown in Fig. 20. Three different practical systems are used, depending on the value of the cross resistance (R_c).

$R_c = \infty$. This system comprises a number of thyristor strings with several thyristors connected in series in order to withstand a high operating voltage. The strings themselves are connected in parallel to increase the rated current.

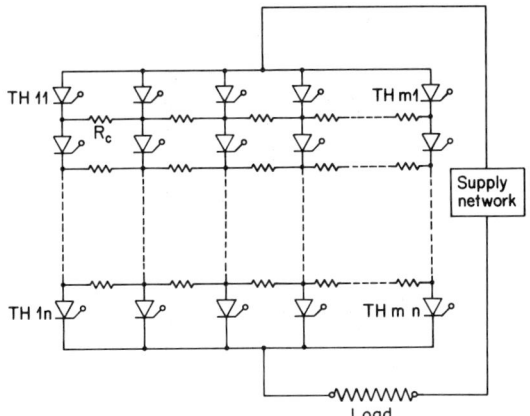

FIG. 20. Thyristor system.

$R_c = 0$. In this case, the system consists of thyristors connected directly in parallel (arrays) and the paralleled units are connected in series.

$0 < R_c < \infty$. The system here comprises indirectly-parallel and series-connected thyristors.

The obvious advantage of the thyristor system is that it offers a higher rated voltage and current, but it also entails various drawbacks which jeopardize its operation.

In a large thyristor system, for instance, thyristors may not turn on and off simultaneously and dangerous overvoltages may occur across individual devices. Furthermore, when all the thyristors in a system are conducting, unequal current distributions may arise which could overload or destroy the individual devices in steady-state conditions. Short circuits present even more dangerous conditions, while overvoltages produced by switching or other sudden changes in the network or load parameters may endanger the thyristor system in the off state.

All of these potential dangers must be taken into consideration at the design stage. Thyristor systems usually function as rectifiers or inverters and, as such, are connected between the load and the supply, as shown in Fig. 20. Their operation can be derived from the operating principles of thyristor strings and arrays.

A. Thyristor Strings

A thyristor string consists of several thyristors connected in series, each shunted by a voltage-grading network (Fig. 21), which comprises (a) a resistance R' connected in parallel with each thyristor to equalize the dc voltage in the off-state condition; (b) a series-connected resistance R and capacitance C

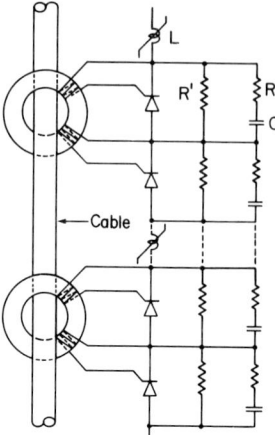

FIG. 21. Electrical connection of a thyristor string with cable firing.

also connected in parallel with each thyristor to reduce the turn-on and turn-off overvoltages.

Furthermore, nonlinear inductances (L) are connected in series with a group of thyristors in order to reduce the dI/dt and turn-on overvoltages. The thyristors are triggered simultaneously from a firing system.

The thyristor string is used for high voltage when the voltage rating of the individual device is insufficient but the current rating adequate.

B. Thyristor Arrays

An array of thyristors consists of several thyristors connected in parallel. A voltage grading network, similar to that used in thyristor strings, is connected in parallel with the array. Uniform current distribution is achieved by using thyristors with matched forward voltage drop characteristics or by inserting a balancing resistance, inductance, or transformers. The array of thyristors is used when the load current is larger than the current rating of the individual thyristors but their voltage rating is adequate.

C. Firing of Thyristor Systems

The firing methods used for individual thyristors have been discussed in detail in several books (5, 6). The firing of a complex thyristor system requires a different technique (30), since any delay in the firing signal produces a nonsimultaneous turn-on of the thyristors, which in turn leads to dangerous overvoltages.

To ensure proper operation, the firing system must fulfill the following requirements (31).

(a) The signal must be sufficiently powerful to turn on the thyristors. Usually, a firing pulse equal to at least three times the minimum gate turn-on current is considered necessary. On the other hand, the firing signal should not exceed the power dissipation capability of the gate-to-cathode junction. The peak current of the firing pulse should generally be between 1 and 10 A.

(b) The rise of the gate-current signal should be less than 0.4–0.5 μsec to minimize the delay between the first and last thyristors turned on in the system.

(c) The time difference or jitter among the gate-current pulses has to be minimized, and should be less than 0.5 μsec.

(d) Although discrete firing pulses of short duration are adequate for normal operation of the system, a higher degree of reliability can be achieved by maintaining a low dc gate current during the conduction period. This dc gate current will restore conduction in the thyristor if a transient decreases its current to zero.

We describe below several firing systems that have been developed to suit the needs of thyristor-system operation.

1. Cable Firing System

This system is shown in Fig. 22. The pulse generator current energizes the pulse transformers threaded on the cable; the secondary current of the pulse transformers fires the thyristors. System performance may be improved by inserting a passive pulse-reshaping circuit at the individual thyristor gate levels to provide a fast rising signal. In the pulse-reshaping circuit, a saturable inductance (L_s) is inserted in series with the gate-to-cathode junction to block the gate current. After a few microseconds, the inductance is saturated and its impedance then falls nearly to zero, suddenly releasing almost the full transformer current through the gate-to-cathode junction of the thyristor. A further development would be to add a capacitance C_s which stores the energy of the transformer pulse and discharges at the saturation of the inductance, thereby increasing the gate current.

A second type of cable firing system contains two cables and an individual pulse generator for each group of thyristors. One of the cables is supplied with a high-frequency current and provides the energy for the pulse generators (dc voltage). The other cable is supplied by a low-level signal impulse which fires the individual pulse generators; these in turn produce the proper gate pulse for each thyristor.

The energy and firing signals are supplied through threaded pulse transformers. Both cables are insulated for the full system voltage.

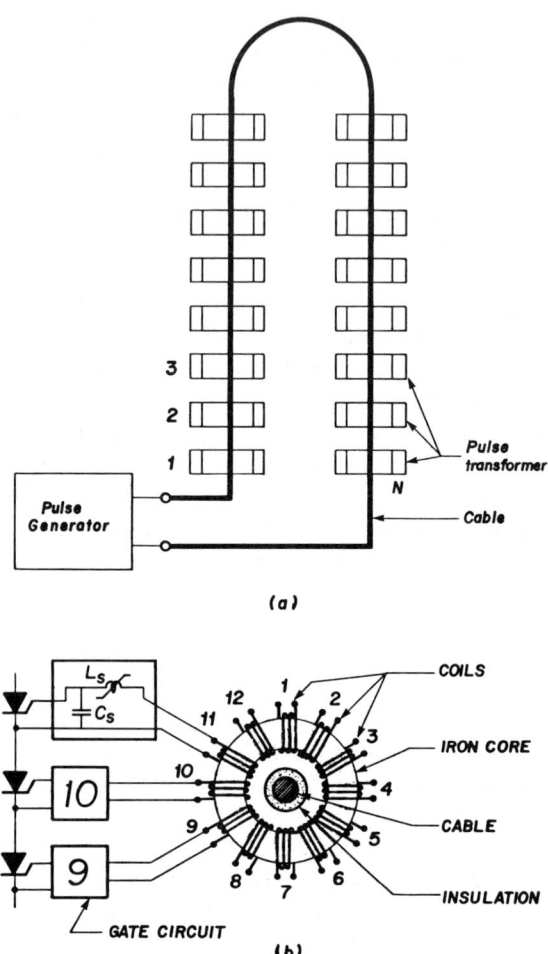

FIG. 22. Cable firing system.

2. Cascade Transformer Firing System

This system comprises a pulse generator for each group of thyristors. Both the energy for the pulse generators and the firing signals are supplied by a series of cascade transformers, each insulated only for the group voltage (20–40 kV).

3. Light Firing System

Among several light firing systems, those described represent the two basic variations.

(a) An insulated pulse generator fires each group of thyristors. The energy required is drawn from the voltage-grading circuit. A distorted ac voltage appears during the blocking period across each thyristor or across each group. This voltage is rectified and the dc voltage obtained supplies the pulse generators, which are fired by a light signal transported from the ground level through light guides. The light signal is generated by a laser or by light-emitting diodes. At the pulse generator end of the light guide the light signal is converted into an electric signal by a photodiode. A variation of this system is obtained if the pulse generators are supplied from cascade transformers.

(b) Connected to each thyristor gate is a simple pulse generator which consists of a capacitor and a small direct light-fired thyristor. The capacitor is charged through a rectifier by the voltage appearing across the thyristor and discharged through the gate-to-cathode junction when the direct light-fired thyristor is turned on by a light signal. The light signal is furnished by light guides.

A comparison of firing systems indicates that the light firing system is advantageous at high voltages (over 150–200 kV) because of the reduced insulation problem and smaller stray capacitance to ground, hence decreasing the overvoltages produced by impulse voltage. The disadvantage of this system is the relatively short lifetime of the light sources.

Recently an interesting new system was suggested and its operation demonstrated by Takahashi *et al.* (*32*). This experimental system consists of an acoustic transmission line made out of piezoelectric material (quartz, $LiNbO_3$, etc.). An interdigitated transducer is installed at each end of the line, using a photolithographic technique. An amplitude-modulated signal is applied to the transducer at one end of the line, where it is converted into an acoustic surface wave that travels along the surface of the piezoelectric material and reaches the second transducer; there it is reconverted into a signal which fires the thyristors. The advantages offered by this system include its compactness, its light weight, and the electrical insulation between the firing signal and the thyristors. It appears that, once fully developed, this system could be promising for high-voltage applications.

IV. Development of Calculation Methods

Today's rapid development of computers is enabling even greater accuracy to be achieved in the design of circuits containing semiconducting elements. This is not without effect in the field of power electronics, where, owing to the size and cost of some of the equipment involved (hvdc thyristor valves with several hundred thyristors, for example), accurate design and optimization of the circuit are extremely important. The mathematical modeling of electronic components has also made great strides, while several computer programs exist for calculating the transient and steady-state currents and voltages in circuits containing transistors, diodes, etc. Among the most widely used programs are the Electronic Circuit Analysis Programs, commonly known as ECAP programs (*33*); already two ECAP versions exist, the second of which permits the simulation of nonlinear elements in addition to standard resistances, capacitances, inductances, mutual inductances, and, furthermore, contains transistor and diode models as well. Network differential equations are solved by the modified Newton–Raphson algorithm. Yet another development is the ASTAP program (*34*), which provides for variation of the elements and produces sensitivity analyses; it can be used, for example, to determine the sensitivity of the output current of a circuit if the value of one of the elements is statistically distributed within a certain limit ($\pm 5\%$). Besides digital calculations, analog and hybrid computation techniques are also used (*35*). Application of these techniques to power electronics calls for accurate modeling of thyristors and thyristor systems, a relatively new field, where the model usually has to be devised by the designers themselves.

A. Thyristor Models

The modeling of a nonlinear device requires (1) formulation of mathematical equations describing the voltage and current characteristics; (2) determination of the discrete components and their interconnection to form an electrical network, comprising linear and nonlinear components, which approximates the device and satisfies its equations; (3) determination of model parameters from measured data; (4) verification of the model under different conditions by comparing measured and calculated current and voltage values. The thyristor is a complicated device to describe mathematically, but the following discrete stages of operation may be distinguished:

(a) *off state*, when only a small leakage current flows through the device; this can be modeled by a resistance R_{off};

(b) *turn-on (a transient state)*. Turn-on can be initiated by the gate pulse, by exceeding the maximum forward blocking voltage, or, third, by a fast change of the anode–cathode voltage (dV/dt). When the thyristor is fired by a gate pulse, turn-on can be represented by a voltage source that provides a collapsing voltage. Because of the finite speed of the carriers, the terminal voltage across the thyristor does not start to collapse until a finite delay time is reached following initiation of the gate pulse, and takes some further time to collapse to the conduction value (36). The phenomenon is shown in Fig. 9.

There is an inverse relationship, which can be determined by measurement, between the gate-pulse magnitude and the delay time. The measured curve can be approximated by an equation using the curve-fitting technique. The typical equation for a 2200-V, 600-A power thyristor is:

$$t_d = \exp(-a \ln I + b), \tag{3}$$

where I is the maximum gate current, $a = 0.42$, and $b = 0.35$, from which it is possible to determine the delay time of a given gate-current pulse. The complete model of the gate-fired thyristor consists of a switch which closes with a delay of t_d in series with a voltage source defined by the following equation:

$$V(t) = V_0 \exp[-\delta(t - t_d)], \tag{4}$$

where V_0 is the voltage across the thyristor before the gate signal is applied, t_d is the delay time, calculated from Eq. (3), and δ is the rise-time constant, whose magnitude can be obtained from Eq. (5) below.

The rate of collapse of the terminal voltage is described by the constant δ, which can be expressed by the following half-empirical equation:

$$\delta = |\ln 9/t_f|, \tag{5}$$

where t_f is as defined in Fig. 9.

Because of manufacturing tolerances, there is always a certain range of delay-time values for any given thyristor type. The delay times are distributed statistically. Figure 23 shows a cumulative delay-time distribution curve based on measurements of a sample quantity of a particular thyristor

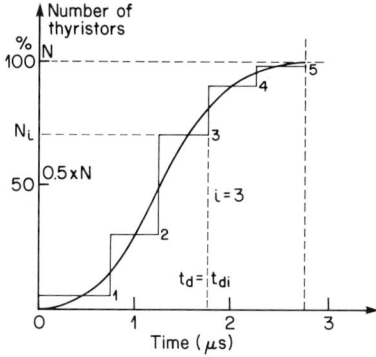

FIG. 23. Statistical distribution of turn-on delay time (36).

type. This distribution curve can be approximated by a Gaussian distribution with the parameters $\bar{t}_d = 1.2$ μsec and $\sigma_d = 0.8$ μsec. When turn-on is initiated by an overvoltage or by dV/dt, the delay time is not well defined, a safe approximation being the highest delay time specified by the manufacturer.

(c) *On state.* The current–voltage characteristics of a typical thyristor are shown in Fig. 24. These may be approximated by straight lines, in which case the thyristor is modeled by a voltage source E_{on} and by a resistance R_{on} (37) or, alternatively, their equation may be determined by the curve-fitting technique.

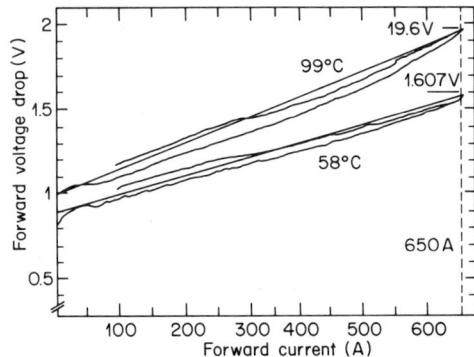

FIG. 24. Forward voltage drop vs. current in a power thyristor.

(d) *Turn-off (transient state).* Figure 14 shows that the thyristor turns off when the reverse current neutralizes the storage charge. For investigating the turn-off process, therefore, the time between the current-zero point and the point when the storage charge is neutralized is an important factor to be considered. The time interval (Δt_{off} in Fig. 14), which may be called the "turn-off delay time" by analogy with the turn-on delay, can be calculated from the following equation:

$$Q_{st} = \int_0^{\Delta t_{off}} I \, dt. \tag{6}$$

In the majority of practical cases, the source voltage is more or less constant during the turn-off process, and the direct voltage source is therefore a satisfactory approximation; as far as the change of current is concerned, it is constant and may be determined from the voltage source and the circuit inductance. A good approximation for the rate of current change is

$$dI/dt = V_0/L, \tag{7}$$

where V_0 is the peak supply voltage, and L_0 is the circuit inductance. The combination of Eqs. (6) and (7) permits the derivation of Δt_{off}:

$$\Delta t_{\text{off}} = \sqrt{2Q_{\text{st}}/(V_0/L)}, \tag{8}$$

where Q_{st} is the measured storage-charge value.

The storage charge of a thyristor is normally measured as a function of the forward current, the rate of change of current through zero, and the junction temperature. Again manufacturing tolerances are such that a range of storage-charge values exists for any type of thyristor. The storage-charge distribution of any given type can be determined from measurements on a sample quantity of thyristors and the distribution curve can be approximated by a Gaussian distribution defined by the mean value and standard deviation.

After the storage charge is neutralized, the thyristor current decreases to zero, as shown in Fig. 14. This current curve can be described by an exponential function and represented by a time-variable current source.

The complete representation of a thyristor during turn-off is a switch connected in series with a time-variable current source. The switch operates when the thyristor storage charge is neutralized. This calls for calculation of the current integral and comparison with the predetermined storage-charge value. Electrically this can be achieved using the circuit shown in Fig. 25, where the thyristor current is integrated with the large capacitance C and where the voltage across the capacitance is proportional to the current integral. This is achieved by comparing the voltage across the capacitance with the source voltage V_{st} which represents the predetermined storage-charge value. The sensing elements used for this purpose are sensitive to the direction of current flow; for example, if the voltage across the capacitance exceeds V_{st}, the direction of current flow reverses, and the sensor operates the switch and turns off the thyristors (38). In addition, the sensors short-

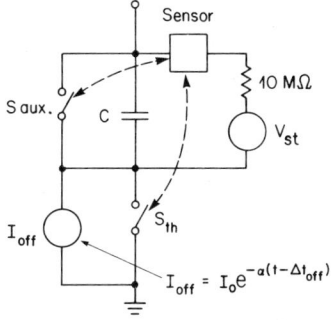

FIG. 25. Electrical model of a thyristor during turn-off.

circuit the capacitance in order to eliminate its influence during the off period.

(e) *Thyristor destruction.* Excessive dV/dt, negative overvoltages exceeding the reverse blocking voltage, and overcurrents are all possible sources of thyristor destruction. Since a fault in the device produces a short circuit, this state may accordingly be represented by a switch operated by sensors which compare the actual dV/dt, I, and V values with the rated ones.

(f) *Gate circuit.* The thyristor gate circuit is essentially a *pn* junction and can be simulated by a diode current generator (*37*). The digital simulation of the thyristor is done by a subprogram (*37*) which examines the variables, determines the state in which the device should be and enters the ON, OFF, TURN-ON, TURN-OFF, or DESTRUCTION model accordingly into the main program, which solves the network equations of the complete circuit (supply, thyristors, load, etc.).

Since the complete model described here is rather complex, simplified models are used for the actual calculation. Development of a program considering all the variables would be rewarding research work.

Harstad (*37*) used the model presented in Fig. 26 in which R_1, R_2, and C_1 simulate the gate delay, J_g the gate–cathode characteristic, and E_n and R_{an} and anode main circuit; the E_n and R_{an} values depend on whether the device is in the on or the off state. This model supposes that the thyristor turns off instantaneously and that the voltage collapses instantaneously across the terminals when the device turns on.

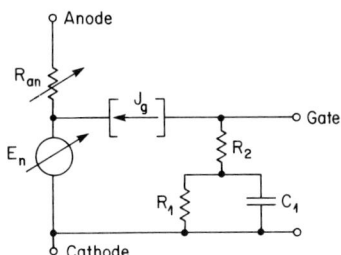

FIG. 26. Simplified thyristor model.

Beattie et al. (*39*) suggest an analog modeling of thyristors where the device is represented by either a short circuit or an open circuit, according to whether it is in the on or the off state. The state of the thyristor is determined by a logic circuit containing AND, NOT, and OR gates and a bistable circuit which produces the open (off state) or short (on state) circuit. This model was modified for digital computer use by the same authors (*40*), replacing the logic circuits by program statements.

An earlier completely different model (5) was recently applied by Kurata (41) for gate turn-off devices. The *pnpn* structure can be subdivided into a *pnp* and an *npn* transistor interconnected to form a regenerative feedback pair (Fig. 27). The transistors are modeled in the usual way using a computerized equivalent circuit. However, determination of the parameters of the component transistors from the measured thyristor data is rather difficult, which makes the model somewhat impractical.

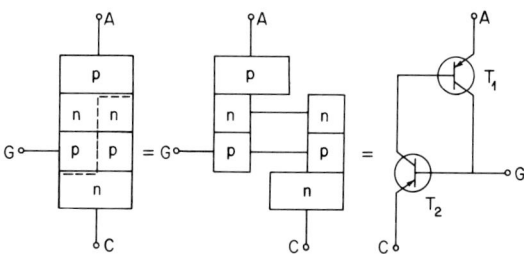

FIG. 27. Thyristor modeling by two interconnected transistors.

B. Thyristor System Models

The operation of a thyristor system can be derived from that of a string or an array of thyristors. An analysis of these two cases is presented in the pages that follow.

1. Thyristor-String Model

The thyristor string consists of several devices connected in series. In the off state, each thyristor is presented by a resistance in the range of 4–20 kΩ. The voltage distribution is determined from the combined impedance of the thyristor and the voltage-grading network.

In the on state, the same current flows through each thyristor in the string, consequently producing overloading in either all or none of the devices. The critical condition occurs during the transient state when the nonsimultaneous turn-on or turn-off of the thyristors may produce dangerous overvoltages.

In order to determine the turn-on overvoltages, a thyristor-string model has been developed (36) in which the N thyristors in the string are divided into n groups. The cumulative turn-on delay-time distribution curve (Fig. 23) is similarly divided into n groups, and it is supposed that all devices within each group turn on simultaneously. Using a previously described thyristor model, the ith group comprises N_i thyristors represented by a switch which closes at time t_{di} in series with a voltage source $V_i = N_i V(t)$

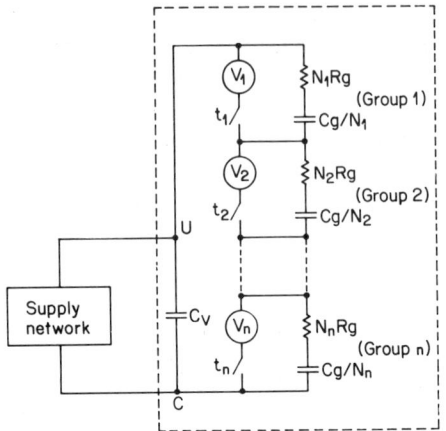

FIG. 28. Equivalent model circuit of a thyristor string.

(Fig. 28). The $t_d = t_{di}$ value is determined from Fig. 23; $V(t)$ is given by Eq. (4), the value of V_0 in the equation being the voltage across each thyristor in group i just before it turns on. The grading circuit for the ith group of thyristors consists of a resistor $N_i R_g$ in series with a capacitor C_g/N_i. The complete model circuit of a thyristor string is shown in Fig. 28.

A program has been developed to operate the switches, monitor the voltage across the groups in the off state, and calculate the value of V_0. The transient produced by the operation of each switch is calculated using an IBM-supplied ECAP program (*33*), which is used as a subroutine. A simplified block diagram of the program is shown in Fig. 29.

The actual voltage appearing across each thyristor can be calculated by the developed program, but a better insight into the physical process may be gained from the considerations below (*30*).

Owing to the presence of the voltage-grading circuit (see Fig. 28), the voltage along the string is distributed uniformly before the turn-on process starts. The thyristor voltage is

$$V_{tH} = V_B/N, \qquad (9)$$

where V_B is the voltage across the string and N is the number of thyristors in the string. The string blocks the voltage V_B until all thyristors have been turned on.

When the first thyristor turns on, the same voltage will be distributed among the remaining $N - 1$ thyristors; the thyristor voltage will thus increase to

$$V_{tH}^1 = V_B/(N - 1). \qquad (10)$$

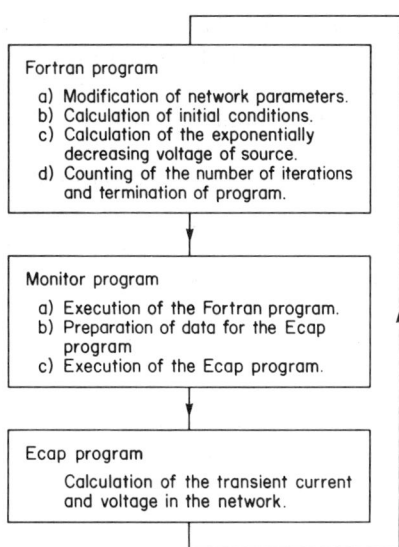

FIG. 29. Simplified block diagram for turn-on calculation program (36).

The thyristor voltage increases gradually because the voltage difference $\Delta V_{tH} = V_{tH}^1 - V_{tH}$ has to charge the capacitor of the voltage-grading circuit for the nonconducting thyristors. As this continues throughout the entire thyristor turn-on process, the highest voltage will appear across the last thyristor turned on.

The physical process during turn-off transients may be described as follows (38). The thyristors will turn off in sequence because of their different storage charges, the first thyristor turning off when its storage charge is neutralized by the negative current, whereupon the voltage across it increases from zero. As current is diverted to the voltage-grading network the current waveform will change, but the current will continue to flow through the thyristors that are on. At some point the current will neutralize the storage charge of the next thyristor, which will also turn off. Voltage will now start to built up on this thyristor as the current is diverted into its grading circuit. This sequential turn-off of the thyristors continues until all thyristors are off. Obviously the highest voltage will appear across the first thyristor turned off and the lowest across the last thyristor turned off. After all the thyristors have turned off, a fast-rising recovery-voltage transient will be generated and superimposed on the storage-charge voltage described above.

Turn-off overvoltages may be calculated by determining the transient current and voltage. An equivalent circuit for the turn-off condition has been developed using a simplified equivalent circuit of the thyristor. This model replaces each of the nonconducting thyristors by a switch closed in the on state; the switch opens when the reverse current neutralizes its storage charge. For the model of the complete circuit, the thyristor string is again divided into n groups. The cumulative distribution curve of the storage charges is approximated by a step function corresponding to the groups. The ith group containing N_i thyristors is represented by a switch Si connected in parallel with a resistance $N_i R_i$ in series with a capacitance C_g/N_i. The complete equivalent circuit of the thyristor string during turn-off is shown in Fig. 30; it should be noted that, for the actual calculation, the supply network must be represented correctly. The operation of this model is outlined below.

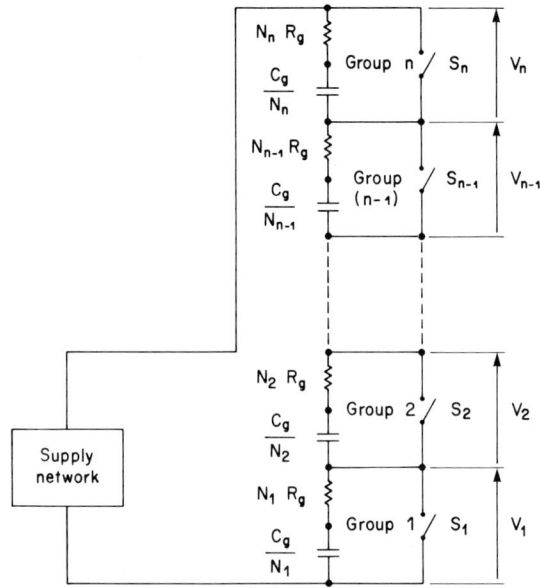

FIG. 30. Equivalent circuit for turn-off calculation.

The turn-off process starts at time $t = T_1$, when the storage charge of the thyristors of group 1 is neutralized, and switch S1 in Fig. 30 opens. The voltage V_1 across group 1 increases from zero as current is diverted into its voltage-grading network.

The current waveform will change, but current will continue to flow through the closed switches S2–Sn. At some time T_2 the current will neutralize the storage charge of the thyristors of group 2 and S2 will open. Voltage will now start to build up on the thyristors of group 2 as the current is diverted into its grading circuit. The sequential operation of switches continues until all thyristors are off.

This sequential calculation process is performed through the use of an IBM-supplied ECAP program. A typical complete equivalent circuit for computer calculation is shown in Fig. 31.

Since the computer calculation starts with the opening of switch S1 at time T_1 the program must be supplied with the initial conditions at T_1. At $t = 0$, the current is decreasing linearly through zero as shown in Fig. 14; all thyristors are conducting and nodes 7 and 0 are short-circuited.

At $t = T_1$, $Q_{min} = Q_1$ has been delivered to group 1 and the thyristors of this group turn off. At this instant, current I_0 flows in the circuit (Fig. 31).

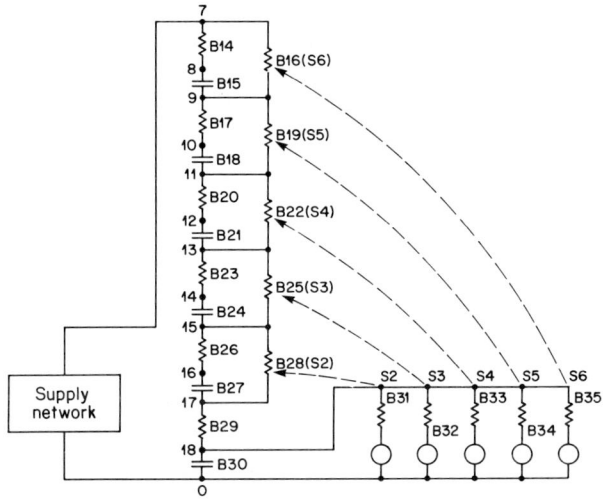

FIG. 31. Modified equivalent circuit for computer simulation of turn-off process.

The initial conditions at time T_1 are calculated by integrating the rate of change of current, given in Eq. 7. From this, the initial current I_0 at $t = T_1$ is

$$I_0 = (V_0/L_t)T_1; \tag{11}$$

$$Q_1 = \int_0^{T_1} I \, dt = (V_0/2L_t)T_1^2. \tag{12}$$

From (11) and (12)

$$I_0 = \sqrt{2V_0 Q_1/L_t} \tag{13}$$

and

$$T_1 = \sqrt{2Q_1/(V_0/L_t)}. \tag{14}$$

In Fig. 31, switches S2–S6 are each represented by stepped resistors having off-state and on-state values of 10 MΩ and 0.1 Ω, respectively. The switches are controlled by branches B31–B35. Each of these branches contains a 10-MΩ resistance and a voltage source connected in series, as well as a sensing element to control the corresponding switches S2–S6. Each voltage source is preset to the storage-charge voltage of the corresponding group of thyristors. The sensing elements are sensitive to the direction of current flow; for example, if the internal voltage of branch B31 is greater than the voltage across capacitor B30, then switch S2 is on; on the other hand, once the voltage across capacitor B30 exceeds that of branch B31, the current direction in B31 reverses and S2 is turned off. New values of current and voltage are computed and the calculation continues until the voltage across B30 exceeds that of the voltage source in B32, at which time S3 is turned off, etc. Since the voltage across capacitor B30 is proportional to the charge delivered by the current, this method permits the operation of switches S2–S6 at the exact instants that the corresponding storage charges are neutralized and is thus an exact simulation of the successive turn-off of groups of thyristors.

The program calculates and prints all branch currents and node voltages, and the thyristor voltages can be derived from the nodal values by elementary methods.

This computational method was used to analyze the operation of thyristor strings, and it was found that (36, 38):

(1) reduction of scattering between the turn-on times considerably reduces the level of overvoltages;

(2) use of thyristors with a longer rise time results in minimum overvoltages, assuming the scattering between the units is small;

(3) because they discharge during the turn-on and inject current into the system, stray capacitances across the string and between the thyristors and ground have a significant effect on turn-on overvoltages;

(4) the current of the stray capacitances and the source current can be reduced by connecting inductances in series with the thyristors; furthermore, this provides a means of controlling turn-on overvoltages. Most manufacturers apply a saturable industance with a saturation time of 2–3 μsec in series with the thyristors or groups of thyristors. This does not reduce the overvoltages up to saturation, but once all the thyristors are turned on and the inductances are saturated they have no effect on conduction;

(5) minimizing the resistance of the voltage-grading circuit decreases both turn-on and turn-off overvoltages. Increasing the value of the capacitance has the same effect;

(6) the turn-off overvoltage is critically dependent on the storage-charge range among the thyristors in the string and this range should therefore be minimized;

(7) The voltage across the thyristor string and across the individual thyristors can be modified during the recovery period by connecting an RC circuit in parallel with the string. The damping circuit can be optimized through use of the computer program described earlier.

2. *Array of Thyristors Connected in Parallel*

The critical operating condition for thyristors connected directly in parallel is the on state. The reasons are as follows:

(1) During the turn-on process, when any one of the thyristors turns on, the voltage collapses across all the devices connected in parallel. Owing to the statistical variation of the turn-on delay times, the delay time expected for parallel arrangements is smaller than for individual thyristors. Consequently, turn-on overvoltages will be lower when several thyristors are connected in arrays, and the arrays connected in series.

(2) Slightly higher overvoltages are expected during turn-off owing to the division of the reverse current which increases the time required to equalize the storage charges. However, the difference is not significant.

In the on state, the current distribution is determined by the thyristors' voltage–current characteristics, which can be approximated by a voltage source in series with a resistance and a diode to avoid the reverse current. The equivalent circuit of an array is shown in Fig. 32. The current distribution in this circuit can be calculated with any computer program that provides the steady-state conditions of a network, e.g., ECAP-I, SPECTRE, etc.

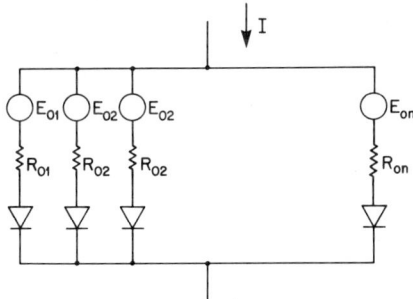

FIG. 32. Equivalent circuit of an array of thyristors connected directly in parallel.

The operation of an array of thyristors connected directly in parallel was analyzed and it was found (*42*) that:

(1) uniform current distribution can be achieved by applying thyristors with matched forward-voltage drops. In a typical case, the forward voltage drops of matched thyristors are within 50 mV for the rated current and temperature;

(2) an identical resistor in series with each thyristor equalizes the current sharing and increases the forward-voltage drop of each thyristor. A resistor that produces a voltage drop of about 0.4 V at rated current keeps the current of the thyristors within 25%. The disadvantage of this method is the additional power loss in the series resistor;

(3) a balancing transformer with a unit turns ratio is connected in such a way that it tends to maintain a net flux of zero by inducing a voltage proportional to the current unbalance in the winding in series with the thyristor having the lower voltage drop;

(4) a balancing inductance can equalize the current of devices connected in parallel, while a further balancing effect can be achieved by arranging the rectifiers and connecting busbars so that the magnetic flux linked by parallel circuits equalizes the current distribution.

V. Fields of Application

In recent years the applications of power electronics, and of thyristors in particular, have expanded dramatically. Table I gives an idea of the major areas in which these devices have come to play such an important role in less than 20 years. Owing to the fast pace of their development, full coverage of these applications has not been attempted here, but only a selection of the applications which the author feels to be more important or representative of the present state of the art.

A. Rectification (ac to dc Converters)

One of the principal applications of power diodes and thyristors is ac to dc conversion. Three-, six-, and twelve-phase circuits are used for this purpose. The current transfers cyclically from one thyristor or diode to the next, the order of commutation (current transfer) being determined by the supply voltage. When the voltage becomes negative across the device, commutation can take place: the device where the voltage is positive takes over the current, while the device where the voltage is negative turns off. In cases where thyristors are used, commutation can be delayed by holding back the gate pulse that fires the incoming device. This phase-control method reduces the rectifier's average direct-voltage output and, in addition, controls the current. Increase of the delay angle to over 90° changes the direction of the dc voltage and allows power to be fed back from the dc into the ac network.

This rectification system presents certain disadvantages, however: harmonic currents are produced in the ac network and harmonic voltage is generated on the dc side. Both of these must be kept within a reasonable

TABLE I
Major Fields of Application of Power Electronics

Rectification	ac to dc controlled rectifiers	Electrochemical industry Smelters and refineries Battery chargers Lighting control Resistance welding Exciters for large generators Transformerless power supplies
	ac to dc converters for motor drives	Pulp and paper industry Plastic industry Textile industry Electric trains Machine tools (milling, grounding, lathes, drills, etc.)
Power control	dc to ac inverters	Power supplies for aircraft Emergency power supplies
	ac to dc converters	Battery-powered vehicles
	ac to dc variable-frequency converters (via dc link)	Ultrasonic crack detector Variable-speed induction and synchronous motor drives Induction heating Pumping station controls
	Cycloconverters	Variable-speed motor drive (low speed) Frequency stabilizers for aircraft
	Voltage control	Low-voltage stabilizers for color TV Voltage control of distribution networks
Special circuits	Solid-state switches	Ignition for cars dc switches ac switches without moving contacts Tap changer for regulating transformers Overvoltage protectors Electronic crowbar circuit Commutator-less dc machines
	hvdc thyristor valves	Long-distance hvdc power transmission (1000 km, ± 600 kV, 1–2000 MVA)

limit by applying auxiliary circuits and filters or increasing the number of pulses per cycle, etc.

One of the most typical uses of rectification is the dc motor drive, where continuous control of the dc supply voltage allows accurate regulation of the motor speed. In fact the solid-state thyristor rectifier has generally replaced the traditional Ward–Leonard system. In these new rectifier systems, the firing delay can be included in a feedback loop to provide a means of control by keeping the speed constant or by following predetermined characteristics. The speed range of the drive system is 10 : 1, although special systems of up to 100 : 1 have been built. System accuracy is high: the control normally keeps the speed within a $\pm 1\%$ limit, but here again specially built circuits are capable of achieving a limit of $\pm 0.1\%$ when the supply voltage fluctuation is $\pm 5\%$, the load variation $\pm 50\%$ and temperature variation $\pm 10\%$ (43).

In the typical dc motor drive shown in Fig. 33, there are two control loops: one for speed, the other for the current. The tachometer (T) produces a signal proportional to the speed of the motor; the signal is processed by

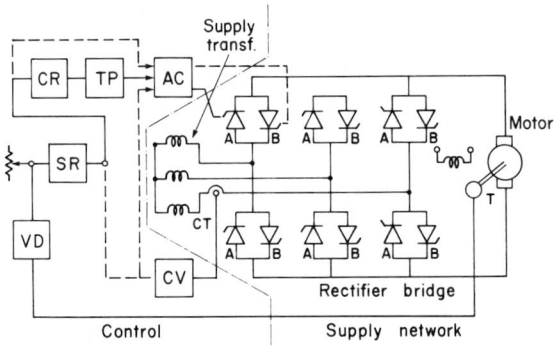

FIG. 33. Basic circuit for thyristor regulators of dc motors (43).

the signal conditioner (VD) and fed to the speed regulator (SR), then through the current regulator (CR) to the trigger pulse generator (TP), which fires the thyristors according to the set value. Additional current regulation is achieved through the current transformer (CT), current value conditioner (CV), and current regulator (CR). The trigger generator fires either thyristors A or thyristors B according to the required direction of motor rotation. The rotation can be reversed by a changeover relay (AC).

Another typical rectifier application is in electrochemical plants. Corby (44) describes a rectifier system for aluminum smelters with a total rating of 840 V, 150,000 A. This is an outdoor installation with rectifier cubicles containing silicon diode stacks, protective fuses, and surge suppression circuits, together with an oil cooling system. General control is achieved by autotransformers equipped with on-load tap changers, while greater control precision is achieved using tranducers driven by thyristor amplifiers from a solid-state closed-loop control system. At the present time, this seems to be more economical than all-thyristor systems for extra high power levels, although thyristors can be expected to replace the diodes in the stacks in the more or less distant future and gate control will be used instead of tap changers.

One of the most important applications of the rectifier is in the static excitation system for large turbogenerators. This normally consists of a step-down transformer, a diode or thyristor rectifier, and a control loop; the dc current output from the rectifier is fed through slip rings and brushes to the generator. The fast control offered by this system is offset, however, by the need for brushes and slip rings, which require frequent inspection and maintenance.

Wright et al. (45) report the development of an experimental brushless thyristor-controlled system (see Fig. 34), which comprises a three-phase bridge rectifier built on the shaft and rotating with the main generator. Each leg of the bridge contains several thyristors connected in parallel and each thyristor is protected by fast fuses. The gate of the thyristor bridge is fired by the control exciter generator which has two field windings on both the direct and the quadrature axes. The windings are supplied independently with controlled current from the automatic voltage regulator circuit, the ratio of

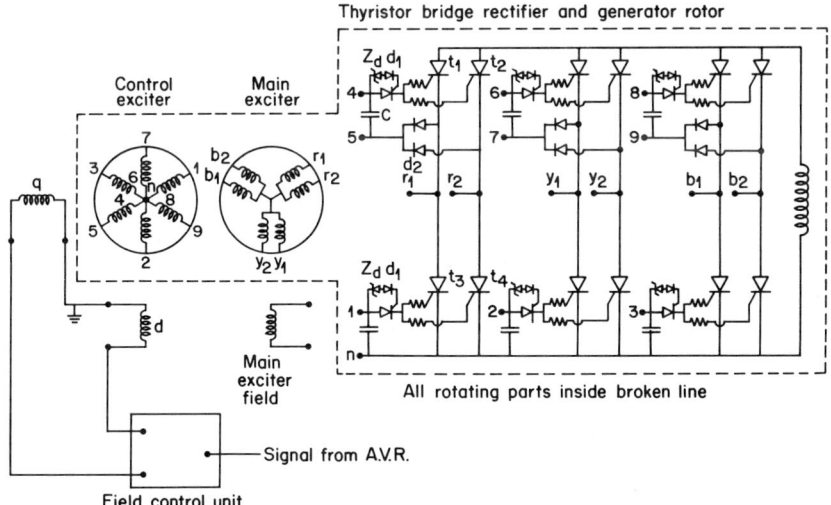

FIG. 34. Brushless thyristor excitation system for large turbogenerators (45).

the currents being controlled in such a way that the net field flux is constant but its position in space can be changed, thus varying the thyristor's firing angle. The rectifier is supplied from a main exciter generator with a constant field. The range of the generator's excitation current can be changed from a positive to a negative value and the time constant is in the range of 10 msec.

The advantage of this system is the direct connection between the generator, thyristor bridge, main exciter, and control exciter, without brushes or slip rings. The excitation current is free from system disturbances and is automatically available when the generator is rotating.

B. Power Control

1. Direct-Current Circuits

The power of a dc circuit (47) can be controlled by a chopper, as shown in Fig. 35. Choppers are used for the control not only of dc traction motors with series windings and battery-powered electric vehicles, but also of line-supplied electric locomotives, streetcars, cablecars, etc. In this circuit, thyristor Th_M is turned on and off with a variable frequency producing a train of

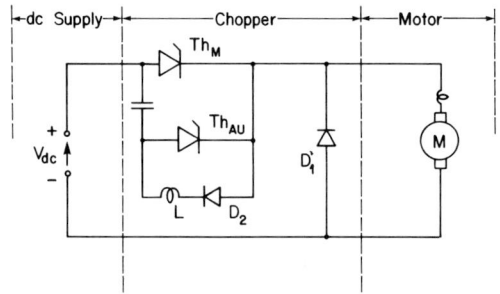

FIG. 35. Thyristor chopper circuit for dc motor control.

voltage pulses to the motor. The average value of the voltage pulses is controlled by the ratio of the on and off times and can be used to control or maintain the speed of the motor as required. Thyristor Th_M is turned off by auxiliary thyristor Th_{AU}, which discharges the charged capacitor, thereby forcing the current of Th_M to zero. The freewheeling diode D_1 allows the inductive load current to circulate during the off period.

2. Alternating-Current Circuits

Back-to-back-connected thyristor pairs and triacs connected in series with the load are equally suitable for controlling power in an ac circuit, since either arrangement is capable of conducting in both directions when the thyristors are fired. Control is effected by delaying the firing gate signal. An alternative control method is to fire the thyristors for several cycles and then remove the gate signals for several cycles; this produces a burst of current pulses, the average value of the current depending on the length of the on time. The latter method is particularly suitable for heating and furnace control and in fact a system regulating a 20-MW induction heating furnace has already been developed. The same principle can be used for high-speed circuit breakers in which the circuit is switched off by removing the gate signal.

A further important application in this field is light control, where a triac or a back-to-back-connected thyristor pair is used. The regulation of filament lamps calls for phase control of the thyristor firing, but in the case of fluorescent lamps a more sophisticated circuit is required with both a constant and a variable voltage output, the former supplying the heater transformers, the latter supplying the lamps and, at the same time, regulating the illumination. The heater transformer and a RC network are needed for initiating and maintaining the lamp arc.

In many cases, the triac is controlled by electronic circuits which vary the light intensity automatically. The sound and light systems used in nightclubs and discotheques have a random control, for example, and the daylight compensation control system measures the light intensity and adjusts the artifical lighting to keep the illumination at a predetermined level regardless of the natural lighting (46).

C. Frequency Conversion

The frequency converter, as its name implies, converts ac power of one frequency (generally 60 or 50 Hz) to another, constant or variable, frequency. High frequencies are used mainly to drive variable-speed squirrel-cage induction motors, dc-excited synchronous motors, or reluctance-type synchronous motors. The variable-speed operation of ac motors has a number of technical advantages over the thyristor-controlled dc motor drive: no commutator, simple motor construction, better power/weight ratio, etc.

In the case of induction motors, frequency and slip control can be combined: the speed of the motor is determined mainly by the supply frequency and the torque of the machine is adjusted by slip control. In this way a full-load torque can be obtained for a wide range of speeds. Furthermore, negative slip permits regenerative braking. In the case of synchronous machines, the speed of the motor again depends on the frequency, which can in fact be kept very accurate by driving the inverter with a precision quartz-crystal-controlled oscillator. For instance, the synchronous operation of several motors can be achieved when they are connected in parallel and supplied with variable frequency from the inverter. Typical applications are rollers in steel mills and various machines in the textile industry.

The two main types of static-frequency converters in present-day use are described below.

1. *Direct-Current Link Frequency Converter*

This circuit comprises a rectifier which converts ac power to dc and an inverter which converts dc power to ac with a different frequency. Figure 36 shows a widely used three-phase frequency converter consisting of a bridge rectifier and a force-commutated inverter circuit also in bridge connection.

With regard to the inverter part of the circuit, several such circuits have been developed to date (47). The operation of a single-phase force-commutated inverter was discussed previously in connection with Fig. 35. When used in the frequency converter, the inverter operates as follows: given certain operating conditions, thyristors Th_1 and Th_6 are conducting and capacitors C_1, C_2, and C_3 are charged. Firing any thyristor with a positive anode voltage discharges the appropriate capacitance and turns off the conducting thyristor in its row: e.g., the firing of Th_2 discharges C_3, which turns off Th_6. Sequential firing of the thyristors provides the proper three-phase output. The output voltage and frequency can be controlled by the firing.

One of the latest developments in the field of inverters is pulse-width modulation, in which the ac voltage waveform is synthesized from a number of pulses of varying width, as shown in Fig. 37. This circuit is characterized by the low harmonic content of its waveform and it requires thyristors with a very short turn-off time.

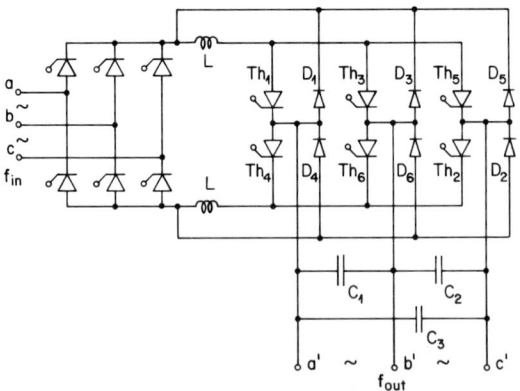

FIG. 36. Three-phase frequency changer.

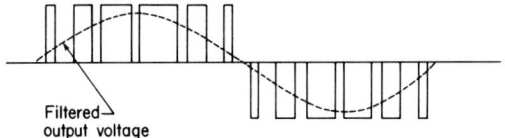

FIG. 37. Voltage waveforms of a pulsewidth modulated inverter.

In some inverter circuits, thyristors are replaced by transistors or gate turn-off thyristors. Since these components do not require forced commutation, this simplifies the inverter circuit, but the present state of development permits only medium-power applications as yet.

2. *Cycloconverters*

The cycloconverter circuit changes an input power at one frequency to an output power at a lower frequency by synthesizing a low-frequency voltage waveform from sections of a high-frequency waveform. A typical cycloconverter circuit and its input waveforms at a resistive load are shown in Fig. 38. The circuit comprises two inversely-connected converters. The thyristors in the circuit are turned on in such a way as to obtain a low-frequency output with a frequency reduction of 3 : 1 between input and output. Between t_1 and t_2, the supply busbar A is positive; this permits the firing of Th_3 and Th_6, at t_1. At t_2, owing to the change of polarity of the supply busbar, Th_3 and Th_6 turn off, and Th_4 and Th_5 are fired. At t_3, the polarity

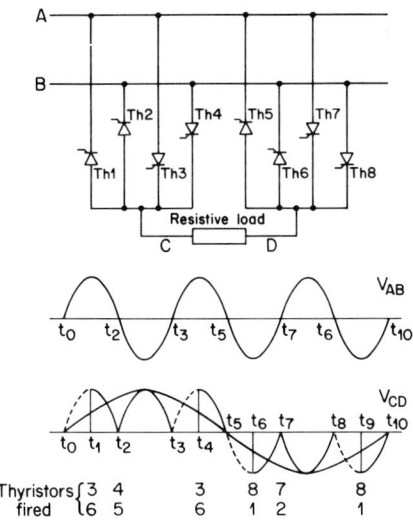

FIG. 38. Typical cycloconverter circuit and its output waveform (47).

of the busbar changes again, which turns off Th_4 and Th_5. In order to obtain the desired waveform, Th_3 and Th_6 will be fired only at t_4. The further sequence of firing and its effect can be followed in Fig. 38. The inductive load modifies the waveform and the firing delay but not the sequence. It can be seen that the firing delay of the thyristors oscillates between zero and 90°, at a frequency determined by the desired output frequency.

The harmonic content of the output voltage is reduced by filters; however, when the converted frequency becomes greater than about half the input frequency, the harmonic content of the output voltage becomes excessive. A major application of the cycloconverter is in aircraft, where it is used to convert into fixed low-frequency power the variable-frequency power produced by the generators driven from the variable-speed main motors of the aircraft. Certain advantages are also offered if the regeneration of power is required, another popular utilization.

D. High-Voltage dc Transmission

One of the most complex applications of thyristors is in high-voltage dc transmission. High-voltage dc transmission systems are economical for long-distance bulk power transfer or in the case of interchange between two large ac energy systems where direct current provides an asynchronous link enabling energy to be transferred economically from one system to the other without incurring stability problems and without increasing the short-circuit current of the two systems. There are several hvdc systems in operation in various parts of the world, the older ones applying mercury arc valves, the new systems applying thyristor valves. The transmitted energy is in the range 1500–2000 MW, the transmission line voltage about ± 400 to ± 500 kV, with a current of 2000 A.

The basic circuit is shown in Fig. 39. Owing to its symmetry, the hvdc system connecting two three-phase ac power systems enables energy to be transferred in either direction. In one operating mode, converter C_1 acts as a rectifier: the energy is transmitted through the dc transmission line to converter C_2 which acts as an inverter, transforms the dc power to ac, and supplies energy to the second ac system. The operating mode of converters C_1 and C_2 depends on the firing delay of the valves. During operations of the dc system the valves in the bridge converter circuit are fired in sequence (1, 2, 3, 4, 5, 6). The output voltage of the converter can be controlled from a positive maximum to a negative maximum value by selecting the appropriate firing angle of the circuit. When the firing angle is between 0 and 90°, the output voltage is positive; between 90 and 180° the output voltage is negative. The current always flows in the same direction through the bridges.

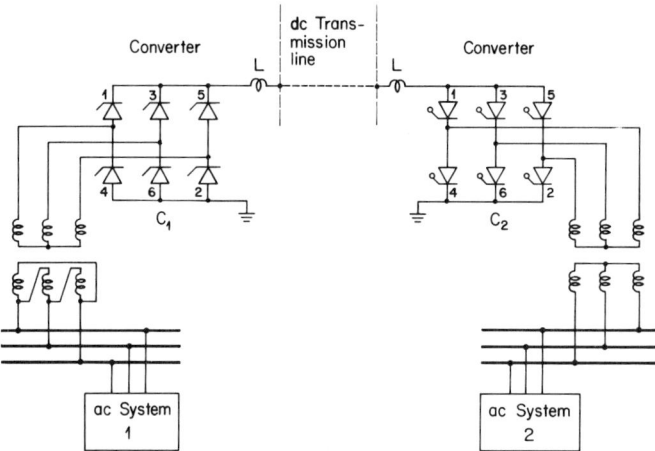

FIG. 39. An hvdc system.

On the other hand, the direction of the power flow is governed by the direction of the voltage; thus for a reversal of power, the direction of the system voltage has to be reversed.

With C_1 operating as an inverter, for instance, the firing angle of C_1 will be in the range of 15–20°, that of C_2 between 160 and about 170°; consequently, the voltage of C_1 will be positive and C_2 will be negative. However, in converter C_2, the valves are positioned in the direction opposite to that of the C_1 valves to allow the flow of current, with the result that the voltage on the dc side of C_2 will balance the voltage of C_1. The small difference in voltage between C_1 and C_2 drives the current through the system. In the case of a power reversal, the firing angle of C_1 will increase to 160–170° and simultaneously, the firing angle of C_2 will be reduced to 15–20°. This changes the direction of the voltages, but the voltage difference can be maintained, thereby allowing the current to be driven in the same direction; however, owing to the change of polarity of the voltages, the direction of power flow is in fact reversed.

The process outlined above and the constant regulation of power dictate the need for a complex electronic control system.

A typical 450-kV system contains three converters in series (Fig. 40), each converter containing six valves with a rating in the range of 150 kV, 2000 A. The valves are complex thyristor systems comprising a large number of devices connected in series and in parallel. A typical electrical arrangement of a thyristor valve was shown in Fig. 21 of this chapter. The mechanical deployment of valve components can be seen in Fig. 41, which gives a top view of one stage of the thyristor valve. The thyristors and associated voltage-grading circuitry are arranged in modules, each of which

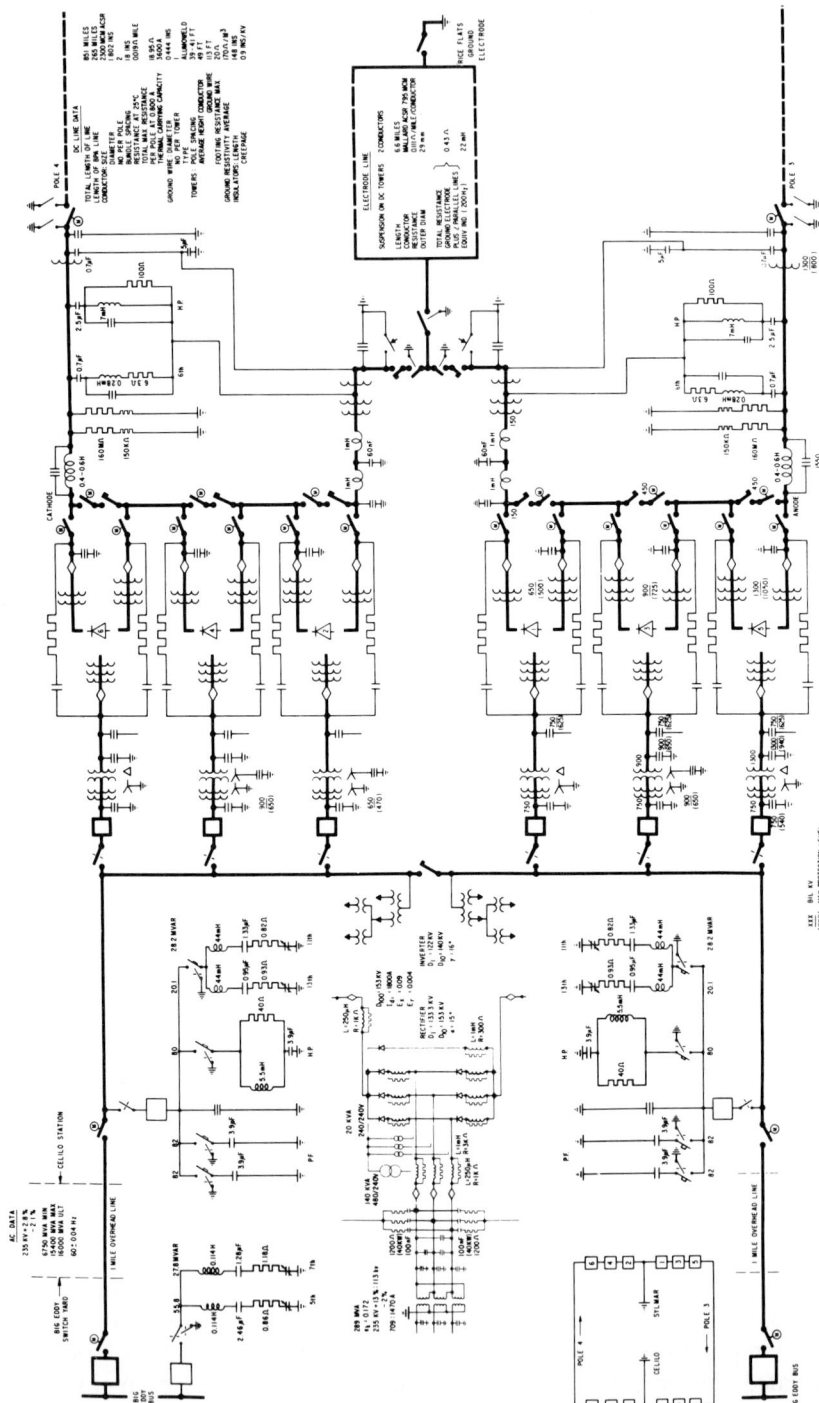

FIG. 40. Circuit diagram of the CELILO hvdc substation of the Bonneville Power Administration.

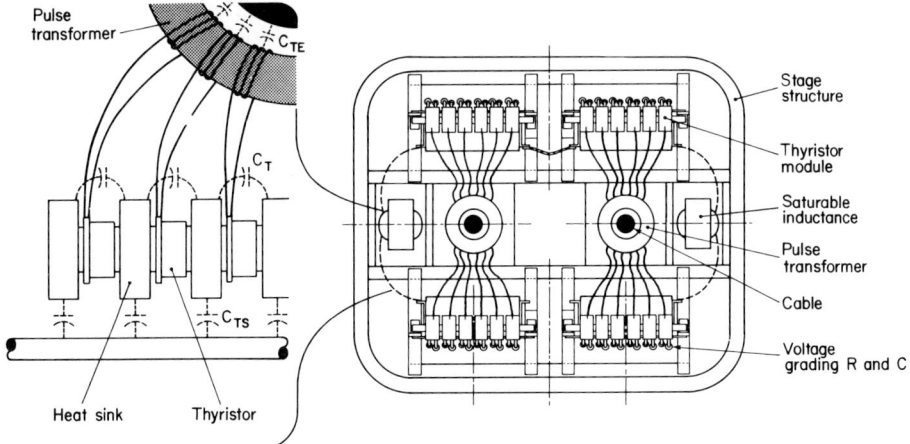

FIG. 41. Top view of one-stage assemblies of a thyristor valve (30).

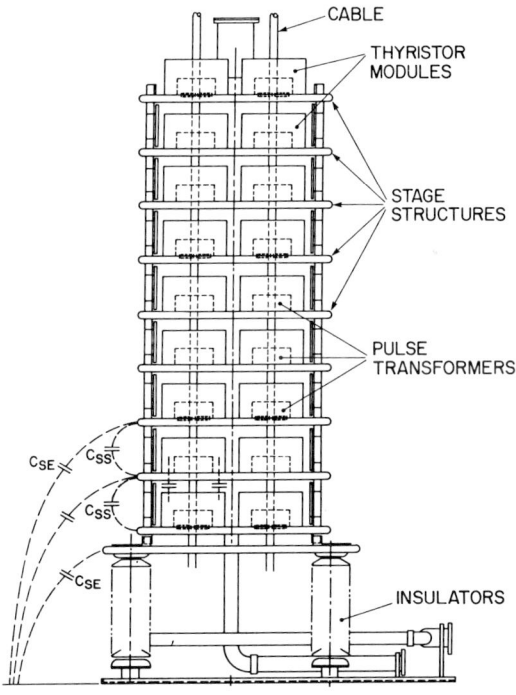

FIG. 42. Diagram of an experimental 150-kV thyristor valve (Westinghouse Canada Ltd.).

comprises several thyristor units connected in series, and cooled by appropriate heat sinks. The modules themselves are arranged in a horizontal stage structure (see Fig. 41); each stage contains ring-type pulse transformers that thread the firing cable as well as saturable inductors in series with the thyristors. Figure 42 shows a complete valve structure in which the stages are arranged in a vertical column. The structure here is supported by porcelain insulators and the valve is shielded by electrodes surrounding each stage.

E. Special Thyristor Applications

Thyristors are successfully utilized for commutator-less dc machines, in which each commutator segment is replaced by a pair of back-to-back-connected thyristors (47). This arrangement requires a large number of devices but, with a little sacrifice in torque smoothness, a four-segment commutator can be obtained. The electrical circuit is shown in Fig. 43. The thyristors here are fired by a rotor-position sensor which provides firing pulses at 0, 90, 180, and 270° positions. The rotor-position sensor is either a magnetic, an optical, or a Hall-effect device. The latter is used in many small motors. Two Hall generators are placed on the rotor at an angle of 90° to each other and the rotor field produces the required firing signal at the proper positions. A typical circuit is shown in Fig. 44, where the thyristors are replaced by transistors. The sinusoidally distributed rotor flux will produce a sinusoidal output voltage from the Hall generators which is amplified by the transistors. The stator-current distribution in the four windings will

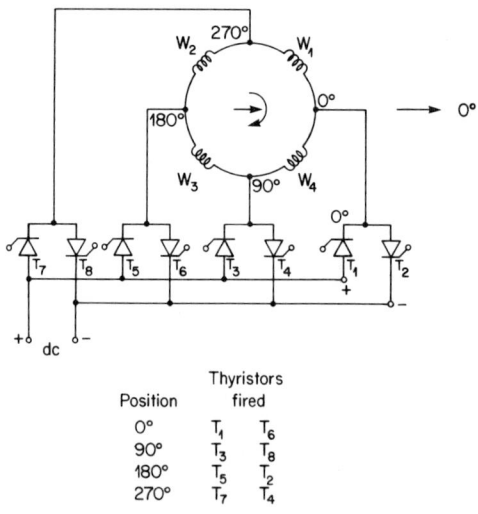

Position	Thyristors fired	
0°	T_1	T_6
90°	T_3	T_8
180°	T_5	T_2
270°	T_7	T_4

FIG. 43. Commutator-less dc machine (47).

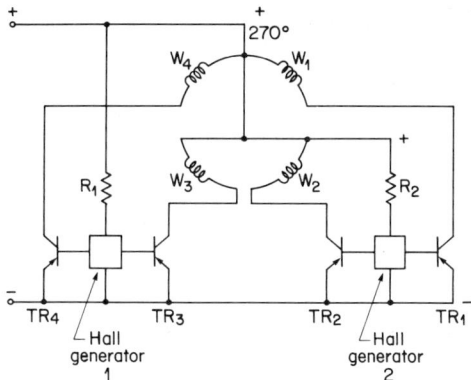

FIG. 44. Winding arrangements of a four-segment Hall sensor for electronic motor drive (47).

be such that the stator flux remains constant. This results in a very smooth operation but, owing to the limited power of the transistors, it can be applied only for small motors. Similar circuits have been developed for thyristors. The advantages of commutator-less motors are: sparkless operations, the reduction of audible noise, and little or no service requirement.

A forced commutation technique, similar to that described in the previous section, was adopted recently to develop a current-limiting circuit breaker, an experimental device suitable for voltage levels of up to 440 V and designed to operate in a circuit with a fault-current limit of 5000 A with an interruption time of less than 250 μsec (49). However, in spite of its advantages, the high cost of this equipment seems to have prevented widespread use to date.

VI. Conclusion

Power electronics has developed very rapidly over recent years and found applications in many areas of our daily lives. This may be due to the fact that the reliability of thyristors and silicon diodes has shown marked improvement: Holloway (11) for example, estimates that present reliability is 15.10^{+6} hours mean time between failures.

This paper has attempted to present the state of the art and development trends in the power electronics field. From this broad survey the following conclusions may be drawn:

(1) The thyristor is the most widely used device in most power electronics circuits. However, the future will probably see thyristors replaced by transistors and special devices in many fields, a trend which will simplify circuits and reduce costs.

(2) Increasing power requirements call for the development of semiconductor devices with a diameter of 102 mm. Present technological limitations (low manufacturing yield) should be solved in the next few years and it should be possible to use these devices up to 2000 A and 4–5 kV.

(3) Besides the development of the devices themselves, the improvement of auxiliary equipment (cooling, mounting, etc.) should result in better performance and reduce the overall size of the circuit. Particularly important here is the development of the cooling method, which could contribute to a significant reduction in the size of the thyristor in the not too distant future.

(4) The application of computers has opened the way to more exact design of the power electronic circuit. In the future, the statistical variation of component values will be taken more into account at the circuit design stage. This, together with sensitivity analyses, cannot fail to improve the reliability of the design methods.

(5) Application of power electronics is spreading. It can be foreseen that its future use in cars, household appliances, etc., will be promoted by lower prices. A further application trend is the growing need for regulation of high-power drives.

Acknowledgments

The author would like to acknowledge the contribution of Ms. L. Régnier of Hydro-Québec for careful editing of the text. Mr. D. Dallaire and Ms. Huguette Martin of IREQ must also be thanked for collecting the literature and preparing the references.

References

1. J. D. Harnden, Jr., and F. G. Golden. *In* "Power Semiconductor Applications (selected reprints of the most important papers)," pp. 1–8, Institute of Electrical and Electronics Engineers, New York (1971).
2. Physics of semiconductor materials. *Advan. Electron. Electron Phys.* **7**, 1 (1955).
3. F. F. Mazda, "Thyristor Control." 381 pages. Halsted Press, New York (1973).
4. Transistors—the first 25 years 1948–1973. *Radio Electron. Eng.* **43** (Jan./Feb. 1973).
5. F. W. Gutzwiller (ed.), "SCR Manual," 4th edition. General Electric Company, Syracuse, New York (1967).
6. K. Heumann and A. C. Stumpe, "Thyristoren," 2nd edition. B. G. Teubnes, Stuttgart (1970).
7. W. W. Reid, Changing views on protecting and packaging thyristors. *Elec. Times*, pp. 9–10 (June 28, 1973).
8. J. De Warga, High power semiconductors. *Eng. Digest.* pp. 29–32 (Nov. 20, 1972).
9. W. Gerbach and A. C. Stumpe, Thyristoren mit Querfeldemitter. *Z. Angew. Phys.* No. 19, pp. 396–400 (1965).
10. F. E. Gentry and J. Moyson, The amplifying gate thyristor. International Electronic Devices Meeting, Washington (1968).
11. L. W. Holloway, Evolving fast switching thyristors for medium and high power industrial applications. *Elec. Times*, pp. 11–13 (June 28, 1973).

12. E. Reimers, Testing of a 470 ampere rms 2000 volt regenerative gate turn-on thyristor. *IEEE Power Process. Electron. Spec. Conf., Detroit, PPESC 71 Records*, pp. 37–43 (1971).
13. P. Ganner and F. Kirschner, Fast high-voltage thyristors (in German). *Siemens-Z.* **46**, 841–843 (Nov. 1972).
14. F. M. Roberts and E. L. G. Wilkinson, The relative merits of thyristors and power transistors for fast power-switching applications. *Int. J. Electron.* **33**(3), 319–341 (Sept. 1972).
15. I. S. Somos and D. E. Piccone, Some observations of static and dynamic plasma spread in conventional and new power thyristors. *Proc. IEEE Conf. Power Thyristors, London*, pp. 1–7 (1969).
16. T. P. Nowalk and J. B. Brewster, High voltage and current, gate assisted, turn-off thyristor development. NASA-CR-121161, Semiconductor Division, University of Cincinnati, Cincinnati, Ohio (1972).
17. E. D. Wolley and R. Yu. Characteristics of a 200 amp gate turn-off thyristor. *In* "8th Annual Meeting of the IEEE Industrial Applications Society, Milwaukee, Wisconsin," pp. 251–257, Institute of Electrical and Electronics Engineers, New York (1973).
18. M. Arai, Magnetosensitive thyristor. *Jap. J. Appl. Phys.* **12**, 1278–1279 (1973).
19. Y. Beauséjour, and G. Karady, Possibility of application of direct light-fired thyristors in hvdc systems. *In* "Digest of the International Electrical Electronics Conference, Toronto," Paper No. 73111, pp. 66–67. Institute of Electrical and Electronics Engineers, New York (1973).
20. F. E. Gentry, R. I. Scace, and J. K. Flowers, Bidirectional triode p-n-p-n switches. *Proc. IEEE* **53**(4), 355–369 (1965).
21. M. Maurer, State of development and future trends of power thyristors and triacs (in German). *Technica* **22**, 711–713 (1973).
22. A. Steimel, Untersuchung über das Einschaltverhalten eines neuartigen gatelosen Thyristors. *ETZ Ausg. A* **95**(5), 282–289 (1974).
23. A. Stamberg, Accent on reducing the cost of thyristor cooling. *Elec. Times* p. 11 (June 28, 1973).
24. J. A. Gardner, Jr., Liquid cooling safeguards high-power semiconductors. *Electronics (USA)* **47**(4), 103–108 (1974).
25. H. Lawatsch, H. and E. Weisshaar, "Ein Silizium Spannungs-Begrenzer zur Beschutzung von Leistungs-Thyristoren. *BBC Nachr.* **59**(9), 476–482 (1972).
26. R. M. McKechnie and S. W. Kessler, Transcalent solid-state power devices. *IEEE Power Process. Electron. Spec. Conf., Atlantic City, N.J., PPESC 72 Records* pp. 128–133 (1972).
27. E. B. Borchert and A. C. Stumpe, Stosspannungsfeste Siliziumzellen. *AEG Mitt.* **54**(5/6), 469–473 (1964).
28. F. B. Golden, A new component—the metal oxide varistor suppressor. *IEEE Power Process. Electron. Spec. Conf., Atlantic City, N.J., PPESC 72 Records* pp. 134–139 (1972).
29. E. T. Schonholzer, Fuse protection for power thyristors. *IEEE Trans. Ind. Appl.* **8**(3), 301–309 (1972).
30. G. Karady and T. Gilsig, The thyristor valve in hvdc transmission. *IEEE Spectrum* **10**(12), 36–43 (1973).
31. G. Karady, G. N. Trinh, and G. Elop, Analysis of the cable firing system for hvdc thyristor valves. *IEEE Trans. Power Appar. Syst.* **93**(2), 571–578 (1974).
32. S. Takahashi, Y. Ebata, and K. Kishi, Application of acoustic surface wave to power electronics. *IEEE Power Proc. Electron. Spec. Conf., PPESC 74 Records* No. 74 CH0863-1-AES, pp. 187–196 (1974).
33. R. W. Jensen, and M. D. Liebermann, "IBM Electronic Circuit Analysis Program Techniques and Application." Prentice-Hall, Englewood Cliffs, N.J. (1968).
34. International Business Machines, "Advanced Statistical Analysis Program—GH 20-1271." IBM, New York.

35. F. Vogt, The simulation of converters (in German). *ETZ, Ausg. A* **94**, 479–482 (1973).
36. G. Karady and T. Gilsig, The calculation of turn-on overvoltages in a high-voltage thyristor valve. *IEEE Trans. Power Appar. Syst.* **90**(6) (1971).
37. D. N. Harstad, SCR model simplifies computer programs. *Electron. Design* **17**(22), 92–95 (1969).
38. G. Karady, and T. Gilsig, The calculation of turn-off overvoltages in a high voltage thyristor valve. *IEEE Trans. Power Appar. Syst.* **91**(2), 565–574 (1972).
39. W. C. Beattie, and W. Monteith, Analogue modelling of a thyristor. *Proc. IEE* **120**(7), 786–788 (1973).
40. W. C. Beattie and W. Monteith, Digital modelling of a thyristor. *Proc. IEE,* **120**(7), 789–790 (1973).
41. M. Kurata, A new CAD-model of a gate turn-off thyristor. *IEEE Power Electron. Spec. Conf., PESC 74 Records* No. 74-CHO-363-1-AES, pp. 125–133 (1974).
42. A. R. Mulica, How to use silicon controlled rectifiers in series or parallel. *Control Eng.* No. 5, 95–99 (1974).
43. H. Winkler, Development of standard thyristor drives. *Elec. Times,* **162**(24), 29–33 (1972).
44. D. B. Corbyn, Rectifiers for electro-chemical plant. *Elec. Times* **162**(16), 3–5 (1972).
45. W. F. Wright, R. Hawley, and J. L. Dinely, Brushless thyristor excitation systems. *IEEE Trans. Power Appar. Syst.,* **72**, 1848–1854 (1972).
46. A. J. D. Sant, Daylight compensation controlled by thyristors. *Elec. Times* pp. 14–15, (June 28, 1973).
47. J. Murphy, "Thyristor Control of ac Motors," p. 345. Pergamon, London.
48. E. W. Kimbark, "Direct Current Transmission, Vol. 1, p. 508. Wiley-Interscience, Toronto (1971).
49. K. Kishi, K. Takigami, M. Moroshohi, and A. Takenaka, Ultra high speed solid state circuit breakers for AC and DC power lines, *IEEE Power Electron. Spec. Conf., PPESC 74 Records,* No. 74-CHO-863-1-AES, pp. 313–320 (1974).
50. P. Atkinson "Thyristors and their applications," p. 126. Mills and Boon, London (1972).

AUTHOR INDEX

Numbers in parentheses are reference numbers and indicate that an author's work is referred to although his name is not cited in the text. Numbers in italics show the page on which the complete reference is listed.

A

Abdallah, J., Jr., 88, 89, *111*
Åberg, T., 96, *108*
Aboaf, J. A., 291, *306*
Albers, V. M., 229, *247*
Albright, N. W., 62, *72*
Alder, H., 144(63), 147, *164*
Aleksandrov, G. N., 42(74), 44(74), 71, *72*
Alfvén, H., 22(33), *71*
Allen, K. R., 39, *72*
Allibone, F. E., 6(16), *71*
Altick, P. L., 87, 92, *108*
Alvarado, S. F., 134(40), 138(48), 140(40), 142(40, 56), 163, *164*
Ambartsumyan, V. A., 22(31), *71*
Ames, I., 144, *164*
Amusia, M. Ya, 83, 92, *108*
Anderson, P. W., 150(78), 159, *164*
Aoi, K., 149(70), 159, *164*, *165*
Aoki, T., 291, *308*
Arai, M., 329(18), 330, *369*
Arnau, C., 92, *109*
Athay, R. G., 25, *71*
Atkinson, P., *370*

B

Backx, C., 106, *108*
Baer, Y., 73, 107, *111*, 118(13), *163*
Baganov, A. B., 144(60), *164*
Bagus, P. S., 83, *108*
Bahr, J. L., 75, *108*
Baker, A. D., 75, *111*
Baker, C., 75, *111*
Balk, P., 278, 290, 291, *306*
Bänninger, V., 144(61), 145(61), 150(76), *164*
Bartelink, D. J., 268, *306*
Basilier, E., *109*
Bassous, E., 291, *306*
Bates, D. R., 76, 84, *108*
Bauer, R. W., 37, *72*
Baum, G., 159(101), *165*
Beattie, W. C., 346(40), *370*
Beauséjour, Y., 330, *369*
Bell, R. L., 152(85), *165*
Benford, G., 37, *72*
Bennemann, K. H., 149(70, 73), *164*
Bennett, M. R., 146, *164*
Berger, H., 303, 304, *306*
Bergman, D. C., *247*
Berger, K., 5, 12, 13, 14, 15, 16, 17, 18, *71*
Bergmark, T., 73, 92, 107, *109*, *111*
Bernhardt, A. F., 36(65), *72*
Berrington, K. A., 90, *108*
Bethe, H. A., 76, 97, 98, 99, 105, *108*, *110*
Bethke, G. W., 35, *72*
Bishop, S. G., 142(51), *164*
Blais, R. N., 32, *72*
Boonstra, E., 259, 303, *306*, *307*
Booth, N. O., 200, 223, 244, *247*, *248*
Borchert, E. B., 335(27), *369*
Bosselaar, C. A., 271, *306*
Boulin, D. M., 292, 293, *307*
Bowles, K. L., 31(56), *72*
Bowman, G. G., 29, *71*
Brash, H. M., 149(71), 155, *164*
Breed, D. J., 297, *306*
Brenden, B. B., 169, *247*
Brewer, J. E., 288, *306*
Brewster, J. B., 326(16), *369*
Brion, C. E., 106, *111*
Brook, M., 17, 19, 20, *71*
Brown, G. A., 275, *306*
Brown, S. C., 34(62), *72*
Bruce, C. E. R., 22, *71*
Brueckner, K. A., 90, *108*
Brundle, C. R., 75, *111*
Bryant, S. B., *247*

AUTHOR INDEX

Bucher, E., 144, *164*
Bulucea, C., 268, *306*
Burke, P. G., 88, 89, 90, 96, *108*, *110*, *111*
Burlaga, L. F., 26, *71*
Busch, G., 115(6), 118(13), 131(36), 134(41), 136, 144(61, 62), 145(61), 147(62), 148(62), 150(76), *162*, *163*, *164*
Butler, S. R., 291, *308*

C

Callaway, J., 89, *108*
Camp, L. W., *247*
Campagna, M., 115(6), 124(25), 131(36), 132(37), 134(41), 136(45), 142(56), 144(61, 62, 63), 145(61, 41), 147(62, 63), 148(62), 150(76), 157(25), 159(94), *163*, *164*, *165*
Campbell, D. M., 149(71), 155, *164*
Camphausen, D. L., 142(50), *164*
Canavan, G. H., 36, *72*
Caner, B., 28, *71*
Capon, J., 203, *247*
Card, H. C., 300, *306*
Carlson, H. G., 275, *306*
Carlson, T. A., 92, *108*
Carlstedt, G., 281, *306*
Carnes, J. E., 260, 290, *306*, *307*
Caroli, C., 117, *163*
Carr, W. N., 258, 259, 261, *307*
Carver, J. C., 75, *109*
Cassaverde, M., 30(54), *72*
Caudano, R., 107, *108*
Chakraverty, B. K., 142(50), *164*
Chang, T. N., 92, *108*
Chase, C. T., 113(1), *162*
Cherepkov, N. A., 83, 92, *108*
Chernysheva, L. V., 83, 92, *108*
Chricchi, J. R., 275, 283, 284, 285, 286, 291, *309*
Christensen, R. L., 144, *164*
Christensson, S., 282, *308*
Clark, M. G., 155, *165*
Clementi, E., 83, *108*
Clinton, J., 149(72), *164*
Coey, J. M. D., 142(50), *164*
Cohen, M. L., 152(88), 153(88), *165*
Cole, B., 303, *306*
Collens, H., 3, 4, 5, 6(7, 3, 4, 5), 8(4, 3), 9(4), 11(4), 13(4), 19(4), 20(4), 70, *71*

Collet, M. G., 260, *306*
Combet Farnoux, F., 82, *108*
Conti, F., 300, *307*
Conti, M., 300, *307*
Cooper, J., 97, 99, *108*, *109*
Cooper, J. W., 80, 81, 82, 87, 88, 97, 98, 106, *109*, *110*
Corbyn, D. B., 356, *370*
Cotti, P., 115(6), 134(43), *162*, *163*
Cox, H., *247*
Crawford, R. H., 256, *307*
Cullen, D. E., 274, *308*
Cuomo, J. J., 255, *308*
Cutler, P. H., 158(96), *165*

D

Dalgarno, A., 83, *109*
Damany, N., 75, 107, *109*
Damburg, R., 89, *109*
Davies, A. J., 65(89), *72*
Davies, A. R., 90, *110*
Davies, C. S., 65, *72*
Davis, J. R., 31, *72*
Davisson, C. J., 113, *162*
Dawson, E. F., 35, *72*
Dawson, G. A., 42(73), *72*
De, B. R., 24, *71*
Deal, B. E., 265, 297, *307*
de Graaff, H. C., 267, 295, *307*
de Maagt, B. J., 266, 267, 268, 273, *309*
de Niet, E., 254, *307*
DeWarga, J., 321(8), *368*
Diatroptov, D. B., 144(60), *164*
Dill, D., 93, 100, 101, 102, 103, 104, *109*, *110*
Dinely, J. L., 357(45), *370*
Donelon, J. J., 122(19), *163*
Dorda, G., 283, 284, *307*
Druyvesteyn, W. F., 254, *307*
Dubau, J., 90, *109*
Duffy, M. T., 284, 290, *307*, *308*
Dunn, P. J., 274, *308*
Dunning, F. B., 158(103), 162, *165*

E

Eastman, D. E., 118, 122(19), 140(49), 147(68), 149, *163*, *164*

Eaton, J. E., 142, *164*
Ebata, Y., 341(32), *369*
Eckstein, W., 115(7), 123(7), 157(7), 159(7), *162*
Edwards, D. M., 149(72), *164*
Edwards, P. J., 30(53), *72*
Eib, W., 130(34), 131(34), 134(40, 42), 138(48), 140(40), 142(40, 56), *163*, *164*
Eimbinder, J., 257, *307*
Einstein, A., 73, 92, *109*
Eisenbud, L., 90, *111*
Ekimov, A. I., 153, *165*
Eland, J. H. D., 75, 107, *109*
Elop, G., 338(31), *369*
El-Sherbini, Th. M., 106, *109*
Endo, H., 146(67), *164*
Esser, L. J. M., 260, *306*
Evans, B. J., 142, *164*
Evans, C. J., 65(89), *72*

F

Fadley, C. S., 107, *109*, 150(77), *164*
Fahlman, A., 73, 92, 107, *111*
Fano, U., 81, 82, 86, 87, 88, 89, 92, 96, 100, 101, 106, *108*, *109*, 156, *165*
Farago, P. S., 114(4), 115, 149(71), 155, *162*, *164*
Farley, D. T., Jr., 31(56), *72*
Feigl, F. J., 291, *308*
Felsch, W., 146, *164*
Fermi, E., 81, *109*
Fernsler, R. F., 63(87), 64(87), *72*
Ferris-Prabhu, A. V., 282, *307*
Fischer, C. F., 83, *109*
Fisher, G. B., 157(93), *165*
Fisher, R. J., 10(19a), 13(19a), *71*
Flowers, J. K., 330(20), *369*
Flowers, J. W., 49, *72*
Fowler, H. A., 144(58), *164*
Fowler, R. G., 1(1), 5, 7(15), 22(34), 26(1), 29(49), 32, 34, 50(1), 55, 56, 57, 70, *70*, *71*
Freeouf, J. L., 118 *163*
Frenkel, J., 269, *307*
Friedel, 130, *163*

Fritzler, D., 223, *247*
Fritzsche, H., 254, *308*
Frohman-Bentchkowsky, D., 276, 295, 296, 297, *307*
Fulde, P., 149(74), *164*
Funk, C. J., *247*

G

Gadzuk, J. W., 158(95), *165*
Ganner, P., 324(13), 326(13), *369*
Gardner, J. A., Jr., 334(24), *369*
Gelius, U., 73, 107, *111*
Geltman, S., 99, *109*
Gentry, F. E., 322(10), 323, 330(20), *368*, *369*
Gerbach, W., 322(9), *368*
Germer, L. H., 113, *162*
Giesecke, A., 30(54), *72*
Gilsig, T., 338(30), 343(36), 345(38), 347(36), 348(30), 349(38), 352(36, 38), 365(30), *369*, *370*
Giovanelli, R. G., 25, *71*
Glassgold, A. E., 92, *108*
Gleich, W., 124, 157, 123(24), *163*
Gluckstern, R. L., 121(18), *163*
Gnadinger, A. P., 291, *307*
Godwin, R. P., 107, *109*
Golden, F. B., 312(1), 335, *369*
Goldstone, J., 90, 91, *109*
Gomer, R., 123(21), 124(26), *163*
Goodenough, J. B., 142(55), *164*
Goodman, A. M., 284, *307*, *308*
Goodman, N. R., 203, *247*
Gordon, N., 281, *307*
Gorin, B. N., 42(74), 44(74), *72*
Goser, K., 274, *307*
Grant, I. P., 85, *109*
Greenberg, J. S., 121, 144, *163*, *164*
Greenfield, R. J., 203, *247*
Grobman, W. D., 118, *163*
Grove, A. S., 297, *307*
Guggenheim, H. J., 118(14), *163*
Gutzwiller, F. W., 316(5), 328(5), 329(5), 335, 338, 347, *368*

AUTHOR INDEX

H

Haberstich, A., 32, *72*
Haensel, R., 107, *110*
Hall, 81, *109*
Hamrin, K., 73, 92, 107, *111*
Hara, H., 300, *307*
Harari, E., 291, *307*
Harger, 203, *247*
Harmuth, H. F., 168, 200, 204, 210, 241, *247*
Harnden, J. D., Jr., 312(1), *368*
Harris, E. P., 255, *308*
Harris, F. E., 88, *109*
Harstad, D. N., 344(37), 346, *370*
Hart, K., 304, *307*
Hartree, D. R., 81, 82, 83, 86, *109*
Hashemi, J., 22(34), *71*
Havlice, J. F., *247*
Hawley, R., 357(45), *370*
Hayashi, Y., 300, 301, 302, *308*
Headrick, J. M., 31, *72*
Healey, R. N., 107, *109*
Heckman, P. J., *247*
Heden, P. F., 73, 107, *111*
Hedman, J., 73, 92, 107, *111*
Hedmann, J., 73, 107, *111*
Heiland, W., 115(7), 123(7), 157(7), 159(7), *162*
Heinzmann, U., 151(82), 155, 156(91), 157(82), *165*
Heitler, W., 75, *109*
Helman, J. S., 132(38), 133(38), *163*
Heno, Y., 82, *108*
Henry, R. J. W., 83, 90, *109*, *110*, *111*
Hercules, D. M., 75, *109*
Herman, F., 82, *109*
Herz, J. A., 159(100), *165*
Heumann, K., 317(6), 338, *368*
Hildebrand, B. P., 169, *247*
Hill, R. D., 13, *71*
Hodges, D. B., 10, *71*
Hofmann, M., 123(23), 124(23), 157(23), *163*
Hofstein, S. R., 297, *307*
Hollander, J. M., 107, *109*
Holloway, L. W., 323(11), 327(11), 367, *368*
Holzberlein, T. M., 37, *72*
Holzwarth, G., 117, *163*
Horninger, K. H., 274, *307*
Hotop, H., 93, *109*
Hüfner, S., 118(14, 17), *163*

Hughes, V. W., 121(18), 144(59), *163*, *164*
Hummert, G. T., 65, 66, *72*
Hutchins, M. T., 137(47), *164*
Huzinaga, S., 92, *109*

I

Igarashi, R., *308*
Iida, K., 290, *307*, *308*
Iizuka, H., 300, *307*, *308*
Inokuti, M., 105, 106, *109*
Ishihara, T., 91, *109*
Ishikawa, M., 300, *307*

J

Jacobs, V. L., 96, 97, 100, *109*, *110*
Jensen, R. W., 342(33), 348(33), *369*
Johansson, G., 73, 92, 107, *111*
Johnson, D. W., Jr., 142(52), *164*
Johnson, W. C., 281, *307*
Joisce, J., 29, *71*
Jones, R. C., 13, *71*
Jost, K., 151(82), 155(82), 156(82), 157(82), *165*

K

Kahng, D., 256, 263, 292, 293, 294, *307*
Kalisvaart, M., 158(103), 162, *165*
Kalter, H., 291, *307*
Kamal, J., 200, *247*
Kamimura, M., 137(46), *164*
Kaminsky, M., 129, *163*
Kamoshida, M., 290, *307*
Karady, G., 330, 338(30, 31), 343(36), 345(38), 347(36), 348(30), 349(38), 352(36, 38), 365(30), *369*, *370*
Karlson, S. E., 73, 92, 107, *111*
Karule, E., 89, *109*
Kasperkovitz, D., 259, *307*
Kato, 30(52), *72*
Keating, P. N., 223, *248*
Keid, J. S., 30(53), *72*
Kelly, H. P., 91, 92, *110*
Kemeny, P. C., 118(16), 142(51), *163*
Kenedy, P. J., 123(23), 124(23), 157(23), *163*

Kennedy, D. J., 82, 83, 84, 88, 93, *110*, *111*
Kennedy, D. P., 255, *307*
Kerr, D. R., 291, *306*
Kessler, J., 114(3), 151(82), 155(82), 156(82), 157(82), 162, *162*, *165*
Kessler, S. W., 334(26), *369*
Kim, Y. K., 106, *110*
Kimbark, E. W., *370*
Kino, G. S., *247*
Kirschner, F., 324(13), 326(13), *369*
Kishi, K., 341(32), 367(49), *369*, *370*
Kisker, E., 159, *165*
Kittagawa, N., 17(28), 19(28), 20, *71*
Kline, L. E., 65, 67, 68, 69, *72*
Klingbeils, R., 63, 64, *72*
Knauer, K., 274, *307*
Kobayashi, K., 284, *307*, *308*
Koch, E. E., 107, *110*
Kohn, W., 88, 89, *110*
Kolker, R. J., 203, *247*
Koo, T. K., 287, 288, 292, *307*
Kooi, E., 291, *307*
Koons, H. C., 26(45), 27, *71*
Koopman, D. W., 42, *72*
Koppelman, R. F., 223, *248*
Koyama, K., 157, *165*
Kramer, R. P., 264, 273, 297, 298, 299, 300, *306*, *309*
Kravse, M. O., 75, 96, 98, 107, *110*
Krider, E. P., 10(19a), 13(19a), 42(73), 43(76), 44(76), 48(76), *71*, *72*
Kritzinger, J. J., 46, *72*
Kuijpers, F. A., 254, *307*
Kunz, C., 107, *110*
Kurata, M., 347, *370*
Kurbatov, B. C., 73, *111*
Kuyatt, C. E., 113(2), 123(2), *162*

L

Laibowitz, R. B., 292, *307*
Lambrechtse, C. W., 259, *307*
Lampel, G., 152, *165*
Latter, R., 82, *110*
Lauer, E. J., 37(68), *72*
Lawatsch, H., 334(25), 335(25), *369*
Lederer-Rosenblatt, D., 117(11), *163*

Lederman, S., 35, *72*
LeDourneuf, M., 90, *108*
Lee, C. M., 89, 100, *109*, *110*
Lee, Q. H., 203, *248*
Lee, S. W., 203, *248*
LeFebvre, R., 86, *110*
Lenzlinger, M., 265, 266, 277, *307*
Leung, L. K., 142, *164*
Levin, K., 149(73), *164*
Lewis, J. B., 203, *248*
Liebermann, M. D., 342(33), 348(33), *369*
Liebermann, L., 149(72), *164*
Liebsch, A., 149(73), *164*
Ligenza, J. R., 292, 293, *307*
Lin, C. D., 89, 92, *110*
Lindberg, B., 73, 92, 107, *111*
Lindgeren, I., 73, 92, 107, *111*
Lintz, P. R., 203, *248*
Lipsky, L., 90, 100, *110*
Livingston, M. S., 105, *110*
Lo, Y. T., 203, *248*
Lockwood, G. C., 284, 287, 288, 292, *307*, *308*
Long, R. L., Jr., 144(59), *164*
Longinotti, L. D., 144, *164*
Lorenz, J., 156(91), *165*
Lu, K. T., 100, *110*
Luecke, G., 258, 259, 261, *307*
Lundkvist, L., 284, 285, 286, *307*
Lundström, K. I., 276, 279, 280, 281, 282, 283, 284, 285, 286, 288, *307*, *308*
Luther, A., 149(74), *164*

M

McCabe, M. K., 25, *71*
McEachron, K. B., 5, *71*
McGuire, E. J., 82, *110*
McKechnie, R. M., 334, *369*
McLain, D. K., 10(19a), 13(19a), *71*
MacPherson, D. A., 26(45, 46), 27, *71*
McVicar, D. D., 90, *108*
Mahan, A. H., 159(101), *165*
Mahr, D., 93, *109*
Mainstone, J. S., 29, *71*
Maison, D., 158(102), 162, *165*
Malan, D. J., 3, 4, 5, 6, 8(4), 9(4), 11(4), 13(4), 19(4), 20(4), *70*, *71*

Malone, D. P., 121(118), *163*
Mann, J. B., 83, *110*
Manne, R., 73, 107, *111*
Manson, S. T., 81, 82, 83, 84, 88, 93, 97, 98, 100, 103, 104, *109*, *110*, *111*
Marom, E., 223, *247*, *248*
Marr, G. V., 75, *110*
Marton, L., 144(58), *164*
Masuoka, F., 300, *307*, *308*
Mathon, J., 149(72), *164*
Mattheis, L. F., 142(54), 144, *164*
Mattis, D. C., 130(31), 131(31), 134, *163*
Maurer, M., 331(21), *369*
Mavor, J., 274, *308*
Mayer, J. W., 290, *307*
Mazda, E. F., 313(3), *368*
Mead, C. A., 265, *307*
Meek, J. M., 6(16), 46, 47, 48, *71*, *72*
Mehta, D. A., 291, *308*
Meier, F., 130(34), 131(34), 134(31, 39, 40, 42), 140(40), 142(40), 151(79, 80, 81), 153(81), 155(81), *163*, *165*
Meister, H. J., 117, *163*
Mellor, P. J. T., 274, *308*
Merz, H., 157, *165*
Meservey, R., 128, 159, *163*
Methfessel, S., 130(31), 131(31), 134(31), *163*
Meyer, N. J., 268, *306*
Meylan, A. A., 25(39), *71*
Michels, H. H., 88, *109*
Mies, F., 86, *110*
Miller, W. F., 105, 106, *110*
Miner, C. E., 107, *109*
Mitchell, I. V., 290, *307*
Mize, J. P., 258, 259, 261, *307*
Moll, J. L., 268, *306*
Mönch, W., 274, *308*
Monteith, W., 346(39, 40), *370*
Moore, C. E., 93, *110*
Moore, E. N., 87, *108*
Moores, D. L., 90, *108*
Moreton, G. E., 25, *71*
Moroshohi, M., 367(49), *370*
Morrish, A., 142, *164*
Moser, C., 86, *110*
Moyson, J., 323(10), *368*
Mueller, R. K., 223, *247*, *248*
Mulica, A. R., 353(42), *370*
Müller, E. W., 123(7, 20, 22), 125(27), *163*
Müller, N., 115(7), 123, 157(7), 158(97), 159, *162*, *165*

Munz, P., 134(43), 142(56), *163*, *164*
Murphy, J., 358(47), 360(47), 361(47), 366(47), 367(47), *370*
Murthy, S. S. R., 200, *247*
Myers, F. E., 113(1), *162*

N

Naber, C. T., 284, 287, 288, 292, *307*, *308*
Nagai, K., 300, 301, 302, *308*
Nakai, Y., 107, *110*
Nakagiri, M., 289, *308*
Nakanvma, S., *308*
Neppell, C. T., 292, 293, *309*
Ness, N. I., 26, *71*
Nicollian, E. H., 256, *307*
Nielsen, R. E., 36, *72*
Ning, T. H., 273, 274, *308*
Nordberg, R., 73, 92, 107, *111*
Nordling, C., 73, 92, 107, *111*
Norwalk, T. P., 326(16), *369*

O

Oberoi, R. S., 89, *108*
Ochs, G. R., 31(56), *72*
Ohnemus, B., 151(82), 155(82), 156(82), 157(82), *165*
Ohta, K., 284, *307*
Ohuchi, K., 300, *307*
O'Neill, M. R., 158(103), *165*
Ong, R. S. B., 60, 61, *72*
Onoda, K., *308*
Ormonde, S., 90, *110*
Orville, R. E., 11, 13(24), 43(76), 44(76), 48(76), 50, *72*
Ovshinsky, S. R., 254, *308*
Owen, T. C., 37, *72*

P

Pan, Y. L., 36, *72*
Pao, H. C., 256, *308*
Parker, E. N., 26, *71*
Penn, D. R., 158(98), *165*
Penney, G. W., 65, 66, *72*
Pepper, M., 272, 273, *308*

Percival, I. C., 88, 89, *110*
Peshkin, M., 97, *110*
Phillips, K., 39, *72*
Piccone, D. E., 326, *369*
Pierce, D. T., 130(34), 131(34), 134(39, 40, 42), 140(40), 142(40), 144(62), 147(62), 148(62), 151(79, 80, 81), 153(81), 155(81), *163, 164, 165*
Platzman, R. L., 105, 106, *110*
Plummer, E. W., 158(98), *165*
Poe, R. T., 91, *109*
Politzer, A. B., 158(96), *165*
Postolache, C., 268, *306*
Prats, F., 87, *109*
Price, W. G., 75, *110*
Prokhorov, V. G., 201, 242, *248*
Pruett, M. L., 11, *71*
Pulver, M., 283, 284, *307*

Q

Quartly, C. J., 251, *308*
Quate, C. F., *247*
Quinn, D. J., 274, *308*

R

Raether, H., 38, 39, *72*
Raith, W., 159(101), *165*
Rau, C., 129, 160(30), *163*
Redkov, V. P., 42(74), 44(74), *72*
Regenfus, G., 123(23), 124(23, 24), 157(23, 24), *163*
Reid, W. W., 321(7), 326(7), 327(7), *368*
Reimers, E., 323, *369*
Remeika, J. P., 134(40), 138(48), 140(40), 142(40, 52, 56), *163, 164*
Reuss, A. D., 35, *72*
Richards, C. N., 42(73), *72*
Richman, D., 290, *307*
Riley, W. B., 259, 262, *308*
Robb, W. D., 89, 90, *108*
Roberts, F. M., 326, *369*
Robinette, W. C., 275, *306*
Roetti, C., 83, *108*
Rolle, A. L., 169, *248*
Romand, J., 75, 107, *109*
Ron, A., 91, *110*
Roquet, J., 28, *71*
Rosenzweig, W., 291, *307*
Ross, E. C., 276, 281, 282, 283, 284, 288, *307, 308*
Roulet, B., 117(11), *163*
Rountree, S. P., 90, *110*
Royce, B. S. H., 291, *307*
Ruchti, P., 130(34), 131(34), *163*
Russell, A., 251, *308*
Russell, G., 37, *72*
Rusu, A., 268, *306*
Rutz, R. F., 255, *308*

S

Safarov, V. I., 153, *165*
Sah, C. T., 256, *308*
Saint-James, D., 117(11), *163*
Salama, C. A. T., 291, *308*
Salpeter, E. E., 75, 96, 98, *108, 110*
Salters, R. H. W., 259, 260, 277, *307*
Saltzer, B. A., 200, 244, *248*
Samson, J. A. R., 96, 97, 98, 107, *110, 111*
Sangster, F. L. J., 260, 303, *306, 308*
Sanmann, E. E., 50(83), 53, 55, 56, 57, *72*
Sant, A. J. D., 359(46), *370*
Sato, S., 288, 291, *308*
Sato, T., 300, *307, 308*
Sattler, K., 130(33), 132(37), 133(33), 134(39, 40), 135(33), 140(40), 142(40), *163*
Saum, K. A., 42, *72*
Saxe, R. F., 46, 47, 48, *72*
Scace, R. I., 330(20), *369*
Schärpf, O., 123(23), 124(23), 157(23), *163*
Schatorjé, J. J. H., 291, *307*
Scheer, J. J., 152, *165*
Schlich, R., 28(47), *71*
Schmidt, V., 97, *111*
Schoenes, J., 135, *163*
Schonholzer, E. T., 336, *369*
Schonland, B. F. J., 3, 4, 5(11), 6(7, 3, 6), 8, 9, 11, 13, 19, 20, *70, 71*
Schröder, K., 159(101), *165*
Schultheiss, P. M., 203, *248*
Schultz, M., 26, *71*
Scott, J. H., 276, *309*
Scott, R. P., 31, 33, 34, *70, 72*
Scudder, J. D., 26, *71*

Seaton, M. J., 83, 88, 89, *108, 110, 111*
Seiler, G. J., 89, *108*
Seligson, C. D., 203, *248*
Selzer, B., 28(47), *71*
Sevier, K. D., 75, *111*
Sewell, F. A., 274, *308*
Sheftel, S. I., 83, *108*
Shelton, Jr., G. A., 7(15), 34, *71, 72*
Shen, Y. R., 152(88), 153(88), *165*
Sheridan, K. V., 25(39), 29, *71*
Shevchik, N. J., 118(16), *163*
Shirley, D. A., 75, 92, 107, *111*
Shkilev, A. V., 42, 44(74), *72*
Shockley, W., 272, *308*
Shreider, E. Ya., 75, 107, *111*
Shull, C. G., 113, *162*
Shuskus, A. J., 274, *308*
Siambis, J. G., 65, 67, 68, 69, *72*
Siegbahn, K., 73, 92, 107, *109, 111*
Siegmann, H. C., 115(6), 130(33), 131(35, 36), 132(37, 38), 133(33, 38), 134(40, 41), 135(33), 136(45), 138(48), 140(40), 142(40, 56), 144(61, 62, 63), 145(61, 41), 147(62, 63), 148(62), 150(76), *162, 163, 164*
Silverstone, H. J., 92, *111*
Simpson, G. C., 22, *71*
Simpson, J. R., 36(65), *72*
Sizmann, R., 124(24), 129, 157(24), 160(30), *163*
Skillman, S., 82, *109*
Sklar, M., 297, *307*
Skudrzyk, E., *248*
Slater, J. C., 81, 82, 83, 86, 89, *111*, 144, *164*
Sletten, A. M., 43(76), 44(76), 48(76), *72*
Slob, A., 304, *307*
Smit, J., 250, 251, *308*
Smith, E. R., 90, *110*
Smith, K. J., 88, 89, 90, *108, 110, 111*
Smith, N. V., 148(69), 157(93), *164, 165*
Smith, R., 31(57), *72*
Smith, R. L., 88, 89, *111*
Snow, E. H., 265, 266, 297, *307*
Somos, I. S., 326, *369*
Sondhi, M. M., *248*
Stamberg, A., 333(23), 334(23), *369*
Starace, A. F., 85, 87, 88, 93, 97, 98, 100, 103, 104, *109, 110, 111*
Steimel, A., 332, *369*
Stekol'nikov, I. S., 42, 44(74), *72*

Stephany, F., 290, 291, *306*
Stewart, A. L., 83, *109*
Stiles, P. J., 292, *307*
Stoner, E. C., 144, *164*
Stumpe, A. C., 317(6), 322(9), 335(27), 338, *368, 369*
Sugano, S., 137, *164*
Sundburg, W. J., 292, 293, *307*
Sutton, J. L., 169, 200, 223, 244, *247, 248*
Svensson, C. M., 276, 279, 280, 281, 282, 283, 284, 285, 286, 288, *306, 307, 308*
Svensson, S., *109*
Sze, S. M., 256, 269, 294, *207, 308*

T

Takahashi, S., 341, *369*
Takei, S., 30(52), *72*
Takeishi, Y., 300, *307*
Takenaka, A., 367(49), *370*
Takigami, K., 367(49), *370*
Tanabashi, K., 284, *308*
Tamura, K., 146(67), *164*
Tanabe, Y., 137(46), *164*
Tarui, Y., 300, 301, 302, *308*
Tedrow, P. M., 128, 159, *163*
Teer, K., 260, *308*
Terenin, A. N., 73, *111*
Thomas, L. H., 81, *111*
Thorn, J. V., 200, 244, *248*
Thornber, K. K., 292, 293, *309*
Tidman, D. A., 62, 63(87), 64(87), *72*
Toburen, L. H., 107, *111*
Torres, B. W., 90, *110*
Traum, M. M., 148(69), *164*
Trinh, G. N., 338(31), *369*
Truhlar, D. G., 88, 89, *111*
Tsong, T. T., 123(22), 125, *163*
Tsujide, T., 290, 291, *307, 308, 309*
Turcotte, D. L., 60, 61, *72*
Turner, D. W., 75, *111*
Tuska, J. W., 287, *309*

U

Udo, T., 44, 45, *72*
Ueberall, H., 116(9), *162*

Uman, M. A., 2(4), 4, 5, 6(2), 10, 11, 13, 21, 42(73), 43, 44(76), 48, 70, 71, 72
Utsumi, T., 124(25), 157(94), 158, 163, 165

V

Vampola, A. L., 26, 71
van den Enden, W. A. M., 254, 307
Vanderkulk, W., 203, 248
van der Wiel, M. J., 106, 108, 109, 111
van Laar, J., 152, 165
Verbist, J., 107, 108
Verhulst, A. G. H., 254, 307
Verwey, J. F., 264, 266, 267, 268, 269, 270, 271, 273, 297, 298, 299, 300, 309
Vilesov, F. I., 73, 111
Vodar, B., 75, 107, 109
Vogt, F., 342(35), 370
Vo Ky Lan, 90, 108
Voorhoeve, R. J. H., 142(52), 164
Voshall, R. E., 21, 71

W

Wachter, P., 136, 163
Wada, T., 289, 308
Wagner, K., 39, 41, 72
Walmark, J. T., 276, 281, 282, 283, 288, 308, 309
Waltar, J. P., 152(88), 153(88), 165
Walters, G. K., 158(103), 162, 165
Ward, A. L., 65, 72
Watson, R. E., 149(74), 164
Webb, T. G., 89, 108
Weiss, A. A., 25, 71
Weiss, A. W., 86, 111
Weisshaar, E., 334(25), 335(25), 369
Weisshaar, H., 334, 335, 369
Wells, J., 90, 109
Wendin, G., 91, 111
Werme, L. O., 73, 107, 111
Wertheim, G. K., 118, 144, 163, 164
Whitaker, W., 90, 110

White, M. H., 275, 283, 284, 285, 286, 291, 309
Wick, G. C., 91, 111
Wiebes, G., 106, 111
Wiedman, S. K., 304, 306
Wight, G. R., 106, 111
Wigner, E., 89, 111
Wijn, H. P. J., 250, 251, 308
Wilde, J. P., 25(39), 71
Wilkinson, E. L. G., 326, 369
Wilson, W. E., 107, 111
Winkler, H., 356(43), 370
Wohlfarth, E. P., 144, 150(77), 164
Wolley, E. D., 328(17), 369
Woods, J. W., 203, 248
Woods, M. H., 287, 309
Workman, E. J., 17(28), 19(28), 20, 71
Worrall, A. G., 300, 306
Wright, J. G., 146, 164
Wright, W. F., 357, 370
Wuilleumier, F., 96, 110

Y

Yamaguchi, T., 288, 291, 308
Yang, C. N., 98, 111
Yaspen, A., 247
Yin, M. L., 92, 111
Yu, H. N., 273, 274, 308
Yu, R., 328(17), 369
Yun, B. H., 283, 309

Z

Zaidel, A. N., 75, 107, 111
Zaidi, M. H., 96, 110
Zare, R. N., 97, 99, 108, 109
Zener, C., 142(53), 164
Zilinskas, G., 223, 248
Zinn, W., 115(7), 123(7), 157(7), 159(7), 162
Zucca, R. R. L., 152, 153, 165
Zürcher, P., 151(79, 80), 165

SUBJECT INDEX

A

Abnormal avalanches, 40–42
Acceleration matrix element, 81
Acoustic imaging in water, 202
Acoustic waves
  electrical processing of, 224
  image generation by, 168
Alias points, in image generation, 241–242
Alkali metals, in polarized electron emission, 155–157
Aluminum oxide layers, in MAOS device, 289–291
Amplifying-gate structure, 324
Amplitude comparator, in image generation, 213–216
Antiforce leader, in long sparks, 43–44
Antiforce waves
  in avalanche process, 41–42
  Class I–III, 50–57
Antiforce wave velocity
  for argon, 34
  for nitrogen, 33
Argon, proforce and antiforce wave velocities for, 33–34
Array size
  for classical limit of resolution, 200–203
  reduction of an image generation, 200–216
Asymmetry parameter
  central field calculation and, 98–100
  general calculation for, 100–104
ATMOS (adjustable threshold MOS transistor memory device, 264, 268, 297–300
  erase operation in, 299
  switching degradation in, 299
Atomic photoelectron spectroscopy, 73–107
  see also Photoelectron spectroscopy; Photoemission
Autoionizing states, 87
Avalanche-injected electron current, 267–268

Avalanche injection of holes, 268–271
Avalanche process, 37–42
  abnormal avalanches in, 40–42
  constant charge density in, 39
  corona in, 38
  electron drift velocity in, 40
  long sparks in, 42
  normal Townsend, 38
  proforce and antiforce waves in, 40–42

B

Band-band tunneling, 278–281, 283
Band-trap tunneling, 281–284
Bistable flip-flop, 257–258
Bohr magneton number, 147
Breakdown waves, in electric fields, 31
Brillouin zone
  transitions in, 154
  wave functions in, 153–154
Bubble memories, 253–254, 302

C

Cable firing system, for thyristors, 339–340
Calculation methods, for high-power electronic devices, 341–354
Cartesian coordinates
  focusing in, 231–234
  storage and distribution circuit for sampling in, 194
Cathode-ray geometry, interdigitated, 323
Central-field calculations
  multiplet structure in, 82
  in photoionization cross-sections, 79–82
Central potential calculation, Hartree self-consistent field potential and, 81
Central processing unit, 249
  compatibility with, 302
  semiconductor circuitry of, 250

SUBJECT INDEX

Cesiated films, spectrum of electronic spin polarization for, 146–149
Channel injection, floating-gate devices and, 300
Charge carrier, nonavalanche injection and, 271–274
Charge injection waves, 36–37
Charge retention, in MNOS memory devices, 284–287
Charge separation, electric fields arising from, 22
Charge-transfer device, 260–261, 303
Class I waves, 50, 53–60
Class II waves, 50, 58–59
Class III waves, 59
Close coupling method
　in photoelectron spectroscopy, 88–89
　in photoionization cross section, 86–92
Clouds, in lightning strokes, 3–4
Collision chamber, in photoelectron spectroscopy, 107
Color images, with optical holography or spatial electrical filters, 228–234
Comparator, in image generation, 213–214
Computer
　memory unit of, 249–250
　subsystems of, 249–250
Computer avalanche simulations, in electron wave theory, 65–68
　see also Avalanche process
Constant charge density, in avalanche process, 39
Continuum configuration interaction, 86–92
Continuum wave functons, 86
Contour effect, in image generation, 242
Corona
　in avalanche process, 38
　long sparks and, 42, 46–47
　photoemitting efficiency, of 66–67
CPU, *see* Central processing unit
Crystal-field splitting, 137
CTD, *see* Charge-transfer device
Cycloconverters, 361–362
Cylinder of focus, 231

D

Dart leaders, in lightning strokes, 4, 17–21
Dc transmission, thyristors in, 362–366

Degradation
　in ATMOS memory device, 299
　in MNOS memory device, 287
Delay principle, in image formation, 176
Density dependence, in electron wave speeds, 51–52
Deuterons, electron capture by, 160–161
Diagrammatic perturbation theory, 91
Diffusion equation theories, in electron acoustic waves, 60–62
Diffusion radius, in electron wave theory, 62
Digital circuits
　implementation of filters by, 216–223
　for inverse Fourier transform, 217–219
Digital delay circuits, 216–217
Dipole approximation, 77
Dipole matrix element
　alternate form of, 77–79
　in photoionization process, 96
Dipole-velocity form
　in Hartree–Fock calculations, 85
　of matrix element, 78
Doppler images, in image generation, 226–227
Doppler resolution, time sharing and, 241

E

Electrical filters
　color images with, 228–231
　spatial, 228–231
Electric fields
　breakdown waves and, 31
　from charge separation, 22
　corona in, 38
　electron growth in, 37
　gas discharges in, 1–2
　wavefront velocity and, 31
Electric spark discharges, lightning as, 2–6
Electric voltages, conversion of sound pressures into, 171
Electromagnetic waves, image generation by, 168
Electron acoustic waves, nonlinear, 1–70
　see also Nonlinear electron acoustic waves
Electron capture, in superconducting tunneling, 128–129
Electron cloud, thermalization of, 40

Electron drift velocity, in normal avalanche, 40
Electron fluid equations
 photoionization models and, 63–65
 time-dependent, 62–63
Electron fluid wave ionizing, Shelton theory of, 50
Electron impact ionization, Shelton-Burgers equations for, 51
Electron ionization rate, avalanche process in, 40
Electron-optical beam, in spin-polarized photoemission, 118–119
Electron spin polarization, 114
 see also Polarized electrons
 detection of by Mott scattering, 116–117
 electron-optical system beam in, 118–119
 experiment in, 114–115
 film thickness dependence and, 149–150
 measurement of, 114, 125–128
 of photoelectrons, 138
 spectrum of from cesiated films, 146–149
Electron waves
 Classes I-III, 50, 53–60
 nonlinear electron, *see* Nonlinear electron acoustic waves
Electronic devices, high-power, *see* High-power electronic devices
 weak perforce waves and, 57–58
Electrons
 avalanche injection of, 266–268
 capture of by deuterons, 160–161
 polarized, *see* Polarized electrons
 spin of, 113
 spin-polarized, *see* Electron spin polarization
Energy distribution curves, 117, 142
ESP, *see* Electron spin polarization
Exchange potential, 82
Expected function, Fourier transform and, 179

F

FAMOS (floating-gate avalanche-injection MOS transistor) memory device, 263–264, 267, 294–296
Fast charged particles, ionization by, 105–106

FEED, *see* Field emission energy distribution
Fermi-Dirac anticommutation relations, 91
Ferrites, in spin-polarized photoemission, 136–142
Ferromagnetic hysteresis loop, 251
Ferromagnetism, band theory of, 145
Ferromagnets, electron capture at surface of, 129
Field emission, of polarized electrons, 123–124, 157–159
Field emission energy distribution, 158
Field evaporation, in spin-polarized field emission, 125
Field ion microscopy, 124–125
Film thickness, ESP dependence on, 149–150
Flash, return strokes in, 19
 see also Lightning
Flip-flop, bistable, 257–258
Floating-gage devices, 294–302
Focus, vertical and horizontal lines, in 231
Focused images, generation of with multipliers, 238
Focusing, for spherical wavefronts, 231
Focusing without approximation, 235–239
Fourier transform, 172 n.
 circuit for two-dimensional processor using, 188–189
 circuits based on, 186
 discrete, 183–185
 inverse transformation by means of, 179–190, 217–218
Fowler-Nordheim tunneling, 265, 280, 283–284, 292
Frequency converters, 359–360

G

Gallium arsenide, spin-oriented electrons and, 151–155
Gap length, for long sparks, 47
Gap overvoltage, in spark discharge, 43
Gap threshold potential, in laboratory sparks, 42
Gas discharges, in electric field, 1–2
 photoionization and, 2
Gases
 laser breakdown of, 34–36
 microwave breakdown of, 34–36
 propagating breakdown of, 31

SUBJECT INDEX

Gateless thyristors, 332
Gate turn-off thyristor, 328–329

H

Hall effect, thyristor and, 366
Hamiltonian
  approximate, 79
  central-field, 88
  single-electron, 80
Hartree-Fock calculations, 83–86
  dipole-velocity form in, 85
  methods beyond, 86–92
Hartree-Fock molecular field, 145
Hartree-Fock orbitals, 93
Hartree self-consistent-field potential, 81–82
Hartree-Slater orbitals, 93
Heisenberg ferromagnets, nonsaturated surface sheet in, 133
Herman-Skillman potentials, 82
HF calculations, *see* Hartree-Fock calculations
High-power electronic devices, 311–368
  see also Thyristors
  accessories in, 332–336
  calculations methods for, 341–354
  cooling of, 334–335
  modeling of, 342–343
  mounting attachments for, 333
  protection for, 335–336
  semiconductor devices and, 313–336
  thyristor systems in, 336–341
Holes, avalanche injection of, 268–271
Holography
  color images with, 228–231
  defined, 223–234
  image generation by, 168–169, 223–224
HS potential, *see* Herman-Skillman potential
Hydrogenic potential, photoionization with, 81
Hydrophones
  electric voltage generation with, 171
  inverse transformation of array in, 181
  super-resolution for one-dimensional array of, 215
  telelens effect for array of, 228

I

Image formation
  delay principle in, 176
  in lens, 176
Image generation
  amplitude comparators in, 213–216
  beyond photography or holography, 223–242
  Doppler images in, 226–227
  experimental equipment for, 242–246
  field of view in, 201
  focusing in, 231–239
  by holography, 168–169
  by linear transformation, 172–197
  observation angle in, 173
  reduction of array size in, 200–216
  signal-to-noise ratio improvement in, 211–212
  with spatial electric filters, 167–246
  spherical wavefront focusing in, 197–200
  super-resolution for, 203–211, 213–216
  synchronous demodulation in, 219–223
  telelens effect in, 227–228
  test results for, 242–244
  time sharing for high resolution in, 239–242
  with two-dimensional sampling filters, 192–193
  field evaporation with, 125
Injection and conduction mechanisms, 264–274
Insulators, 4f, 130–131
Integrodifferential equation formalism, in photoionization cross sections, 90
Interchannel coupling, in photoelectron spectroscopy, 88
International Conference on Electron Spectroscopy, 107
International Conference on Vacuum Ultraviolet Radiation Physics, 107
Intrachannel coupling, in photoelectron spectroscopy, 88
Inverse Fourier transform, digital circuits for, 217–219
Inverse transformation, with sampled storage circuits, 190–197
Ion getter pump, in photoemission, 122
Ionic wave functions, 94–95

Ionospheric effects, 26–31
  electrostatic waves in, 26–27
  Starfish high-altitude hydrogen bomb explosion and, 27–31

J

Johnston Island "Starfish" explosion, 27–31

K

Kirchhoff equation, electron acoustic waves and, 62–63
Kronecker delta, 86

L

Laser breakdown of gases, 34–36
Leader strokes
  in lightning discharge, 3–4, 44
  in long sparks, 46–47
Light-emitting diode displays, 242–245
Lightning, 2–22
  see also Long sparks
  in Alps, 5
  blazing out velocity in, 8
  clouds, in 3–4
  dart leaders in, 4, 17–21
  Empire State Building studies in, 4–5
  flash in, 4, 19
  general description of, 2–6
  hook process in, 21
  in laboratory, 2, 46–47
  leaders in, 3–4, 7–10, 44, 46–47
  luminosity in, 3–4, 21
  M components in, 21
  multiple strokes in, 19
  pilot leaders in, 5–8
  propagation directions, in 13
  return strokes in, 3, 10–17
  stepping in, 3, 5, 7–10
  stroke in, 4, 19
  stroke length in, 19–21
  strokes per flash in, 17
  stroke types in, 3
Linear transformations, image generation of 172–197

Liquid cooling, of high-power electronic devices, 334–335
Locus of focused points, 231
Long sparks
  see also Lightning
  antiforce leaders in, 43–44
  breakdown potential of, 45
  gap length and, 47
  gap overvoltage in, 43
  gas pressure reduction in, 48
  leaders in, 44, 46–47
  luminosity in, 44
  main strokes in, 46–47
  in nonuniform fields, 42
  return stroke in, 48–50
Luminosity
  in lightning, 3–4, 21
  in long sparks, 44

M

Magnetic core, in memory devices, 250–252
Magnetic fields, in magnetohydrodynamics, 22
Magnetite, in spin-polarized photoemission, 136–142
Magnetohydrodynamics
  plasmas and magnetic fields in, 22
  solar wind shocks in, 26
Many-body perturbation theory, 86–92
  MAOS (metal-alumina-oxide-silicon) devices, 263, 274, 288–291
Marx circuit, in primary breakdown wave study, 32
Matrix element, dipole acceleration form of, 78
MPBT, see Many-body perturbation theory
M components, in lightning, 21
Metals, 3d, 144–145
Memory, of computer, 249–250
Memory cell, static and dynamic semiconductor types of, 259–260
Memory devices
  bubbles in, 253–254
  magnetic cores as, 250–252
  nonvolatile, 250–255
  ovonic devices in, 254–255
  semiconductor, 255–262

SUBJECT INDEX

Memory transistor
  double-junction type, 301
  plane-injection type, 301
Metal-nitride-oxide semiconductor, *see* MNOS
Microwave breakdown of gases, 34–36
MIOS (metal dielectric oxide semiconductor) devices, 262, 274
  with interfacial doping, 292–294
MNOS (metal-nitride-oxide semiconductor) devices, 263, 266
  band-trap tunneling of, 281–284
  charge retention and, 284–287
  degradation in, 287–288, 293
  double gate-insulator thickness in, 288
  switching phenomena in, 274–278
  tunneling in, 283
MOS (metal-oxide semiconductor) transistors
  p-channel type, 256
  read-only memories and, 262
  in semiconductor memory devices, 255–257
  in static semiconductor memory cell, 303
Mott scattering, 116–118
  spin-polarized field emission and, 124
Multipliers, in focused image generation, 238

N

Natural phenomena
  gas discharge in, 2–31
  lightning as, 2–22
NEA (negative electron affinity), 152–154
Neglect of retardation, 79
Nitrogen, proforce and antiforce wave velocities for, 33
Nonavalanche injection, of charge carriers, 271–274
Nonlinear electron acoustic waves, 1–70
  see also Electron fluid equations; Electron waves
  avalanche processes in, 37–42
  charge injection waves of, 36–37
  Class I–III waves and, 56–50
  computer avalanche simulations and, 65–68
  density vs. wave speeds in 51

diffusion equation theories in, 60–62
ionospheric effects in, 26–31
laboratory experimentation in, 31–50
long sparks in, 42–50
microwave and laser breakdown in, 34–36
natural phenomena and, 2–31
primary and secondary classifications in, 50
primary breakdown waves in, 31–34
in solar phenomena, 22–26
theories of, 50–68
time-dependent solutions in, 62–65
weak proforce waves and, 57–58
Nonvolatile semiconductor memories, 249–304
  avalanche injection of electrons in, 266–268
  band-band tunneling in, 278–281
  dynamic semiconductor memory cell and, 257–258
  floating-gate devices in, 294–302
  injection and conduction mechanisms in, 264–274
  nonavalanche injection in, 271–274
  semiconductors for, 255–262
  static semiconductors and, 257–258
npn silicon wafer, 315
npnp devices, 330

O

Observation angle, in image generation, 173
One-electron approximation, in satellite lines, 95
Optically magnetized solids, polarized electron emission and, 150–157
Optical photography, focusing methods in, 200
Ovonic memory devices, 254–255

P

Paschen curve, for primary breakdown waves, 32
p-Channel MOS transistor, 256, 264
PEA, *see* Positive electron affinity

Perturbation theory, in photoionization cross sections, 76
Photoelectric magnetization curves, 131–134, 145
Photoelectron angular distributions, 74, 97–104
  asymmetry parameter calculation of, 100–104
  basic theory in, 97–98
  central field calculations for, 98–100
  dipole approximations for, 98
Photoelectron lines
  intensity of, 75
  in photoionization cross section, 93
Photoelectrons
  angular distribution of, see Photoelectron angular distribution
  electron spin polarization of, see Electron
  spin polarization
  energies of, 74
  intensity of, 117
  time-of-flight method for, 107
  from unpolarized atoms, 100
Photoelectron spectrometers, 107
Photoelectron spectroscopy, 73–107
  close-coupling equations in, 88
  defined, 73–74
  experimental techniques in, 106–107
  two categories of, 74–75
Photoelectron spin polarization, see Spin-polarized electrons; see also Electron spin polarization
Photoemission
  energy distribution curves in, 117
  models of, 117–118
  negative electron affinity and, 152
  of polarized electrons, 115–118
  positive electron affinity in, 154
  spin-polarized, see Spin-polarized photoemission
Photography, image generation by, 223
Photoionization
  asymmetry parameter for, 104
  in computer avalanche simulations, 67
  defined, 73
  dipole matrix element in, 96
  equation for, 73
  final state in, 94
  in gas discharges, 2
  hydrodynamic potential and, 81
  and ionization by fast charged particles, 105–106
  photoelectron angular distributions in, 97–104
  theoretical description of, 75–106
Photoionization channel, transitions in, 103
Photoionization cross sections, 75, 96
  autoionizing states in, 87
  central field calculations for, 79–82
  close coupling, 86–92
  continuous configuration interaction in, 86–92
  general theory in, 76–79
  Hartree-Fock calculations in, 83–86
  integrodifferential equation formalism in, 90
  many-body perturbation theory and, 86
  perturbation theory and, 76
  photoelectron lines in, 93
  pseudostates in, 89
  satellite lines in, 92–96
Photoionization models, in electron fluid equations, 63–65
Photoionization theory, fast charged particle ionization in, 105–106
Pilot leaders
  in lightning, 5–7
  speed of, 7–8
Plasmas
  in magnetohydrodynamics, 22
  ultraviolet emission from, 107
PMCs, see Photoelectric magnetization curves
pn junction, 313–314
pnp transistor, 315
$pnpn$ structure, 316, 330
  in thyristor, 319
Poisson's equation, in computer avalanche simulations, 65
Polarization detector, calibration of, 120–121
Polarization vector, defined, 114
Polarized electron transmission
  see also Electron spin polarization; Spin-polarized electron emission
  catalyst $La_{1-x}Pb_xMNO_3$ in, 142–144
  field emission and, 157–159
  optically magnetized solids and, 150–157
  superconducting tunneling and, 159–160

SUBJECT INDEX

Polarized electrons
  alkali metals and, 155–157
  emission of from solids, 113–162
  existence of, 113
  field emission of, 123–124
  photoemission of, 115–118
Polarized light, photoelectron angular distributions in, 97–98
Polaroid linear polarized, 152
Poole-Frenkel conduction, 269–270
Positive electron affinity, in polarized electron emissions, 154
Power diodes, semiconductor, 314
Power electronics
  future applications of, 368
  major fields of, 355
  new devices in, 328–332
Primary breakdown waves, 31–34
Proforce return strike, for long sparks, 49–50
Proforce waves
  in avalanche process, 41–42
  Shelton-Fowler theory of, 51–52
  velocity of for nitrogen, 33
  wave speeds in, 51
  weak, 57–58
PROM (programmable read-only memories), 261–264
Pseudo-color images, in image generation, 228–234
Pseudostates, method of, 89

R

RAM (random-access memory), 259–260
Range images, 223–226
  Doppler imaging and, 227
  three-dimensional images from, 225
Read-only memories, 261–264
Relative Sherman function, 121
RePROM (reprogrammable read-only memories), 261, 304
Return stroke
  brightness of, 19
  in lightning discharge, 3–4, 10–17
  for long sparks, 48–50
ROM, *see* Read-only memories
Rydberg atomic units, 76

S

Sampled storage circuits, inverse transformation by means of, 190–197
Satellite lines, in photoionization cross sections, 92–96
SCR, *see* Silicon controlled rectifier
Semiconductor devices
  see also Semiconductor memory devices;
    Transistors
  development of, 312
  high-power, 331–336
  nonvolatile, *see* Nonvolatile semiconductor memories
Semiconductor memory cells, static and dynamic, 257–262
Semiconductor memory devices, 255–262
  charge-transfer device and, 260–261
  MOS transistor in, 256–257
  nonvolatile memory devices and, 261–262
  random-access memory and, 259–260
  3d, 136–144
  4f, 130–131
Semimetals, 3d, 136–144
Shelton-Burgers equations, 51
Shelton-Fowler proforce wave theory, 51
SICF, *see* Single ion in crystal field
Signal-to-noise ratio
  improving of, 211–212
  time sharing and, 241
Silicon, tunneling from, 264–266
Silicon-controlled rectifier, in electric power control, 311
Silicon dioxide
  band-band tunneling in, 279
  tunnel current in, 265
Single ion in crystal field, 118, 136–137
Slater approximation, 82
Solar atmosphere, low electrical conductivity layers in, 22
  see also Sun
Solar flares, lateral waves and, 25
Solar noise, sunbursts as, 25
Solar phenomena, 22–26
  bursts in, 25
  lateral waves in, 25
  surges in, 25
Solar prominences, 23–24
Solar wind shocks, 26

SUBJECT INDEX

Sound pressures, conversion of to electric voltages, 171
Spark discharge, electrode configuration and gap overvoltage in, 43
Spatial electrical filters
  color images with, 228–231
  generation of images by, 167–246
  two-dimensional, 169
Spectroscopy, photoelectron, *see* Photoelectron spectroscopy
Spectrum of spin polarization, 142
Sphere of focus, in image generation, 197
Spherical wavefronts, focusing for, 197–200, 231–234
Spin, of electron, 113
Spin-orbit splitting, of valence band, 151–152
Spin-oriented electrons, gallium arsenide in, 151–155
Spin polarization, spectra of, 134–136, 140, 156
  see also Electron spin polarization; Spin polarized photoemission
Spin-polarized electron emission
  future of, 162
  ion getter pump for, 122
  measurements in, 144
  turbomolecular pump in, 122
Spin-polarized electrons
  see also Electron spin polarization
  defined, 114
  field emission of, 123–128
Spin-polarized photoemission
  see also Electron spin polarization; Spin-polarized electron emission
  apparatus for, 118–123
  electronic levels in, 130
  ferrites in, 136–137
  4f semiconductors and insulators in, 130–131
  light source in, 122
  magnetite in, 136–142
  photoelectric magnetization curves in, 131–134
  results of, 130–157
  3d semiconductors and semimetals in, 136–144
SSP, *see* Spectrum of spin polarization
Static semiconductor memory cell, 257–260, 303

Stepped leaders, in lightning discharge, 7–10
Stroke, in lightning, 4, 19–20
Sun
  see also Solar phenomena; Solar prominences
  atmosphere of, 22–23
  electrohydrodynamic phenomena in, 25
  lateral waves across, 25
  magnetic field of, 22
  surges in, 25
  Sweet's mechanism in, 25
Sunbursts, 25
Superconductor tunneling, 159–160
  electron emission in, 128–129
Super-resulution
  for hydrophone array, 215
  in image generation, 213–216
Sweet's mechanism, in solar wind shocks, 26
Synchronous demodulation, in image generation, 219–222

T

Telelens effect, in image generation, 227–228
Thomas-Fermi potential, 81
Three-dimensional image
  in photography and holography, 223–226
  from range images, 225
Three-phase surface charge transfer device, 260
Thyristors, 316–328
  in ac-to-dc rectification, 354
  in alternating current control, 358–359
  applications of, 354–367
  bidirectional, 330–331
  cooling of, 334–335
  current rating of, 321
  as cycloconverters, 361–362
  in direct-current link frequency converter, 360–361
  in electric power control, 311
  in frequency conversion, 359–362
  gateless, 332
  gate turn-off, 325, 328–329, 344
  as Hall-effect device, 366

SUBJECT INDEX

high-frequency applications of, 321–322
in high-voltage dc transmission, 362–366
laser firing and, 330
in motor drive control, 356
parallel array of, 353–354
in power control, 358
protective devices for, 335–336
radiation-sensitive, 329–330
regenerative gate structure in, 324
special applications of, 366–367
steps in manufacture of, 318
structure and principle of operation in, 316–319
triac, 330–331
turn-off of, 325, 328–329, 344
turn-on calculation program and, 349–350
voltage rating of, 320
Thyristor arrays, 338
Thyristor cooling, 334
Thyristor destruction, 346
Thyristor models, 347–356
  calculation methods for, 342–347
Thyristor strings, 337–338, 347–353
Thyristor systems, 336–341
  cable firing system for, 339–340
  cascade transformer firing system for, 340–341
Thyristor technology, development trends in, 320–328
Time-dependent electron fluid equations, 62–63
Time-of-flight method, for low-energy photoelectrons, 107
Time sharing, for high-resolution image generation, 239–242
Timing circuits, in inverse transformation, 196–197
Townsend avalanche, 38
Transformation, inverse, 190–197
  Fourier transform and, 217–219

Transistors
  see also MNOS devices
  MOS type, 256, 262, 303
  power applications of, 314–316
Triac thyristor, 330–332
Tunneling, superconducting, 128–129
  see also Band-band tunneling; Band-trap tunneling; Fowler-Nordheim tunneling
Tunneling experiment, electron emission in, 128
Turbomolecular pump, in photoemission, 122
Turn-off, of thyristors, 325–326, 344, 351
Two-dimensional sampling filter, 194–195

U

Ultraviolet emission, from hot plasmas, 107
Unpolarized atom, photoelectron ejection from, 100

V

Valence band, spin-orbit splitting of, 151–152
Visible range layer, in image generation, 225

W

Ward-Leonard system of motor control, 356
Wavefront velocity, electric field and, 31–32
Wave functions, simplicity of, 83
Wave speeds, density dependence of, 51–52

ENGINEERING

RETURN EN
TO →